SEA POWER ASHORE AND IN THE AIR

edited by

David Stevens and John Reeve

HALSTEAD PRESS

Sydney, 2007
First published by
Halstead Press
66 / 89 Jones Street
Ultimo, New South Wales 2007

ISBN 978 1 920831 45 5

Disclaimer
The views expressed are those of each contributor and not necessarily those of the Department of Defence. The Commonwealth of Australia will not be legally responsible in contract, tort or otherwise for any statement made in this publication.

Sea Power Centre – Australia
The Sea Power Centre–Australia (SPC-A) was established to undertake activities to promote the study, discussion and awareness of maritime issues and strategy within the RAN and the defence and civil communities at large. The aims of the SPC-A are:

- to promote understanding of Sea Power and its application to the security of Australia's national interests,
- to manage the development of RAN doctrine and facilitate its incorporation into ADF joint doctrine,
- to contribute to regional engagement,
- within the higher Defence organisation, contribute to the development of maritime strategic concepts and strategic and operational level doctrine, and facilitate informed force structure decisions, and
- to preserve, develop and promote Australian naval history.

Internet site: www.navy.gov.au/spc

Comment on this book and any inquiry related to the activities of the Sea Power Centre – Australia should be directed to:

Director Sea Power Centre – Australia
Department of Defence
CANBERRA ACT 2600
Australia
Telephone: +61 2 6127 6512
Facsimile: +61 2 6127 6519

Contents

Foreword

I think this fascinating collection of essays, the fourth publication in the King-Hall conference series, clearly demonstrates the Navy's growing maturity in dealing with its history, while at the same time raising community awareness of the Navy's critical role in national defence. Selfishly, as the Chief of Navy, I hope that it will also increase our knowledge of both the strengths and weaknesses of our service to help us build a more secure and successful future. We are, after all, the product of those who have come before us.

In the context of today's uncertain and unstable strategic environment, the multi-dimensional theme of this volume is very appropriate. As a maritime nation, we must always be prepared to defend our interests on, over and under the open ocean. At the same time, we must keep in mind that whatever happens at sea is inextricably linked with events ashore. The Navy routinely operates across the length, breadth and depth of our national security environment. We play a key enabling role for other elements of the Australian Defence Force, which is inherently joint in all that it does.

Today, there is almost no part of the globe remote from naval power. Yet, in truth, this is nothing new. As early as 1804, HMS *Calcutta* landed 150 sailors at Sydney to help quell an uprising by Irish convicts. Fifty years later, naval detachments helped restore order during the Eureka uprising at the Ballarat goldfields. It is little wonder that British naval power seemed ever present to the Empire's enemies in countless small and not so small wars.

When Australia felt ready to make a self confident contribution on the world stage, she likewise used her naval men. From New Zealand and the Maori Land War in 1860, through to the great wars of the twentieth century, Australian sailors were routinely employed on a myriad tasks ashore. This was not necessarily because they were the most capable tool in the defence inventory, but rather, because sailors were invariably professional, disciplined and innovative. Their ships could rapidly self-deploy to the scene of crisis or conflict and they possessed inherent skills sufficient to cope with almost any eventuality. Most importantly, they took their logistic tail with them – and were ready to operate immediately upon arrival.

By way of an example, I recall the story of three Aussie gunners, who, in August 1940, were landed by the light cruiser HMAS *Hobart* to help defend British Somaliland against an advancing Italian army. Responding to an urgent request for an anti-tank weapon, *Hobart* provided a 3-pounder saluting gun on an improvised mounting made from a 40-gallon drum. The three Australian sailors had only 63 rounds of ammunition, but they remained with the British rearguard until it was overrun by the enemy. The three *Hobart* sailors spent eight months as prisoners of war, but all were recognised for their bravery, not least because their presence and conduct were judged to have been of utmost value to the morale of the retreating British forces.

Sailors have a fierce sense of identity, and the habit of using their talents to the best effect, wherever they might be employed. In the last decade the men and women of the Royal Australian Navy have served on the front line in East Africa, Asia, the Pacific and the Middle East. Whether for sea patrol, blockade enforcement, sealift, ordnance disposal, humanitarian support, logistics or communications their unique and varied skills have been essential to a succession of ADF operations. I remain immensely proud of their contribution and I am confident that the performance of our men and women at sea will continue to inspire Australians well into the future.

Vice Admiral Russ Shalders
Chief of Navy

A member of the Victorian Naval Brigade in the late 19th Century. (RAN)

Contributors

Dr J.G. Barlow

Jeffrey Barlow has been a historian with the US Naval Historical Center's Contemporary History Branch since 1987. His service in the US Army from 1967 to 1970 included a tour in South Korea and an assignment with the US Army Military History Research Collection at Carlisle Barracks, Pennsylvania. A 1972 graduate of Westminster College in Pennsylvania, he received his MA and PhD degrees in International Studies from the University of South Carolina. He has written chapters for more than a dozen books dealing with the events of World War II and the postwar period and is the author of *Revolt of the Admirals: The Fight for Naval Aviation, 1945–1950* (1994). Awarded the 1995 John Lyman Prize for US Naval History, this book was republished by Brassey's in 1998. He recently completed a manuscript entitled *From Hot War to Cold: The US Navy and National Security Affairs, 1945–1955*, the first book of a projected multi-volume history of the US Navy in the Cold War.

Dr C. Clark

Chris Clark is the RAAF Historian and heads the Office of Air Force History from his position on the staff of the Air Power Development Centre. He has written or edited more than twenty books in the field of Australian defence history from colonial times to the present. Several of these works have been on naval subjects, including the battles of the Coral Sea and Savo Island. His biography of Vice Admiral Sir William Clarkson appeared in 2002 and continues previous work on naval industry in Australia in the twentieth century.

Dr J. Connor

Dr John Connor is a Senior Historian at the Australian War Memorial, Canberra. A graduate of the University of New South Wales (MA, PhD), he later taught history at the Menzies Centre for Australian Studies, King's College, London. His book *The Australian Frontier Wars, 1788–1838* (2002) was shortlisted for the Westminster Medal for Military Literature awarded by the Royal United Services Institute for Defence Studies. He has just completed a biography of Australia's longest-serving defence minister, Senator Sir George Pearce, as well as writing a volume of the *Official History of Australian Peacekeeping and Post-Cold War Operations*.

Dr A. Dorman

Andrew Dorman is a Lecturer in Defence Studies at the Joint Services Command and Staff College, a Senior Research Fellow at the Centre for Defence Studies and

Deputy Editor of *World Defence Systems*. He has previously taught at the Royal Naval Staff College, Greenwich, and at the University of Birmingham where he completed his masters and doctoral degrees. He was recently seconded to the United Kingdom's Ministry of Defence where he conducted a study with the Secretary of State for Defence on contemporary operations. His most recent publications include: *Defence under Thatcher, 1979–89* (2002); (co-ed.) *The Changing Face of Military Power* (2002); *Britain and Defence 1945–2000: A Policy Re-evaluation* (2001); and, *European Adaptation to Expeditionary Warfare: Implications for the US Army* (2002).

Associate Professor B.A. Elleman

Bruce Elleman is an Associate Professor in the Maritime History Department in the Center for Naval Warfare Studies at the US Naval War College. His fields of specialisation are Chinese, Japanese and Russian history, East Asian international relations, Sino-Soviet and Soviet-Japanese diplomatic history and Chinese military history. He has taught at the Cheng Chih University, Taipei, and Texas Christian University, Fort Worth. He lectured at Hunter College, New York, and was a research assistant at Columbia University. His most recent published works include: *Modern Chinese Warfare, 1795–1989* (2001); *Wilson and China: A Revised History of the Shandong Question* (2002); (co-ed.) *Naval Mutinies of the Twentieth Century: An International Perspective* (2003); and (co-ed.) *Naval Blockades and Seapower: Strategies and Counter-strategies, 1805–2005* (2006).

Commodore D.D. Farthing, DSC, RAN (rtd)

David Farthing was admitted as a barrister in the Supreme Court of New South Wales in 1986 and began full-time practice at the Sydney Bar on 11 December 1989. He came to the bar after a long and varied career in the Australian Navy, spanning 35 years of service. As a naval pilot, he qualified in both fixed-wing aircraft and helicopters and completed over 300 deck landings on board the aircraft carrier HMAS *Melbourne*. He commanded the RAN Helicopter Flight Vietnam in 1969–70, when this unit was integrated into the 135th Assault Helicopter Company of the US Army. On return from Vietnam, he commanded 723 Helicopter Squadron. In his final command at sea, HMAS *Hobart*, he led the international saluting line for the Bicentennial Naval Review on 26 January 1988 as the senior Commanding Officer of the Australian Fleet. On completion of this command, *Hobart* was awarded the 'Gloucester Cup' as the most efficient ship in the RAN. His final posting, prior to retirement, was as the Deputy Maritime Commander Australia.

Commander D. Hobbs, MBE, RN (rtd)

David Hobbs joined the Royal Navy in 1964. As a naval pilot, he flew Gannets, Hunter and Canberra aircraft as well as Wessex Commando Helicopters. He served in several Royal Navy aircraft carriers. He retired from the active list in 1997 and until recently was Curator and Deputy Director of the Fleet Air Arm Museum at RNAS Yeovilton, United Kingdom. His publications include *Aircraft Carriers of the Royal and Commonwealth Navies* (1996), *Aircraft of the Royal Navy since 1945* (1982), and *Royal Navy Escort Carriers* (2003). He has also read papers at major historical conferences around the world and established a reputation as a broadcaster on naval aviation matters on radio and television in the UK, Canada and New Zealand. In June 2005 he received the Lockheed Martin Award at the Paris Air Show, having been judged the Aviation Writer of the Year in the defence sector.

Professor A. Lambert

Andrew Lambert is Laughton Professor of Naval History in the Department of War Studies at King's College, London. He gained his PhD from the University of London in 1982, and taught previously at Bristol Polytechnic, the Royal Naval Staff College, Greenwich, and the Royal Military Academy, Sandhurst. His books include *The Crimean War: British Grand Strategy against Russia 1853–1856* (1990), *The Foundations of Naval History: Sir John Laughton, the Royal Navy and the Historical Profession* (1997), *War at Sea in the Age of Sail (2000), Trincomalee: The Last of Nelson's Frigates* (2002), and *Nelson: Britannia's God of War* (2004). He wrote and produced *War at Sea*, a three part series for BBC2, screened in 2004. He is a Fellow of the Royal Historical Society, Honorary Secretary of the Navy Records Society, Vice President of the British Commission for Maritime History, a Councillor of the S.S. *Great Britain* Project, Honorary Research Consultant to the *Oxford Dictionary of National Biography* and Vice President of the HMS *Trincomalee* Trust.

R. Mallett

Ross Mallett hails from Melbourne. He has a Bachelor of Science degree from the University of Melbourne, a Master of Business Administration from Monash University and a Master of Arts (Hons) from the Australian Defence Force Academy in Canberra, submitting a thesis on Military Technology and the Australian Army in World War One. He now lives in Canberra where he works as a software engineer. He is currently working on a PhD thesis on the logistics of the South West Pacific campaigns of 1943–45.

Commodore J.H. McCaffrie, AM, CSM, RANR

Jack McCaffrie is the Visiting Naval Fellow at the Sea Power Centre - Australia. As an aviator, most of his flying career was spent in Grumman Trackers, embarked and ashore. Most of his later career was spent in a succession of jobs in Canberra. He retired from the RAN in 2003 on returning from his final posting as Australian Naval Attache, Washington.

Captain C.L.W. Page, RN (rtd)

Christopher Page took up the post of Head of the Naval Historical Branch of the Naval Staff of the Ministry of Defence in 1999 on his retirement from the Royal Navy after 37 years service. He specialised in mechanical engineering, and his sea service included aircraft carriers, frigates, a minesweeper, HMS *Bristol*, and the Royal Yacht *Britannia*. Ashore, his service included postings to Portsmouth Dockyard, the Engineering Training School HMS *Sultan*, Head of Defence Studies in the MOD, and on the staff of the UK Delegation to NATO as the Deputy Military Delegate to the Western European Union. In 1992 he gained his MA in War Studies from King's College, London. He is the author of *Command in the Royal Naval Division* (1999), and many other talks, articles and chapters on various topics of naval and military history in books and journals. He is a member of the Western Front and Gallipoli Associations, the British Commission for Military History, is a Councillor of the Navy Records Society, and an Honorary Research Fellow of Birmingham University.

Associate Professor S.C.M. Paine

Sarah Paine is an Associate Professor in the Strategy and Policy Department at the US Naval War College. Before coming to the Naval War College, she was a research assistant at Columbia University, and later lectured at Texas Women's University and at the University of Dallas. She is author of *Imperial Rivals: China, Russia and their Disputed Frontiers* (1996) and *The Sino-Japanese War of 1894–1895: Perceptions, Power and Primacy* (2002), and is also co-editor of *Naval Blockades and Seapower: Strategies and Counter-strategies, 1805–2005* (2006). Her current research interests include a book project on Japanese-Soviet rivalries over northeast Asia from 1931 to 1949.

Dr. S. Papadopoulos

Sarandis (Randy) Papadopoulos is a historian in the Contemporary History Branch, US Naval Historical Center; and studied at the University of Toronto (BA 1985), the University of Alabama (MA 1991), and received his PhD in 1999 from the George Washington University. Since 1999 he has been a Lecturer in History at George Washington and, more recently, at the University of Maryland, College Park. He joined the Naval Historical Center in June 2000, and is the author of a forthcoming monograph *'The Steel Grey Stabilizer': The Sixth Fleet in Three Crises, 1946–1958*. Another project, written under the sponsorship of the Historian, Office of the Secretary of Defense, is provisionally entitled *Pentagon, 9/11: Anatomy of the Response*. He is also the Region III Coordinator for the Society for Military History and a Trustee of the US Commission of Military History. Other publications include entries in *History in Dispute*, and *The Oxford Companion to American Military History*. His principal fields of interest are German military and naval history and the US Navy.

S. Prince

Stephen Prince is a historian in the Naval Historical Branch of the Royal Navy's Naval Staff. He has particular responsibility for operational records and the historical input into current policy, and is the Policy Advisor for the UK's Maritime Battlestaff. He is a graduate of Warwick University and King's College, London, where he received the Russell Prize. He has been Sir Robert Menzies Scholar at the Australian War Memorial, a lecturer at Warwick University and Britannia Royal Naval College and Senior Lecturer at the Joint Services Command and Staff College. His publications include *The Royal Navy and the Raids on St Nazaire and Dieppe* (2006) and articles in *The Journal of Strategic Studies and Defence Analysis*.

Professor R. Prior

Robin Prior is the former head of the School of Humanities and Social Sciences, University of New South Wales at the Australian Defence Force Academy in Canberra. His principal research area is the military history of the First World War. He is concerned about the competence or otherwise of command, the impact of technology on tactics and the economics of war. He has written: *Churchill's 'World Crisis' as History* (1983), and with Trevor Wilson *Command on the Western Front* (1992) and *Passchendaele: the Untold Story* (1996). He has also contributed to *The Oxford Companion to Australian Military History* and to *The Oxford Dictionary of National Biography*. His second interest is in the military, political, social and economic history of the Second World War.

Dr J. Reeve

John Reeve is Senior Lecturer and Osborne Fellow in Naval History at the University of New South Wales, Australian Defence Force Academy. A graduate of Melbourne University (MA) and Cambridge (PhD), he has taught at Cambridge, Yale (as a Fulbright Fellow), Hong Kong, and Sydney Universities, has held visiting fellowships at Cambridge and London Universities, and is a Fellow of the Royal Historical Society and a Member of the International Institute for Strategic Studies. He began his career as a political and diplomatic historian and for fifteen years has specialised in international, naval and strategic history and contemporary maritime strategic affairs. His publications include *Charles I and the Road to Personal Rule* (1989); (co-ed.) *Southern Trident: Strategy, History and the Rise of Australian Naval Power* (2001); (co-ed.) *The Face of Naval Battle: the Human Experience of Modern War at Sea* (2003); (co-ed.) *The Navy and the Nation: the Influence of the Navy on Modern Australia* (2005) and many articles, essays, chapters, working papers, and dictionary entries. He is writing a study of early modern diplomacy and strategy and is an Associate Editor of *The Oxford Dictionary of National Biography* (2004).

M. Schweikert

Mark Schweikert joined the Australian Department of Defence in February 1999. He currently works in the Directorate of Future Warfighting where he manages the day-to-day running of Defence's Multi-National Experimentation engagement program involving Australia, Canada, Finland, France, Germany, the UK, the USA, Sweden and NATO. He has a Bachelor of Arts Degree in modern history and a Masters Degree in defence studies. He was an Army Reserve Officer for 12 years specialising in Artillery Intelligence with some full time work as his unit's Regimental Adjutant. He has written for a number of Australian and overseas defence publications since 1995 resulting in his appointment in 1998 as editor of the national magazine *The Navy*, produced by the Navy League of Australia.

Professor R.H. Spector

Ronald Spector is Professor of the History of International Relations, Elliott School of International Affairs, the George Washington University, a position he has held since 1990. Previous academic positions were as Visiting Professor of History, Princeton University, 1991, Professor of History, the University of Alabama, 1986–90, Director of Naval History and Curator for the Navy Department 1986–89, Associate Professor of History, the University of Alabama, 1984–86, Historian, Southeast Asia Branch, U.S. Army Center of Military History, 1971–84 and Acting Chief, Southeast Asia Branch, 1979–80, 1982–83. Recent Books include *At War At Sea: Sailors and Naval Combat in the Twentieth Century* (2001) which received the History Book Club Distinguished Book Award in 2002, and was listed among the *Washington Post*'s 'Best Books of 2001', as well as (co-ed.) *The Oxford Companion to American Military History* (1999) which received the Society for Military History's Distinguished Reference Book Award, and *After Tet: The Bloodiest Year in Vietnam* (1993).

Dr D. Stevens

David Stevens is a former naval officer, a graduate of the University of New South Wales (PhD) and the Australian National University (MA), and currently Director of Strategic and Historical Studies within the Sea Power Centre - Australia. He has contributed articles and essays to many publications and his work has been translated into several languages. His most recent books include the third volume in the Australian Centenary History of Defence: *The Royal Australian Navy* (2001); (co-ed.) *The Face of Naval Battle: the Human Experience of Modern War at Sea* (2003); *The Royal Australian Navy in World War II* (2nd edn. 2004); (co-ed.) *The Navy and the Nation* (2005); *A Critical Vulnerability: the Impact of the Submarine Threat on Australia's Maritime Defence 1915–54* (2005) and *Australia's Navy in the Gulf* (2006).

Dr G.E. Weir

Gary Weir is head of the Contemporary History Branch of the US Naval Historical Center in Washington DC and a Guest Investigator at the Woods Hole Oceanographic Institution. He specialises in the history of submarines, undersea warfare, and the ocean sciences. He also teaches at the University of Maryland University College and is currently working on an official history of the US Navy's Cold War deep ocean surveillance system, SOSUS. His study, *Forged in War: the Naval-Industrial Complex and American Submarine Construction, 1940–1961* received the 1993 Theodore and Franklin D. Roosevelt Naval History Prize as the best book in naval history published in that year. More recently, his book *An Ocean in Common: Naval Officers, Scientists, and the Ocean Environment, 1919–1961*, a study of the US Navy's role as participant and patron in oceanographic research, was selected by the Organization of American Historians as a recipient of the Richard W. Leopold Prize for 2002.

Illustrations, figures and tables

Illustrations

Figures

Tables

Abbreviations

ADAWS	Action Data Automation Weapons System
ADF	Australian Defence Force
ADM	Admiralty Files, Public Record Office, Kew, London
AIF	Australian Imperial Force
AIR	Air Ministry Files, Public Record Office, Kew, London
APA	Assault Transports
ARVN	Army of the Republic of Vietnam
AWACS	Airborne Warning and Control System
AWD	Air Warfare Destroyer
AWM	Australian War Memorial
BR	Book of Reference
CAG	Carrier Air Group
CAP	Combat Air Patrol
CBAL	Carrier Borne Army Liaison
CID	Committee of Imperial Defence
CinC	Commander-in-Chief
CINCPAC	Commander-in-Chief Pacific
CINCPOA	Commander-in-Chief Pacific Ocean Areas
CNF	Commonwealth Naval Forces
CNS	Chief of Naval Staff
CHINCOM	China Committee
COCOM	Coordinating Committee for Multilateral Export Controls
COMAFV	Commander Australian Forces Vietnam
COMINCH	Commander-in-Chief, US Fleet
DDG	Guided Missile Destroyer
DDL	Light Destroyer
DMS	High-Speed Minesweeper
DUKW	Amphibious Truck
DSC	Distinguished Service Cross
DSO	Distinguished Service Order
ECC	Expeditionary Combat Command
EMU	Experimental Military Unit
EWTGPAC	Expeditionary Warfare Training Group Pacific
FAA	Fleet Air Arm
FAC	Fast Attack Craft
FFG	Guided Missile Frigate
FOAC	Flag Officer Aircraft carriers
FO2FES	Flag Officer Second-in-Command Far East Station

FOME	Flag Officer Middle East
FPB	Fast Patrol Boats
GHQ	General Headquarters
GIUK	Greenland-Iceland-UK
GRT	Gross Registered Tons
HMAS	His/Her Majesty's Australian Ship
HMNZS	His/Her Majesty's New Zealand Ship
HMS	His/Her Majesty's Ship
IJA	Imperial Japanese Army
IJN	Imperial Japanese Navy
INTERFET	International Force East Timor
ISAF	International Security Assistance Force
ISTDC	Inter-Services Training and Development Centre
IWM	Imperial War Museum
JANI	Journal of the Australian Naval Institute
JOI	Joint Operational Instructions
JOOTS	Joint Overseas Operational Training School
JSF	Joint Strike Fighter
LCA	Landing Craft Assault
LCI	Landing Craft Infantry
LCM	Landing Craft Mechanised
LCS	Littoral Combat Ship
LCS	Landing Craft Support
LCT	Landing Craft Tank
LCVP	Landing Craft Vehicle Personnel
LRAPP	Long Range Acoustic Propagation Program
LSH	Landing Ship Heavy
LSI	Landing Ship Infantry
LST	Landing Ship Tank
MAD	Mutually Assured Destruction
MFO	Multinational Force and Observers
MPF	Maritime Prepositioned Force
MTI	Moving Target Indicator
NAA	National Archives of Australia
NATO	North Atlantic Treaty Organisation
NCO	Non-Commissioned Officer
NEC	Network-Enabled Capability
NHC	Naval Historical Center, Washington
NHS	Naval Historical Section, Canberra
NMM(G)	National Maritime Museum, Greenwich, UK
NOIC	Naval Officer in Charge
NOAA	National Oceanographic and Atmospheric Administration

MAL	*Marine-Artillerie-Leichter*
MFP	*Marine-Fähr-Prähme*
MTB	Motor Torpedo Boats
OIC	Officer in Command
OKW	German Armed Forces High Command
ONR	Office of Naval Research
OODA	Observe-Orientate-Decide-Act
1 PARA	1st Battalion Parachute Regiment
PRC	People's Republic of China
POA	Pacific Ocean Areas
PNF	Permanent Naval Force
RAAF	Royal Australian Air Force
RAF	Royal Air Force
RAN	Royal Australian Navy
RANBU	RAN Beach Unit
RANHFV	RAN Helicopter Flight Vietnam
RANR	RAN Reserve
RCN	Royal Canadian Navy
RFA	Royal Fleet Auxiliary
RMA	Revolution in Military Affairs
RN	Royal Navy
RNAS	Royal Naval Air Service
RND	Royal Naval Division
RNVR	Royal Navy Volunteer Reserve
ROP	Report of Proceedings
RRF	Rapid Reaction Force
RNZN	Royal New Zealand Navy
ROC	Republic of China
SAS	Special Air Service
SDSC	Strategic and Defence Studies Centre
SMP	Self-Maintenance Period
SNOBU	Senior Naval Officer Beach Unit
SSBN	Nuclear-powered Ballistic Missile Submarine
SSN	Nuclear-powered Submarine
SURTASS	Surface Towed-Array Sonar System
SWPA	South West Pacific Area
TF	Task Force
TG	Task Group
3 RAR	3rd Battalion Royal Australian Regiment
TNA	The National Archives, London (Kew)
UAR	United Arab Republic
UDT	Underwater Demolition Team

UN	United Nations
UNITAF	Unified Task Force – Somalia
UNSOM	UN Operation in Somalia
UNSCR	UN Security Council Resolution
US	United States
USAAF	US Army Air Force
USAF	US Air Force
USMC	US Marine Corps
USN	US Navy
USS	United States Ship
USSR	Union of Soviet Socialist Republics
VC	Victoria Cross
VC	Viet Cong
V/STOL	Vertical/Short Take Off and Landing
WMD	Weapons of Mass Destruction

Acknowledgements

The essays in this book began as papers presented at the fourth King-Hall Naval History Conference held in Canberra in July 2005. The conference was sponsored by the Royal Australian Navy in conjunction with the University of New South Wales at the Australian Defence Force Academy. This book could not have been produced without the support, not only of the contributors, but also of a variety of individuals and organisations. Particular thanks are due to Tom Lewis and Desmond Woods who also provided papers to the conference; to John Perryman, Brett Mitchell, Linda Parslow, and Graham Dunbar of the Sea Power - Centre Australia; and to Navy Health, the Australian Defence Credit Union, DEFCREDIT, and THALES for their sponsorship. Finally, mention must be made of Peter Hore, whose 2001 collection, *Seapower Ashore: 200 years of Royal Navy Operations on Land*, served as an inspiration for this book.

The analysis, opinions and conclusions expressed or implied in this volume are those of the authors and do not necessarily represent the views of any national military service, department, or other government agency.

Introduction

David Stevens and John Reeve

AUSTRALIANS GENERALLY THINK THAT they understand their past, and it is often remarked that the nation has been shaped by its experience of war. But however good we might be at tradition, we are generally bad at history. We tend to focus on what we think we know and, all too frequently, we examine our defining historical moments while wearing horse-blinkers. Our national culture is indelibly coloured by mythologies, in particular beginning with the struggle of the first European settlers with the harsh Australian bush. Such myths tend naturally to be far more concerned with looking inwards than outwards. The problem for a maritime nation such as Australia is that in this context, the sea that surrounds us is seen as a moat and not a dual carriage highway.

Similar images carry over into our collective wartime stories where, in the main, a single perspective on the historical narrative selectively masks what were invariably complex events. If asked to list some famous battles most Australians would probably find it difficult to get past Gallipoli and Kokoda, and even then they would be hard pressed to flesh out the context of these operations. How our troops and their equipment were safely convoyed over the sea, and then sustained and protected (and often redeployed or evacuated) for months or years at a time and often thousands of miles from home, is not generally the stuff of popular military history. Considered attempts to analyse how sea and air power have shaped our military campaigns are almost unheard of outside staff colleges or strategic policy conferences.

Even our most recent joint service operations, carried out under the full glare of the modern media, have often been either misinterpreted or viewed through a single service lens. We rarely hear in the media, or even from academia, for example, that the Australian Defence Force (ADF) as a whole has a capability gap. Too often the claim is simply made that one service or another has for too long been the 'Cinderella' of the force and needs boosting. Notwithstanding the ADF's progress towards becoming a seamless joint force, there remains a dangerous tendency to view the unique contributions of the three armed services as separate and disconnected endeavours. Few outside the ADF seem to understand that a balanced and effective force requires the provision of complementary capabilities which can work together as part of a mutually supportive system of systems. Without this mutual support, a single capability will never see its potential maximised, and in the worst case may eventually be sidelined as a strategic irrelevance. On the surface this is not a difficult concept to grasp, but in the public arguments over future defence force acquisitions and force transformation the need to procure inherently flexible and complementary assets is regularly ignored. There appears to be a fundamental community misunderstanding of the nature of modern joint combat power.

The INTERFET (International Force East Timor) deployment to East Timor in 1999 provides a useful example. This was largely seen at the time, and has been since, as a situation in which the key to victory was Australian 'boots on the ground', with maritime support either absent or fairly minimal. In fact INTERFET was necessarily very much a joint force and a combined (that is multi-national) force. As its commander, the then Major General Peter Cosgrove unhesitatingly acknowledged, the naval presence was not 'an incidental, nice to have "add on"'. Rather, he continued, 'it was an important indicator of international resolve and most reassuring to all of us who relied on sea lifelines.'[1] The capability, responsiveness and flexibility of coalition naval assets ensured that INTERFET could operate in a secure environment, allowed in-theatre mobility and then offered continuing sustainment. 'Mass', as Norman Friedman underlines in *Seapower as Strategy*, 'has to come by sea',[2] and in terms of logistical support, sea power provided INTERFET's functioning foundation.

Similarly, some have found it difficult to comprehend the role played by the Coalition's air defence vessels during the INTERFET operation, as our fighter aircraft should have quickly cleared the skies of any air threat which might have appeared.[3] But even in relatively nearby East Timor, aircraft from Australian bases would have been hard pressed to remain on station for long, and would have worked best as part of systems already and permanently on the spot. Royal Australian Air Force FA-18 fighters, or their future joint strike fighter replacements, would have succeeded only if properly integrated with other ADF capabilities. In East Timor this role would have fallen almost exclusively to the Coalition's air defence ships. These vessels, which operate with their own organic sensors, weapons, and command systems, possessed the ability to control, direct, and as necessary offer third party protection, and not just in relation to friendly aircraft. The likelihood of interference is not the issue. Rather, it is that INTERFET's success ashore was underpinned by the possession of flexible maritime assets able to cope fully with a multi-dimensional threat.

The overall point here is that highly capable naval forces provide essential elements in littoral operations, and that even when not obviously or actively engaged, sea power has invariably been an operational enabler of the first order. Notwithstanding differences of scale, similar conclusions may be drawn from the 2006 Australian deployment to East Timor and recent operations in the Solomons, the Persian Gulf and elsewhere. Indeed, almost every military operation Australia has ever undertaken has been fundamentally joint – if not always jointly Australian – simply because every Australian military operation overseas has been expeditionary.

Such observations on sea power are hardly new, and as the contributors to this volume make clear, the Navy's history of 'combat operations from the sea' is as long and as varied as its history of 'combat operations at sea'. This collection serves as a timely reminder that a balanced maritime force not only works well in partnership with other military elements, but also provides options for decision makers that no other form of military power has yet been able to match. These derive from its inherent characteristics in terms of mobility, readiness, access, flexibility, adaptability, persistence and resilience.[4]

It is a valuable lesson, as applicable to our understanding of the past as it is in terms of planning for the future, and even if the reader takes away no other, then the editors will be well pleased.

PART I

Naval power projection at the beginning of the twentieth century

1. Sea Power Ashore and in the Air

Andrew Lambert

HOW DO ARMED FORCES deal with strategic change? While strategy will always be subject to influences beyond their control, armed forces are better equipped if they develop a clear understanding of the principles upon which the application of power is based rather than simply responding to the latest issue to hit the headlines. National strategy should be a realistic and coherent response to security needs, not a contested ground on which the individual services try to sell narrowly focused programs.

Education is the key. It is also an environment in which armed forces can relax the rules of subordination and encourage new thinking. After all, what harm can come from a seminar run by an academic? Academics can provide a better grasp of long term trends and context, but only if they are asked the right questions. Too often the military relationship with the academic community takes the form of Captain Thomas Jackson's infamous question to the denizens of Room 40 on 31 May 1916: a closed request for confirmation, rather than an opportunity for an exchange of ideas and information.[1] Jackson asked the civilian intelligence personnel service where the call sign of the German flagship was, and was told in harbour. In fact, as the intelligence operatives knew, the flagship changed call sign on leaving harbour, and they knew she was at sea, but Jackson did not provide the occasion for that information to be passed. As a result Jellicoe did not know the German Fleet was at sea until Beatty came under fire.

Context is essential, because only with context can we see the ideas of key thinkers from Clausewitz and Mahan to Corbett and Richmond as being of their day, addressed to specific communities and specific politico-strategic situations. These were practical men: they were not trying to create timeless wisdom, but advanced contemporary education. Consequently their concern with specific types of naval strategy should not lead us to conclude that these were the only, or even the major issues, just the most pressing. Colomb took a different view to Mahan and Corbett because he was both British and writing in the late 1880s. He is more relevant today than he was for much of the twentieth century because he provided the best approach to sea power in asymmetric warfare.

This suggests that we need to reconsider how we use history. In 2005 the Annapolis Naval History Symposium took as its theme 'Expeditionary Warfare'. Over two days there was much discussion of this critical new tool in the American strategic arsenal. With all the enthusiasm of a convert, the US military has not only taken up the expeditionary theme, but it has re-created its own past along suitably 'expeditionary' lines. As an outsider I was struck by the religious fervour of the newly expeditionary USA, and the extent to which all were expected to conform to the new mantra. While the American passion to become expeditionary is understandable, experience suggests that such choices should never be absolute. Successful strategy builds in flexibility. By contrast, at Annapolis history was once again being rewritten, or at least re-packaged to support the latest agendas. While this is perfectly understandable, it is dangerous, misleading, and a damaging devaluation of the historical profession. History is too important to be left to the dictates of fashion, or the politics of pressing agendas.

That is not to say that history should ignore contemporary context. History has always been written to address current agendas, but good historians recognise this, and correct the bias. In 19th century Germany the development of modern historical methods was intimately linked to aggressive nationalism. Historians affected to see in Prussian military success the completion of a pre-ordained purpose, writing their histories to support this contemporary agenda, imposing a linear inevitability on the chaos of the past that separated Arminius from Bismarck.[2] This prompted a typically explosive response from Friederich Nietzsche. His 1873 essay 'The Use and Abuse of History' recognised the central role of the present in historical writing and stressed: 'You can explain the past only by what is most powerful in the present'.[3] We must remember that History is both a record and a process, and that the process is the more important of the two. How we think is more important than what we know.

The questions addressed in this volume should not be closed questions, with simple prescriptive answers: they must be open questions, prompting debate. Historians who do not recognise the way in which the world is changing are condemned to be ignored. We must read our past masters in context, and not as isolated, timeless paragons of intellectual virtue. While the subject discussed in this chapter shares something with the subject of the Annapolis Symposium I am confident that it will not be expected to support the party line, not least because if there is one no-one has bothered to tell me what it is!

There is no doubt that the present and future prospects for naval forces are very different from those that faced our predecessors. In reality classic naval warfare, fleet against fleet for command of the sea, has long ceased to be a serious possibility. For all their fascination, Trafalgar, Tsushima and Midway are far less relevant to us than the way in which these landmarks of the naval operational art were translated into war-winning strategies. Currently unquestioned naval dominance is in the hands of the United States and its allies without a battle. While the overall picture is one of absolute American control, there will always be regional conflicts where the situation may be less clear cut. In East Timor in 1999 INTERFET (International Force East Timor) forces had to remain aware of the possibility that unstable elements in the Indonesian Armed Forces might decide to contest Coalition sea

In 1999, the Coalition's massed naval might in the approaches to Dili helped convince Indonesian authorities that the International community had in fact ranged itself in full support of an independent East Timor in a way that land forces could not. (RAN)

control. While this would have been suicidal, it would have required a relatively high level of anti-submarine, anti-air and anti-surface capability to respond.

The speed with which the Australian-led 'East Timor Expeditionary Group' arrived off Dili, and the powerful presence it established before any shore footprint had been established, played a key part in overawing any potential recidivist elements within the Indonesian military. The presence of an American Aegis cruiser secured the operation against any air threat. All told, the naval presence gave the ground commander the leverage to establish his bridgehead and take control far faster than the local Indonesian forces had anticipated. Speed was enhanced by the use of the INCAT vessel, HMAS *Jervis Bay*, whose impressive 40 knot speed has influenced the US decision to provide the Littoral Combat Ship with a 50 knot top speed.

As the Western world faces no serious threat at sea, the majority of new naval construction is taken up with expeditionary and air warfare assets, be they carriers or air warfare destroyers, to control the battlespace above any future expeditionary warfare operations. Naval warfare may be dead, but navies have never been more important, because their ability to impact on the land has never been in greater demand. The key requirement is, to quote the Australian Minister of Defence speaking in June 2005, the provision of 'sufficient flexibility to provide military response options to a wide range of contingencies'. He spoke of the sea as a highway, not a moat, and jargon aside, could have been explaining Imperial defence in London a hundred years ago.[4] It is worth stressing that his words were addressed to national defence needs, not naval procurement issues.

Nor is Australia alone. On a recent visit to Paris I was struck by the very public enthusiasm of the French Navy for the 22,000 ton 'projection and command ships' *Mistral* and *Tonerre*, currently completing for service. They are capable of deploying 28 helicopters, 60 armoured vehicles and 700 troops.[5] With new expeditionary assets in France, Spain, the Netherlands, Italy and the United Kingdom, the

European Union will shortly have a significant expeditionary capability, while Britain and France are acquiring the naval aviation platforms to support such operations. If the European Union ever became a serious political organisation it would have a major capability to project power from the sea, second only to the United States.

It is not my business to pre-empt the other contributors, or to provide conclusions. Instead I would like assess the contemporary context, and offer a historical example from the nineteenth century, when the British acquired a wealth of experience of expeditionary warfare only to find that when they went to war in 1914 they needed a very different strategy. In the process we might find some useful experience of the role of history and the dangers of unduly prescriptive approaches to strategic thought.

Theme

Expeditionary warfare currently occupies centre stage because the main politico-strategic issues facing Western nations after the end of the Cold War are terrorism, rogue states, failed states, the threat of weapons of mass destruction (WMD), resource dependency and population movement. At heart it is Western culture that is under threat.[6] While we have been told that there are WMD deployable at short notice in some of the rogue states, and that terrorist groups might have the capability to use crude biological or radiological weapons, the terrorist threat is real, and it will not be defeated by any amount of homeland defence. While the West is safe from invasion by sophisticated hostile nations, staying at home is not an option. Because Western interests are truly global the solution to the next crisis may be military force, it may be humanitarian aid, but it will certainly be delivered in an expeditionary fashion. In the 21st century there is no purpose in Western nations maintaining sophisticated and effective forces that cannot be easily deployed over long distances, and sustained there for significant periods. The mantra of the modern age is 'From the Sea'. The vital naval role is that of enabling land forces, by providing afloat basing, logistics, firepower and sustainment. While there is little chance of classic naval warfare, navies have never been more important.

The key driver of this process is globalisation. The rapid growth of world trade, 90 per cent of which by weight and volume is carried by sea, means that the Western world, whether defined by geography or culture, cannot ignore what is happening beyond its borders. That this trade includes economically essential and increasingly scarce resources from regions of strategic instability only reinforces the key point. Globalisation means that Fortress America was not an option in September 2001, nor was Fortress Europe, nor yet Fortress Australia. The West cannot function without global trade, and if that global trade is to continue the West must be prepared to intervene to deter aggression, stabilise allies and defeat enemies. The Tanker War (1980–88) was an obvious example: the illegal use of naval and air forces against neutral shipping in the Persian Gulf drew in the world's four largest navies to ensure the safe and timely arrival of vital oil supplies. The problem was resolved by naval expeditions. This action was entirely consistent with the British experience in the 19th century, when naval force was used to open and secure trade routes, break down tariff barriers and spearhead international efforts to suppress illegal maritime activities like piracy and slave trading. That both were essentially economic crimes has been largely forgotten amid clouds of romantic fiction and moral outrage. A serious historical analysis of the period reveals a very different situation, and one that makes far more sense. Britain occupied a dominant position in global maritime affairs between 1815 and 1914, and used that position to secure and extend her economic interests.

As the world's population grows and shifts towards the most economically dynamic areas on the coast, most of the problems the world will face will be within range of sea-based expeditionary forces. The Western world must shape the world to promote security, stability and prosperity in an era of rapid change. It will need an expeditionary strategy to achieve this. The United States has taken a clear lead in this respect. This is hardly surprising. The US Navy and Marine Corps possessed the assets to support this strategy at the end of the Cold War, carrier battle groups with powerful strike and air defence assets, with amphibious lift and landing forces. Other Western nations that wish to take an active role in shaping the future, whether in partnership or alone, will need similar assets. Few observers have stressed the centrality of the carrier to this strategy, or the extent to which it remains the key to any expeditionary strategy. Carriers came of age because they were more flexible than battleships, not more powerful. American carriers provided the tactical air power for the initial operations in Afghanistan. Without air control expeditionary warfare is no more than an interesting idea.

Since the collapse of the Soviet Union in 1991 the possibility of significant, large scale, naval warfare, of combat between fleets of ships for control of the oceans, which was never serious even at the height of the Cold War, has receded toward the irrational, if not the impossible. The *Kursk* incident of 2000 revealed a Russian Navy that, for all its size and weaponry, is unable to send many ships to sea, and has very few effective units.[7] The shortage of trained manpower revealed by the accident makes sustained Russian naval activity impossible. President Putin's very Petrine dream of a return to the open ocean will remain, for a long time to come, just that. China, the new bogey in the American budget process, the only other nation with a large fleet that is not allied to the United States, is a pasteboard threat, although useful for long-term planning. Despite significant American assistance in the 1980s,[8] the Chinese Navy (PLAN) does not possess the weapons, electronics and operational capability to engage the US, and lacks the logistics and integral air cover for open ocean warfare. While possessing some dangerous assets for littoral combat, Kilo class submarines and Sunburn missiles, these are at best local area denial systems. China is still years away from producing an effective nuclear-powered attack and ballistic missile submarine fleet.[9] The real significance of the

PLAN is to remind the United States that its current unquestioned maritime preponderance may not last forever, and that naval combat capabilities like strategic anti-submarine warfare and anti-surface warfare should not be ignored.[10] Quite what the PLAN is doing with the rusty hulk of the former Russian carrier *Varyag* is open to speculation, but many years will pass before a Chinese Carrier Battle Group sorties into the Pacific, although there is a short term agenda for more localised naval aggression.[11] There is no other navy of any consequence that is not linked to the USN. With no realistic naval or military threat in prospect Western nations can afford to concentrate on expeditionary warfare, in contrast to the Cold War, when the focus was necessarily on the threat of total war and nuclear destruction.

Expeditionary warfare is an offensive strategy intended to exploit superior skills, assets, and fighting power. In the asymmetric conflicts of the 21st century it will be essential to the success of the West that it uses its advantages to take the initiative, forcing the rogue states and non-state actors to engage in forms of warfare that exploit their weaknesses.

The shift to an expeditionary model has prompted considerable fresh thinking, and this is now being translated into capabilities by the more astute, flexible powers. The nature of 21st century naval force will shift from single role assets to those built with flexibility in mind, from small naval platforms to large, inter-operable assets that can sustain inter-service forces afloat or those operating ashore. Western states cannot afford the cost, let alone the confusion, of allowing each service to operate a distinct, divergent strategic model. Similarly, if they wish to be effective partners in crisis and conflict, the preparations for joint action must be established before the event. This common approach must be ensured throughout the education, training, exercise and procurement processes; and it must be directed from the centre. It is not the business of armed forces to second guess the political system; but they are entitled to expect a clear understanding of national policy, and to exercise a degree of professional judgement as to how to meet those needs. For example, any current front line defence asset that cannot be deployed effectively at a distance, or perform a vital role in supporting such a deployment, needs to be re-assessed.

The ability to use the sea for strategic purposes will remain a major advantage for the West into the medium term future. Sea-based systems are harder to locate and strike than those on land, and are only vulnerable to small powers and terrorists when they enter the littoral. (The obvious examples are the American casualties in the Tanker War and first Gulf War, along with the 2000 attack on the USS *Cole*). Sound strategy exploits such advantages to maximum effect. Furthermore the current threat is susceptible to the usual strategic solutions. Non-state actors and rogue states are just as much political entities as Western democracies, and they have aims, interests, assets and points of vulnerability. The task for the medium term is to recognise these points and to strike them, to seize the initiative, forcing the enemy to conform to our movements. Maritime powers have been dealing with non-state actors, rogue states, illegal immigrants, people traffickers, smugglers and religious fanatics since the dawn of time. They are an economic fact of life.[12] Against such diffuse, amorphous threats modernised Maginot lines are both useless and wasteful. The most economical solution has always been to seek out and destroy the threat at source, while removing the opportunity or the economic causation, to ensure long term stability. The offensive component of maritime dominance is expeditionary warfare. For example, early in the 19th century Britain was drawn into a series of expeditions against piratical communities and Wahhabist religious zealots in the Persian Gulf region.[13] These expeditions secured economic interests, friendly local regimes, and free use of the seas for trade.

British policy and strategy have shifted significantly since 1989, and this has been reflected in procurement and force profile. The 1998 *Strategic Defence Review* explicitly adopted an 'Expeditionary Strategy' making the armed forces more flexible, deployable and inter-operable. In 2001 the Review was updated to reflect the growth of the terrorist threat.[14] The new strategy is essentially the 'Maritime Strategy' developed by Julian Corbett a century back, although renamed to avoid suggestions of a naval take over. This should prompt a close study of how Britain developed and used the strategy Corbett summarised.[15] His ideas came back into the mainstream because they conform to the reality of post-1990 operational deployments of British forces outside the United Kingdom.[16]

Aside from the acronyms, current British defence policy would not surprise a late Victorian strategist. Nor should it: the logic of expeditionary warfare in the 21st century is essentially what it was in the past, to pose a threat to the enemy's territory, assets or interests that forces the enemy to conform to our strategy at times and places of our choice. Whatever their agenda, the opponents of the West have values and interests that can be targeted. In combination with tightened domestic security, expeditions to Afghanistan and Iraq have largely achieved this end. Iraq might be a mess, but the mess is in Iraq, and the hostiles have been drawn in by the fundamental threat which liberal democracy poses to their aims and values. If such people are willing to die for their beliefs it is better that they do so in places and at times of our choice. This option will only be open while the West has effective naval assets, maritime logistics and offshore basing. There is nothing unusual in this, for the greater part of naval history the command of the sea, the control of maritime communications, has not been in doubt. The role of navies has been to project power against weaker or non-naval rivals, through economic blockade, the seizure of overseas territory and attacks on or across the littoral. The difference today is that the amount of power that can be deployed from the sea is far greater, and the strike range of sea-based assets is measured in thousands of miles. Afghanistan was subjected to sustained conventional bombardment by naval assets in the Indian Ocean, firing cruise missiles and launching air strikes. Nowhere on earth is safe from maritime power projection; and very few states could do anything about the platforms from which this attack is launched.

American developments suggest future expeditionary operations may be on a smaller scale, although they require more sophisticated methods of transferring from ship to shore and ever greater intelligence sophistication to arrive in time. The rationale behind the decision to name the first US Navy Littoral Combat Ship (LCS) USS *Freedom* could not be clearer: the current terrorist threat is a fundamental challenge to Western cultural values. If we wish to live in free societies then we will have to deal with those who threaten those values at a distance. The LCS combines high speed, network enabled weapons systems and embarked expeditionary force in a shallow hull intended for littoral operations in hostile zones under the strategic umbrella of American intelligence and sea dominance. These vessels demonstrate that nations which think seriously about the subject procure mission specific assets that enhance their expeditionary warfare capability. In the 21st century there will be no time to acquire them after war begins.

Expeditionary warfare is the current concern for good reason. The problems facing Western nations are over there, and sound strategy dictates that they should be kept over there, by expeditionary operations, and enhanced domestic port security. However, while the United States rushes into Expeditionary Warfare Training Group Pacific (EWTGPAC)[17] and the LCS[18] to enhance combined expeditionary capability, it can do so without sacrificing core naval missions. Other navies and nations are not so fortunate, and will face harder choices. As the 1982 Falklands War demonstrated, expeditionary capability is asset critical, but few of those assets are unique. The future Royal Navy surface fleet will be based around two large deck aviation platforms, together with three large and several smaller amphibious platforms. But as with most Western forces these new assets are seen as multi-role joint service assets, focused on power projection from the sea. The cost of change is covered by reduced destroyer, frigate and mine countermeasure vessel numbers. The submarine force will be sustained.[19] Western navies must be careful to build a degree of flexibility into their long-term procurement, because nothing is forever, certainly not strategic concepts.

History

With the contemporary context clear it may be time to consider the effect that context has on our perception of the past. Expeditionary warfare relies on mobility and sustainability, using high quality intelligence-led forces to seize and maintain the initiative in all quarters of the globe. Curiously enough, that is precisely how the British Empire worked for a century. The British experience of expeditionary warfare before 1914 offers both an example of the problems armed forces face in reducing experience to understanding, and a background to one of this volume's opening case studies, Gallipoli.

History allows contemporary armed forces opportunities to access real experience without war – but to paraphrase Clausewitz, while history is simple stuff, even the simplest history is complex. History is written for contemporary audiences, and addresses the issues of moment. This is why every generation has to reconsider its own past. I began my academic career with a study of the Crimean War, but found it hard to reconcile the evidence that emerged from the archives with existing histories. How could a conflict in which the Anglo-French armies were never more than a day's march from the sea, drew their supplies and reinforcements from the sea, and were invariably attacking fortified naval bases be anything other than an example of large scale maritime expeditionary warfare? And yet the same old story of the Charge of the Light Brigade and of nursing is repeated down to the present day. In the 20th century the Crimean War was re-written to serve a variety of strategic and diplomatic contexts.[20] If we accept that the Crimea was a limited expeditionary war, at least for the 'West', then the relevance is obvious. But we must be careful in balancing context and relevance. The past that matters to us is not the total wars, hot and cold, of the 20th century but the more diffuse problems facing the 'West' in the 19th century.

For example, it is essential to know that the most brilliant of maritime strategists Sir Julian Corbett wrote in 1911:

> *Since men live upon the land and not upon the sea, great issues between nations at war have always been decided – except in the rarest cases – either by what your army can do against your enemy's territory and national life, or else by the fear of what the fleet makes it possible for your army to do.'*[21]

We should also know that his comments were directed at those who followed a simplified version of the argument advanced by Alfred T. Mahan, namely that securing command of the sea was an end in itself, and would ensure victory in war. His mastery of 18th century history persuaded Corbett that *naval warfare* was concerned with control of oceanic communications, so that a *maritime strategy* could be developed to project power against the enemy, without putting at risk the territory or supplies of the successful naval power. It is equally important to know that he did not have the opportunity to examine the 19th century in the same detail.

The British experience of expeditionary warfare in the 19th century could be examined at the tactical or operational level, but it is only comprehensible in the context of national grand strategy. Unique among major powers, Britain's strategy was expeditionary warfare. It sustained Britain's empire of trade and capital across the globe, and enabled the British to take a powerful position in European affairs without a significant field army. From Drake to Nelson British naval officers were heavily involved in amphibious operations, while British policy makers understood that naval mastery was only useful when it could be employed to harm the enemy ashore. The question was how.

As was typical of the British, however, their expeditionary doctrine remained entirely unwritten before the 20th century, passed from generation to generation by word of mouth and hands on experience. This approach had worked well in the

17th and 18th centuries: the Royal Navy, regularly defeated rivals with stronger written traditions. This system would ultimately fail, as Andrew Gordon has shown, not in war but in the absence of war.[22] Decades of peace broke the human connection, while the belated attempt to develop historically based educational and doctrinal models was overtaken by events.

For our purposes we need not trace British practice back much further than the latter stages of the Napoleonic Wars. After Trafalgar the Royal Navy was largely reconfigured as a power projection force: the natural role of a navy. With no rival left on the ocean the British shifted the balance of effort from naval combatants to amphibious craft. Troopships, flotilla craft and flat boats were used by amphibious task forces that ranged across the globe. From Copenhagen and Cape Town to Buenos Aires and Baltimore, British forces were landed with their stores and horses, and supported by naval gunfire and naval working parties. Officers who served in these campaigns led the Royal Navy from 1815 to the Crimean War and they did not forget their experience. Vice Admiral Philip Colomb, the first theorist of British expeditionary warfare, observed that Vice Admiral Sir Charles Napier's operation at Bomarsund in 1854 'demonstrated all the principles which for a century and a half governed the successful attacks upon territory from the sea.'[23]

After 1815 the British emphasised expeditionary warfare because it conformed to their strategic needs. A technologically advanced offensive maritime strategy was the only reliable method of securing global political and economic aims. These were the preservation of European peace and stability, and the security and expansion of British extra-European trade. The British relied on command of the sea to secure their home islands, overseas possessions, and vital oceanic trade routes from attack. To lose command of the sea would be fatal: Britain could not feed her people let alone function economically without access to overseas food and raw materials, while her prosperity depended on international commerce.

British strategy was anything but defensive, however. Rather than create a system to provide local protection for all their possessions the British preferred to seize the initiative in any conflict. Attacking the naval assets and coastal cities of their enemies, when combined with the ability to clear hostile commerce from the ocean and set up a close economic blockade, gave Britain the strategic power to inflict serious damage on any rival.[24] The threat posed fell short of annihilation, but involved inevitable defeat after a long drawn out limited war of economic attrition that only Britain could win. Furthermore, any state that went to war with Britain had to recognise that her other rivals would soon exploit the opportunity to secure political or territorial concessions.[25] Even in the total wars of the Napoleonic era Britain not only survived, but prospered, playing a major part in defeating France through economic warfare and peripheral strategic attacks.

Unlike other major powers Britain had the strategic flexibility to apply her power where her rivals were most vulnerable, rather than trying to confront them at their chosen point of attack. The British did not defend the point attacked, rather they exploited the mobility of expeditionary forces to launch a counter-stroke at the most vulnerable point of a potential enemy. India and Turkey were defended by threats against Cronstadt and St. Petersburg, while Canada was secured at New York and Washington. For Britain the front line of her defence began at the high water mark on an enemy coast, and the forts placed there were the first target. After 1815 the ability to strike at enemy coasts was refined, as new technologies like steam propulsion, explosive shells and rifled artillery enhanced the offensive power of naval forces. In the 1840s the threat posed by the new French naval base at Cherbourg prompted the development of a new approach, a 'Cherbourg Strategy' which used steam-powered flotilla craft to isolate and bombard hostile naval bases. This operational development underpinned the British reliance on a strategy of deterrence. By 1854 the British had developed dedicated coastal warfare assets (Littoral Combat Ships if you like) for the mission. The ubiquitous steam gunboat was designed to attack Cronstadt. It was equally effective in other littoral battlespaces.

After refining their art in a series of small wars with minor powers, rogue states and non-state actors the British declared war on Russia in 1854 in alliance with France and Turkey. The Crimean War (1854–1856) was waged as a large scale expeditionary campaign. Allied troops landed on the Russian coast laid siege to naval bases at Bomarsund and Sevastopol, while other fortresses were attacked by naval or amphibious forces. At no time were allied troops ever more than one day's march from the sea. The Allies' sea-based logistics and mobility gave them the initiative, and the combat power to defeat the world's largest army in Russia. The attempt to seize Sevastopol with a grand raid came unstuck for a number of reasons: inter-allied friction, poor command structures and lack of resolve being key among them, but it should be stressed that it was a British plan, and aimed at British objectives: the destruction of the Russian Black Sea Fleet.[26] Later in the war, operations at Kertch and Kinburn reprised the success of Bomarsund, while the three day bombardment of Sweaborg proved that the other littoral warfare method, using massive firepower without troop landings to destroy naval facilities, was equally effective.[27]

The British celebrated their success in the Russian War with a massive fleet review and mock coastal bombardment at Spithead. The reality of British power in the 19th century was expeditionary power projection, not fleet battle.[28] The Russians quickly assimilated the strategic lessons. Small isolated fortresses that could not resist British amphibious power were destroyed, while efforts to improve the fortress at Cronstadt were redoubled. That said, coastal defences had served the Russian state well. They delayed a British attack on Cronstadt for two years. As Russian strategy traded space for time, the purchase of one or two years was worth almost any price.[29]

Although the post-Crimean War Royal Navy had the experience to plan and conduct extensive expeditionary operations, this living legacy gradually faded away. Yet the Crimean War remained critical to the development of British strategy. While the military aspects of the Crimean campaign were the primary focus of study, the Baltic operations of 1854 and 1855 provided important elements for the development of naval theory. Many key figures witnessed the campaigns at first hand: historian John Laughton served both seasons, as did Philip Colomb, while John Fisher never forgot the Baltic.[30]

While lecturing on history at the Royal Naval College in the 1880s Colomb realised that the Navy had not given much time to strategic issues: it had spent the years between the Crimea and the naval revival of the 1890s mastering the complexities of new and ever changing technology rather then seeking the eternal verities of the naval art. In an effort to reverse this trend Colomb would 'attempt to comprehend laws that governed Naval Warfare, and assess how far recent events may have modified them'. He observed that many of his contemporaries believed recent developments had rendered the experience of the past irrelevant, which was not the case with military studies. For this he blamed the lamentable state of naval history, most of which he condemned as either a simple chronicle, or unalloyed self promotion. He was convinced that 'no state of mind could well be more dangerous for this empire.'[31] Excluded from this excoriation of naval history was the work of John Knox Laughton, pioneer of modern naval history, whom Colomb quoted extensively.[32]

Colomb's strategic concerns were obvious, and the second half of his 1891 book *Naval Warfare* was devoted to expeditionary and coastal warfare.[33] For Colomb the struggle for command of the sea was merely the precursor to the projection of power, either to wipe out the remnants of the enemy fleet, or to use the Army to strike deeper inland. The impact of this book outside Britain was particularly revealing. The Russian and later the Soviet navies based much of their thinking on Colomb because he, unlike Mahan, demonstrated the offensive dynamic of British naval power.[34] The leading pre-1914 Japanese naval theorist shared this preference for the Colomb model, as distinct from that of Mahan.[35] The war between Russia and Japan was fought using expeditionary concepts, with the Japanese Army being used to drive the Russians away from the coast, and to destroy the Russian First Pacific Squadron, before the Second Pacific Squadron (the Baltic Fleet) could arrive in theatre. Colomb concluded his pioneering study with the typically verbose, but entirely correct assessment that:

> *under certain conditions – command of the sea, sufficient and well-handled land forces, landings either away from the batteries, or after their fire has been temporarily silenced – proper appliances and small vessels have always been necessary to secure the success of territorial attacks, and that there is at least nothing in recent times, to show that the rule has in any way changed.*[36]

This was not accidental. For Colomb the ultimate aim of naval warfare was the projection of power ashore and, for that, mission specific assets were essential. This analysis would be validated time and again in the 20th century when attacks conducted without specialist equipment, from landing craft to inshore fire support capabilities, were driven off, or only succeeded at fearful cost. This was all too clear at Gallipoli. Colomb's book had been inspired by the absence of any examination of the strategic significance of naval warfare, which he contrasted with Edward Hamley's *The Operations of War* of 1866. Hamley served on the teaching staff at the new post-war Army Staff College at Camberley, where his book became the recognised text book. The success of the well organised, educated and historically aware Prussian Army in 1870–71 had focused the attention of the British Army on strategy. While Hamley recognised the value of sea power in providing the Army with various options for the basing of overseas operations, and while for him retaining the initiative was of 'especial importance for a maritime power'[37], his remarks were brief, and his colleagues preferred to dwell on Continental models.

After the salutary experience of amateur command in the Crimea the Army acquired new professional models from the Continent. How far this reflected the lack of a modern 'British' military literature, and how far a desire to emulate Continental military systems is unclear. Whatever the cause, the result was an increasing tendency among educated soldiers to look to the Continent, and especially Germany, as an ideal. This approach ignored or at best undervalued the critical naval contribution to British expeditionary campaigns, the 'Small Wars' of Queen Victoria's reign.[38] Without the Royal Navy it is hard to imagine how the British Army would have found half its war zones on a map, let alone completed its campaigns. In 1897 British strategy remained maritime and expeditionary, as Major Charles Callwell's pioneering study of expeditionary warfare for soldiers recognised.[39] While Callwell had distilled experience into common prudence for his army colleagues, he was ignored. By contrast his earlier text *Small Wars* had been a best seller. It took the harsh lessons of the Boer War (1899–1902), to interrupt the Army's continentalist dreams. In 1904 a Royal Marine Officer observed that the post-war Staff College course:

> *had an air of reality, which was missing altogether from the Staff College in the days when I was a student there (1889–1891), and it used to be assumed in all our schemes and Staff Tours that armies numbering hundreds of thousands were fighting in the South of England, our navy being either absent or non-existent; that our nation could exist with our sea communications insecure; and that we had a field army on the Continental scale for home defence, which in those days we had no prospect, or intention of maintaining.*[40]

Evidently the pre-1899 British Army had not troubled itself to think about the strategic needs of a global, maritime trading empire with extensive extra-European possessions and interests which it had been frequently called upon to defend, secure, or extend. It preferred to study strategy as a universal Jominian scientific system. There was one significant exception to this attitude: Colonel G.F.R. Henderson (1854–1903), Staff College educator and historian.[41] Henderson was not an uncritical admirer of German thinking: he demonstrated that British ideas on tactics and strategy were often superior, and certainly better suited to national needs. Having seen the British strategic system in action Henderson emphasised the maritime core of British strategy, notably the subordination of the Army to maritime considerations. He preferred to study the American Civil War rather than the German Wars because it had been heavily influenced by sea power, giving it 'very special value' for British students.[42] He was equally clear that while the armies of Continental Europe could calculate to the last train journey where their next war would unfold, British strategy was almost entirely reactive. It would be impossible and unwise to guess where the next war would break out. 'Except for the defence of the United Kingdom and of India much remains to be provided when the Cabinet declares that war is imminent.' Furthermore: 'Each expedition demands special equipment, special methods and special tactical devices, and sometimes special armaments.'[43]

Britain needed historically derived strategic doctrine which addressed all possible options, rather than a prescriptive campaign plan with detailed mobilisation plans. The 19th century Royal Navy had kept mobilisation and planning distinct for this reason. Suitably equipped British forces would derive 'enormous strategic advantages' from a sea base, and Henderson believed the advantage had been greatly enhanced by modern technology.[44] In British strategy, Henderson argued: 'the naval strength of the enemy should be the first objective of the forces of the maritime power, both by land and sea.'[45] It is unnecessary to look any further for the origins of John Fisher's bon mot: 'the British Army is a projectile to be fired by the Navy'[46], or for his strategic assessment of the British Empire. Whether he read Henderson's book, or relied on his strategic advisors to provide such nuggets is less clear. He had certainly read Mahan's second *Sea Power* work[47] by 1903 when he advocated that the Army should direct its energies to perfecting rapid embarkation in the fleet 'for the only purpose for which the army will ever be required in war, which is for rapid military expeditions for the capture of the enemy's strategic points and Colonies.'[48] In 1908 he declared 'We want a sea-going army that we can launch forth anywhere AT AN HOUR'S NOTICE! *Not 6 months!*'[49] This vision was not shared by the Army, which wanted no part in such maritime campaigns. It was bad enough to be humiliated by Boer farmers without being subordinated to John Fisher's oceanic world view.

The Boer War and the influence of Henderson's teaching opened a brief golden age of inter-service co-operation, with naval officers lecturing at the Staff College from 1901, while Henderson's successor as Professor of Military Art and History, Lieutenant Colonel E.S. May, RA,[50] visited Greenwich to expand on the roles of the two services in Imperial Defence. The Navy then had no equivalent to Camberley. In 1904 Lieutenant Colonel George Aston, RM, with his old friend Staff College Commandant Henry Rawlinson, arranged a lunch with Colonel Henry Wilson of War Office Staff Duties Division and Admiral Edmond Slade, Director of the new Naval War Course.

They agreed to establish officer exchanges between the two courses, which began in 1906. They also set up annual combined staff tours, which necessarily included amphibious practice, the obvious key element in expeditionary warfare.[51] The Annual Manoeuvres of 1904, attended by many officers from the Staff College and the Naval War Course, involved the landing of a hostile division between Clacton and Colchester. The Manoeuvres were analysed by a high-powered joint War Office-Admiralty Conference. Their report of 1905 provided clear principles for a new age of truly combined operations. The report set out the conceptual framework: a sufficient command of the sea was requisite, and such expeditions were only to be planned against colonial territories, not the European mainland. Expeditions would comprise two divisions and a cavalry brigade. The force would have to take all the necessary landing craft, some of which should be procured in peacetime, in a heavily escorted convoy. Opposed landings were effectively impossible against modern weapons, however.[52] Despite the downbeat conclusion and limiting concept, the Clacton Report contained much practical insight, some history and the most recent experience of large scale troop landings. It offered the basis for a serious doctrinal primer on expeditionary warfare. Yet it had no significant impact within either service. The Army was already looking to the Continent, and John Fisher's Admiralty was seeking naval solutions to the strategic conundrums of using sea power to defeat a state like Germany that had no worthwhile overseas interests. In truth the report arrived a year after the rules had been changed by the Anglo-French Entente. At a stroke the new relationship with France removed all the worthwhile strategic targets for expeditionary warfare, mainly French cruiser bases, replacing them with the slight and useless fabric of the German colonial empire. It appeared that expeditionary warfare had become irrelevant.

Despite this the Secretary of the new Committee of Imperial Defence (CID), the navalist soldier Sir George Sydenham Clarke, tried to create a national strategy that combined the two services in a maritime program, avoiding the twin dangers of 'blue water navalism' or conscriptionist homeland defence. Clarke's approach influenced the thinking of Julian Corbett, who had already declared 'The real importance of maritime power is its influence on military operations'.[53] Corbett's next book *England in the Mediterranean* of 1904 was developed from his Naval War Course lecture series of 1902, and credited Clarke with the key idea of showing how the Navy had exerted influence in a major theatre across a long period, neatly combining the twin strands of Corbett's career, historical scholarship and higher education. Corbett's 1903 lecture notes for 'England in the Mediterranean'[54] and the War of the Spanish Succession stressed 'object of course to direct & assist private study', because 'practical advantage of historical study is the chief value of a method of studying strategy.'[55]

Two years after Corbett began lecturing to the Naval Course Admiral Slade passed on a request that he lecture at Camberley, 'as I think it very important that sound ideas should be inculcated in the Staff College. They are quite ready to receive them.'[56] Colonel Rawlinson proposed he study 'the Function of the Army in relation to gaining command of the sea, and in bringing war with a Continental Power to a successful conclusion.' The course delivered in November 1905 examined 'how we can confine enemy's strategy if we are acting with an ally as in 7 years war'.[57] Corbett's approach was heavily influenced by Henderson's posthumously published essays, and these introduced Corbett to Clausewitz. In his 1905 lectures to both naval and military courses, published as *England in the Seven Year's War* in 1907, Corbett used a Clausewitzian analysis of a major conflict as a template for the development of contemporary strategy. He followed Henderson in explicitly stressing the primary importance of naval objectives in British combined operations. Corbett lectured on combined operations at Camberley or Aldershot in 1905, 1906 and 1907, and kept up his annual series with the Naval War Course down to 1914. These lecture series and the related monographs met the educational agendas set by Clausewitz and Henderson. Their creation required a sustained research program, however, much of it drawn from original archives.[58] This originality was vital, for old research reflects old circumstances and old questions. Only a fresh examination of the evidence can hope to produce contemporary insight.[59]

Consequently while the considered development of Corbett's ideas, *Some Principles of Maritime Strategy*, followed Colomb in placing combined operations at the end of the study, the treatment was altogether different. In 1911 Corbett was anxious to stem the tide of Army General Staff driven alarmism, which made much of the danger of invasion. Corbett used his work to deprecate the possibility. In a work prompted by Fisher, Corbett stressed the risk any invader would face when trying to launch a full scale invasion in the presence of a superior Royal Navy. Consequently he devoted little attention to the positive aspects of combined operations.[60] A rather better grasp of Corbett's views can be found in his lectures at Camberley and to the Naval War Course between 1902 and 1914, which emphasised the amphibious core of British strategy. These lectures formed the basis of his monographs, and his work on naval war plans. He had always seen that the object of sea power was to apply pressure ashore, and recognised that navies facing no challenge at sea, as in the Crimean War, could do far more in this respect than those tied to a close blockade of a powerful enemy fleet.

Rawlinson had added naval problems to the Staff College curriculum, but by 1907 the Army was focused on the political and military situation in North-West Europe. Everyone at the Staff College knew where the next war would be fought, and many had visited the area to check the German railway preparations.[61] The inevitable war was being prepared for, but the army that was required for such a war was very far from the one that the Liberal Government wished to acquire.[62] The Staff College syllabus still included a broadly Imperial program, and in Julian Corbett had the ideal visiting lecturer on the subject, but the centre of the British world view had shifted even as Aston, Rawlinson and Slade were building their expeditionary bridges in 1904. The Anglo-French Entente of 1904 and the humiliating defeat of Russia by Japan removed the pre-eminent threats to the British Empire, and replaced them with Germany, which had no power beyond Europe. Britain needed to make a gesture of solidarity with France in the face of German bluster during the First Moroccan Crisis of 1904–05, and used the promise to dispatch an 'expeditionary' force for that purpose. The Army seized on this opportunity to establish an independent mission, and whatever the Staff College was doing between 1904 and 1914 it did not produce a clear grasp of expeditionary warfare.

When Prime Minister Arthur Balfour established the CID in late 1902 to 'co-ordinate' departmental planning he was merely confirming naval dominance. The diplomacy of the Entente required a revised strategy, however, one relevant to a continent dominated by alliances, rather than a single-handed Imperial struggle. Against Germany something more direct would be required. In the autumn of 1904 the First Sea Lord, Admiral Fisher, pressed for the establishment of a CID sub-committee to consider offensive operations against major powers. Military co-operation was limited to the figure of 15,000 men left over from the previous round of joint planning, which had been directed against the French colonies. Fisher hoped to secure Admiralty control of national strategy. This was ambitious, but not unreasonable.[63] In September 1905 the Army even proposed landing 120,000 men in Schleswig, to divert German forces from the French frontier. Within weeks the idea was dropped; the Army claimed it would not divert enough German troops.[64] When Callwell, now a colonel in the Military Operations Branch at the War Office, repeated his expeditionary message of 1897 in 1905, his plea for an amphibious army and a navy with a dedicated coastal support and amphibious fleet following the successful Japanese model was ignored.[65] His case was probably not helped by a hasty, repetitive and often inaccurate text, but more significantly he was now fundamentally out of step with the Army High Command. Callwell resigned his commission in 1909. In a period of only eight years national policy had undergone a major change.

The establishment of the General Staff in early 1906 formalised military opposition to a maritime national strategy by stressing the 'Continental' role. The CID endorsed the plan, but only as a short-term necessity. While the 160,000 man Expeditionary Force was organised to fight in France, this role was never made public, nor was it realistic, with only 74 battalions, a limit that reflected arbitrary financial limits and political realities. Consequently, 'there was a yawning chasm between Britain's chosen strategy and her actual military organisation or long-term planning.'[66] The 'Continental Commitment' was adopted because it was cheap and met a temporary policy need.[67] Cost determined the Government decision to adopt the Entente policy, which deprived the state of its strategic independence. Despite being an 'Expeditionary Force' the pre-1914 British Army did not want to think about expeditionary warfare, only about one specific expedition. This unduly prescriptive approach became, as was intended, self fulfilling. Following Michael

Howard's persuasive book it has been customary to see the 'Continental Commitment' as necessary and inescapable.[68] Whether or not this was the case, the General Staff failed to consider any alternative scenario. This proved to be a crucial error. As Corbett observed in late 1914 'it fettered our freedom of action'.[69] The General Staff had subverted the existing maritime strategy, without being able to provide an alternative. In 1914 the British Army was, in German eyes, 'contemptible' and the Navy, shorn of its expeditionary assets, irrelevant. There may be something for the future in this example. In 1914 the consequence of narrow vision and false assumptions proved catastrophic.

Principles

As an insular power with a global empire, wholly dependent upon seaborne trade for food, wealth and security, the British *had* to make littoral warfare a principle strategic option. Naval dominance was the pre-requisite for any successful British strategy, while the British could not use their army against an enemy without going overseas.

Littoral operations were a key element in British strategy. Their significance went far beyond the littoral area, their purpose invariably being related to the full range of strategic missions that the British might require. Where other powers used expeditionary operations at the margins of their strategic planning, they provided solutions for most of the strategic problems that the British faced, notably securing or sustaining battlefleet sea control, securing insular bases to improve strategic posture, intervening in European conflicts, acting in support of friendly powers or commercial communities (notably at Lisbon and Istanbul), recovering debts, destroying pirate strongholds, and opening commercial arteries.

The primary aim of littoral warfare in a purely British strategy, when used against a major power, was invariably the destruction of enemy naval resources, fleets, naval bases and dockyards. The object was to improve or perfect command of the sea, to facilitate ever more effective power projection operations, and to bring home to the enemy the cost of the conflict. The method of attack could be varied, however, according to circumstance, resources and opportunity. There were always two approaches to expeditionary littoral warfare in British strategy.

Expeditionary operations involved the use of significant bodies of troops, marines or naval detachments for extended service ashore. These were normally raid or siege operations aimed at coastal towns or arsenals. The use of troops preserved the fleet to meet an enemy fleet. Where the enemy fleet was largely or wholly ineffective, as in the Russian War, naval resources could be employed ashore to a considerable extent, to assist the siege. The 'Naval Brigade' was a classic demonstration of the flexibility of British strategy, recognition that littoral warfare often enabled the vagaries of naval bombardment to be replaced by the more stable platforms of land batteries.

Coastal warfare relied on specialist coastal service vessels or surplus warships, to attack coastal positions and arsenals without risking the fleet. Specialist craft have a long, chequered history, but the gunboats and floating batteries of the Crimean War, the monitors of World War I, and the Landing Craft Gun and Rocket of World War II were all critical instruments for effective littoral warfare. While landing parties could be used, to secure positions vacated by the enemy or to complete the work of destruction, they were optional. This method was preferable where the enemy had an overwhelming local advantage in military manpower, and the objects did not require the occupation of ground, however temporary. The modern case would replace guns and mortars with aircraft and cruise missiles. Significantly George Aston was among the first to see the potential of aircraft to solve the age old problem of forcing an inferior fleet out to sea, before World War I.[70]

In both cases the fundamental aim of the British was to draw out the enemy fleet by threatening to destroy it, or to seize territory of such value that the enemy would risk its fleet in an attempt to save the target. These two methods might be differentiated as direct and indirect. In the absence of hostile naval assets other high value assets would be targeted to draw out the enemy.

While British practice down to 1854, or even 1867, was based on wide experience, and case studies of the relevant campaigns were available, there was a serious failing at the level of doctrine. No one wrote down and refined the basic elements of coastal and expeditionary warfare in the systematic form that would make the lessons available to naval and military officers who had not specialised in this work. During the 1882 occupation of Egypt serious mistakes were made in stowing the shipping. While much of this experience was recalled in 1899 for the

A British naval landing party. The skills possessed by sailors proved invaluable ashore on many occasions during the 19th and 20th centuries. (D.J. PERRYMAN)

Boer War,[71] it appears to have been forgotten by 1915. The failure to study 19th century experience systematically, and to reduce it to order in a clear doctrinal primer, deprived the Army and Navy of the practical insights and ideas necessary to make a success of the first major amphibious operation of World War I. Ample evidence for this argument is provided by the complete success of the second major amphibious operation – the evacuation of the Gallipoli bridgehead. Failure at Gallipoli[72] left a legacy of despair that would not be fully reversed until 6 June 1944.

Conclusion

The collapse of joint service planning for expeditionary warfare in 1905 ensured that there was no specialist heavy naval ammunition for shore bombardment, no gunnery training, nor amphibious equipment ready in 1914, while the marines who understood the issue, like George Aston, were used as substitutes for regular soldiers.[73] At heart the failure to develop a single, coherent and flexible imperial strategy left the British with neither a continental army, nor an expeditionary force capable of beach, let alone opposed, landings. The world's most powerful navy went to war without the military assets needed to secure key shore positions, be they enemy naval bases, choke points or cities. In the Crimean War the entire strategy had been dominated by the use of the Army to translate command of the sea into absolute maritime dominance, thereby enhancing the impact of the key maritime assets: economic blockade, coastal operations, and concentrated power projection. The failure to record this reality has distorted the history of that war down to the modern age,[74] and finds echoes in the histories of the two services in most major wars.[75] This historical weakness reflects both parochial attitudes and the failure to open the 19th century archive to scholars while the subjects were still fresh. Corbett belatedly recognised that the Crimean War had been very instructive. Without access to the archives and adequate opportunities for research, however, that experience remained locked away, and the critical doctrine primers unwritten.[76]

Throughout the nineteenth century the British Empire was secured against all manner of threats by a sophisticated expeditionary strategy. Because the British had been successful they did not share the need of Prussians and Frenchmen, who had to analyse a recent past filled with defeats and disasters. Consequently British strategy was not written down, as doctrine or history. Instead it was transmitted by experienced officers down to the late 1850s. In 1914 this system broke down. Both the open-ended commitment of the Army to France and the botched Dardanelles-Gallipoli offensive demonstrated a failure to understand the past, or to align strategy with contemporary political realities.

Between 1906 and 1918 Britain needed a national strategy that could sustain sea control and support the balance of power in Europe. The Government elected to solve this conundrum with the Entente system, bolstering France and Russia against Germany, to save money. However, the failure of the Government and the Committee of Imperial Defence to impose a single coherent policy on the Admiralty and the War Office left strategy to be settled at service level – by default. The Army provided a key diplomatic asset, but the failure to explain to Paris and St. Petersburg the limits of Britain's military commitment, allied to the ambitions of the Army to exercise a long matured 'Continental' role led to chaos and confusion in 1914. Britain needed to influence events in Europe, but not at the cost of dismantling the long-term Imperial strategy that had served the national interest since Trafalgar.

This became clear after 1918, when the British Empire returned to a maritime strategy, only to be forced back on to the Continent by Nazi Germany and Soviet Russia, as it had been by Habsburg Spain, Bourbon and Bonapartist France. This was short-term realism, not the profound structural inevitability that some recent historians have professed to see. In fact their arguments were a short-sighted response to the particular circumstances of the late 1960s and 1970s.[77] The emasculation of the Royal Navy, the growing focus on NATO, the European Union and the inner German border persuaded prominent historians to mark the end of Britain as a world power by stressing the irresistible logic of the 'Continental Commitment', and adopt the Army perspective. Not only were they speaking to their own age but they were wrong on both counts. We must not repeat that mistake.

Narrow prescriptionist approaches to national strategy do not work. Wise nations know their interests, and are prepared to defend them. For most Western nations that defence will require mobile, deployable and sustainable expeditionary forces for the foreseeable future. The biggest danger we face today is that we allow the 'expeditionary' impulse to lapse into the same sort of strategic monoculture that held sway in the Cold War, an excuse not to think. Given the lead times and service lives of modern defence hardware it would be wise to procure assets with a degree of inherent flexibility.

Expeditionary warfare is not a universal panacea: it is merely the best answer to a variety of strategic problems that may face states or alliances at certain times. While Britain as nation and empire has a far longer history of 'expeditionary' strategy than any other modern state, expeditionary warfare is not British strategy, it is only a British strategy. There will be times when alternatives are preferable. We might reflect on that before we rush into an overly restricted vision of the future.

2. The Weihaiwei Blockade and Japanese Hegemony in the Far East

Sarah Paine

FOR THOUSANDS OF YEARS, China was the dominant power of the Far East. In the Chinese mind, the Mongols, who created the greatest land empire of all time, had not overturned this tradition; rather, the Chinese claimed as their own Yuan dynasty (1279–1368) the allegedly Sinified branch of the Mongol ruling house. In 1895, however, Japan toppled China from this position of primacy by military defeat in the Sino-Japanese War. Unlike the Mongol conquest, a contest between land forces and cavalry, Japan's victory relied upon its Navy to deliver troops and joint operations to eliminate the Chinese Navy and to sustain forces on land. With the Anglo-Japanese alliance of 1902 – Britain's only long-term alliance between the Napoleonic War and World War I – Britain recognised this reversal in the traditional Far Eastern balance of power and Japan's admission into the select group of great powers.[1] Japan's victory over Russia in the Russo-Japanese War of 1904–05 then confirmed the verdict of the Sino-Japanese War of Japan being number one in the Far East. Even Japan's defeat in World War II did not restore Chinese primacy.

The Japanese blockade in 1895 of the Chinese port of Weihaiwei on the Shandong Peninsula, the ensuing joint attack by Japanese land and naval forces, and the resulting destruction of the Chinese fleet trapped inside had enormous long-term repercussions. Prior to the war, the quality and quantity of Chinese and Japanese naval vessels put both navies among the top ten of their day.[2] Japan's elimination of the modern vessels of the Chinese Navy coupled with extraction of a large war indemnity not only removed China from the list but, more importantly, marked the end of any Chinese attempt to become an important naval power until the 1990s, that is for a century. The war also cost China sovereignty over Taiwan and the Pescadore Islands, marking the beginnings of the two China problem. When Japan acquired the southern half of Sakhalin Island in the Russo-Japanese War, it controlled the island chains, containing China militarily from Sakhalin Island in the north all the way to Taiwan in the south. Only in our own day are we witnessing a serious Chinese attempt to rebuild a world-class navy to reverse the verdict of the Sino-Japanese War. This attempt includes breaking out of its naval containment by claiming not only Taiwan, but also a host of lesser island chains from the South China Sea northward through the Senkaku Islands, claimed by Japan as part of the Ryuku Archipelago, which also includes the key US military bases on Okinawa.

	Imperial Japanese Navy	Beiyang Fleet
Capital ships	4 (all old)	2
Armoured cruiser	1	–
Protected cruisers (including unprotected frigates and corvettes)	7	3
Unprotected cruisers	12	5
Gunboats	7	10
Torpedo boats	26	5–16

Table 2.1. Relative Strengths of the Imperial Japanese Navy and China's Beiyang Fleet September 1894 (Source Evans and Peattie, *Kaigun*, p. 39)

Prior to the Sino-Japanese War, the European powers hardly thought of Japan at all. When they did, they envisaged a backward and remote island most notable for its appealing pottery, alluring hostesses, and exotic horticulture. In the Far East, the Qing Empire encompassing China, Korea, Mongolia, Turkestan, Tibet, and Annam (Vietnam) was considered to be the only power of note. For thousands of years China had held its many neighbours in its thrall, militarily, economically, politically, and culturally. For Western Europeans, the importance of the Far East lay in commerce and the most lucrative part of that commerce was with the Qing Empire. Until 1894 Japan did not figure in Western power calculations.

In 1903, that is, after the Sino-Japanese War and just before the Russo-Japanese War, an Englishman described the change in perceptions: 'One used to look upon Japan as a kind of dolls' house, or, at the best, as a sort of Feast of Girls, where they had wooden images in gorgeous mediæval Oriental dresses, and wonderful miniature sets of every kind of curious furniture and utensils used in an odd country.' He observed: 'The idea that Japan would ever be a factor in the world's politics was too absurd to contemplate – their role was to be absurd, and supply the suburbs with cheap decorations.'[3] William S. Gilbert and Arthur Sullivan encapsulated the prevailing pre-war European views of Japan in their famous operetta of 1885, *The Mikado* or 'The Emperor of Japan,' which showcased Japan as an example of incompetent Oriental despotism. The Sino-Japanese War shattered these Western preconceptions so that Japan has been integral to Western power calculations ever since.

The Course of the War

Japan's blockade of Weihaiwei and annihilation of the Beiyang Fleet at anchor was the last major engagement of the Sino-Japanese War. Japan and China had both intervened militarily to stem the increasing political instability in Korea. Whereas the Korean King had called upon China for help within the framework of the two countries' traditional suzerain-tributary relationship, Japan came to pre-empt

Russia, whose 1891 announcement of the construction of the Trans-Siberian Railway implied a forward policy in the Far East. The Sino-Japanese War set in motion half a century of rivalry between Russia and Japan for empire in northeast Asia that ended only with Japan's defeat in World War II.

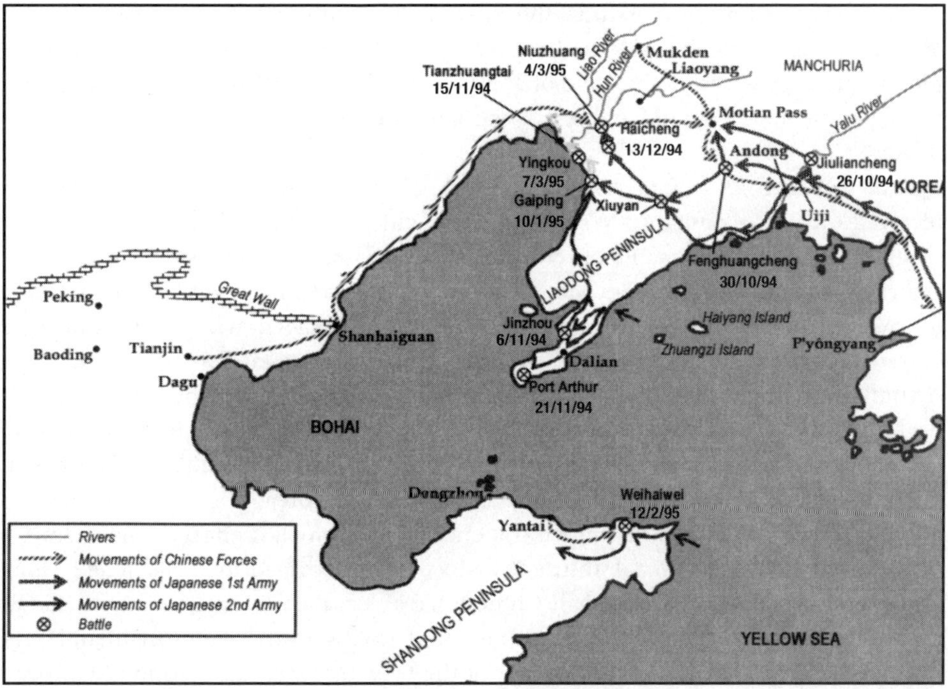

The Sino-Japanese War, 1894–95

The Sino-Japanese War comprised four key battles. During the first pair, the Japanese expelled the Chinese from the Korean Peninsula and cowed the Chinese Navy, while in the second pair, they focused on the land and sea approaches to Beijing. In mid-September 1894 the Japanese rapidly took control of the entire Korean Peninsula with the Battle of P'yongyang (15–16 September 1894) and also seized and retained the initiative at sea with the Battle of the Yalu (17 September 1894).[4] Afterward, the Beiyang Fleet, the modern part of China's Navy, made avoiding the loss of ships its top priority.[5]

The second pair of key battles ensued during the winter of 1894–95. On 21 November the state-of-the-art naval refitting facilities and fortress of Port Arthur (Lüshun), located on the Liaodong Peninsula, fell to the Imperial Japanese Army with hardly a fight. The Han Chinese soldiers filling the ranks of the Qing armies abandoned their positions in droves, unwilling to sacrifice their lives to maintain the minority rule of the Manchus, who constituted perhaps one per cent of the population of the Qing Empire.

The fall of Port Arthur put the Japanese in possession of the northern approaches to Beijing, while control of Weihaiwei would put them in control of the southern approaches. Together the two naval bases on the northern and southern peninsulas defining the mouth of the Bohai controlled the sea approaches to the capital. The Imperial Japanese Army planned to execute a pincer movement on Beijing. After taking Port Arthur, the First Army continued the land campaign in Manchuria to clear the way to Beijing, while the Second Army simultaneously planned to attack the home base of the Beiyang Fleet at Weihaiwei. If successful, the Japanese forces would obliterate the Chinese Navy and have Beijing at their mercy.

The Beiyang Fleet had arrived at Port Arthur in early November 1894, but received orders to return to Weihaiwei. So it did not participate in the defence of China's only facilities capable of refitting severely damaged ships. Worse yet, on the return to Weihaiwei, the *Zhenyuan* (*Chen Yuen*), one of China's two large battleships for which Japan had no counterpart, had been damaged while navigating the entrance to Weihaiwei harbour. In the absence of docking facilities – those at Port Arthur were under imminent attack – the ship had to be beached, rendering it useless for the remainder of the war. Western observers could not understand why the Chinese had bought modern battleships if they never intended to use them. With the best of the Chinese Army destroyed in Korea, the Navy offered an essential means to defend Beijing.[6] But it could do so only if it engaged the enemy.

Before the Battle of Port Arthur a British colonel informed the *London and China Express* that 'a comparatively small Chinese naval force' could make it very difficult for the Japanese to transport large numbers of troops to the Asian mainland.[7] Yet Admiral Ding Ruchang, the commander of the Beiyang Fleet, did nothing to impede the Japanese troop build up in preparation for the assault on Port Arthur. The British editors of *The Japan Weekly Mail* were dumbfounded:

> *When we begin to think what the loss of Port Arthur would signify for the Chinese Fleet, and what the abandonment of the place to its fate would imply under the circumstances, we can not but marvel at China's apparent inaction. Port Arthur is the only dock in north China. Did it come into Japanese possession, the Chinese war-ships would have no place to go for repairs and consequently dare not risk an engagement. Moreover, Port Arthur alone is not invested. The Japanese are holding the entrance to Pechili [Bohai] Gulf...Yet despite its easy accessibility for purposes of relief, and despite the crippling consequences involved in its capture, the Chinese seem resolved to leave it to its fate.*[8]

It was incomprehensible to foreigners how China could possess a modern fleet and then not use it to prevent Japan from supplying its forces. The sea lanes were the lifeline for the Imperial Japanese Army, yet 'ordinary unarmed merchantmen, have been regularly plying to and fro without any escort, and they could have been waylaid and sent to the bottom time after time had China but risen to the occasion', wrote a reporter for the British-owned *North-China Herald*.[9] 'The movements of the Chinese fleet have throughout the war been...utterly and incomprehensibly imbecile...[T]he Chinese fleet has not attempted to meet the Japanese fleet in the open sea, or weighed a single anchor to hinder and debar the unprotected transports of Japan passing to and fro with their freight of eager invaders.'[10]

Soon after the war, Hilary A. Herbert, the US Secretary of the Navy, concluded that, given the well-known and endemic weaknesses of the Chinese Army, 'China had in this war a chance, and only one chance to win, and that lay in her fleet.'[11] To seize this chance required aggressive and daring use of that navy. Instead, China had 'entered upon a losing game of transporting troops to Korea, the battle ground Japan had chosen, in competition with an enemy, whose lines by sea were shorter and whose transports were as three to one.' This resulted in China's defeat at P'yongyang. Herbert continued, 'Japan, having beaten China in transporting troops to Korea, was then allowed to choose her own time and place for a sea fight in the battle off the Yalu.' He excoriated China for 'the untoward results of the unfortunate policy of scattering her war ships.'[12]

The Chinese Rationale for Divided Command

The Qing court hamstrung its officers out of fear that they might turn on their Manchu overlords. Fear of mutiny precluded unified command in the Army or Navy, let alone careful coordination between the two services. Manchu banner forces operated independently of the more numerous Han land forces. Manchu, not Han forces, were deployed to defend the capital. The Beiyang Fleet was the most modern of China's four navies. The other three were the Canton Fleet based at Canton (Guangzhou), the Nanyang Fleet based on the Yangzi River, and the Fuzhou Fleet based at Fuzhou Naval Yard in southeast China. Neither South China armies nor navies came to the aid of the North China forces engaged in the hostilities. Nor did North China armies coordinate with each other let alone with the Beiyang Fleet. This allowed the Japanese to defeat Chinese forces at their convenience and in detail.

The Manchus had deliberately prevented the creation of a unified national army, not out of stupidity but out of sound domestic political calculation. Keeping their subjects divided among themselves was an integral part of the Manchu strategy to maintain their minority rule over a vast land teaming with Han Chinese. Segregating banner forces and preserving Manchu privileges were prerequisites for keeping the Manchu status distinct.[13] Fracturing the command of the far more numerous Han Chinese forces was necessary to prevent the Han from unifying to depose the dynasty.

On the eve of the war, an article in the German press focused on the lack of a unified command. Each of the provincial armies was the personal creation of that province's governor. 'It is naturally in the interest of each [provincial] Viceroy to retain the fruit of his exertions for himself; in no case is he inclined to come to the assistance of a neighbour who is worse provided, and incur the danger of denuding his own province, for whose safety he is responsible with his head.' The same system of individual responsibility applied down through the military ranks. It squelched initiative and promoted narrow-minded rather than creative action. 'By this system, common action is virtually excluded.'[14]

This article came closest to pinpointing the predicament of Chinese officers in the field. The Qing legal code devoted much attention to regulating the Army. The code required officers to depart for expeditions on schedule and to engage only in previously authorised troop movements. The punishments for failure to do so were severe, ranging from military exile on a remote frontier to enough cudgel blows to kill an ordinary man.[15] Those who failed to hold their positions against enemy attack could be decapitated.[16] Destruction of arms issued by the government would also incur a potentially lethal number of cudgel blows.[17] In an era of slow communications, such restrictions made it difficult for commanders in the field to adjust tactics to reflect changing circumstances. Moreover, the regulations precluded an orderly retreat since commanders could not destroy weapons to prevent their falling into enemy hands, nor could they abandon a position without fear of decapitation.

For Admiral Ding, it must have become clear that if he intended to keep his head, his priority would have to be convoy duty rather than offensive action or troop transports. In this he was following common sense and orders. After the Battle of the Yalu, he and his superior, Viceroy Li Hongzhang, the creator of China's modern naval forces and key Han moderniser, both made avoiding further losses of ships their top priority.[18] Ding's instructions throughout the war remained to defend the coast of the Bohai from Weihaiwei to the Yalu River – in other words, to protect Beijing and the Manchu government ensconced there. This strategy wasted the Beiyang Fleet on convoy duty instead of interrupting the transport of Japanese troops to the war theatre.[19] From the Manchu point of view, priority number one had to be the protection of the dynasty. Since they could not assume the loyalty of their subjects, they had to deploy troops defensively. Their most dangerous enemy was not necessarily the Japanese. Rather, the most menacing danger lay in any incipient Han Chinese nationalism.

Preparations for the Battle of Weihaiwei

In December 1894, immediately after the fall of Port Arthur, the Imperial Japanese Navy sent a naval detachment to ascertain the whereabouts of the Beiyang Fleet. This reconnaissance mission confirmed that the fleet remained anchored at Weihaiwei, provided the complete inventory of the ships there, discovered that the harbour entrance had been mined, and concluded that the land defences would have to be neutralised before the Imperial Japanese Navy could engage the Beiyang Fleet.[20] The Japanese continued to monitor the Chinese telegraph traffic, whose code they had broken before the war.[21] From this and other intelligence they understood that Shandong province had minimal land defences. The governor had made no preparations to defend the coastline, only the port of Weihaiwei.[22] In China, provincial governors, not the central government, and civil, not military, officials were ultimately responsible for the defence of their province.

In January 1895, as the Japanese continued their march from the Korean border through Manchuria toward Beijing, they prepared for a joint attack on Weihaiwei. The Second Army divided its forces. The Imperial Japanese Navy was very busy during the third week of January. It simultaneously blockaded the Beiyang Fleet in port, transported the entire Second Division and most of the Sixth Division from the Liaodong Peninsula in Manchuria across the Yellow Sea to the Shandong Peninsula, and launched a diversionary attack on Dengzhou.[23] Dengzhou was located to the west of the treaty port of Chefoo (Yantai), which in turn was located to the west of Weihaiwei, while Japan's amphibious landings were taking place thirty miles east of Weihaiwei on the easternmost tip of the Shandong Peninsula.[24]

Unlike at Port Arthur, where the goal had been the capture of the naval facilities for their present and future use by Japan, at Weihaiwei the goal would be 'the destruction of the fleet inside'. This would give Japan the naval supremacy after the war to dominate the Far East.[25] Japan's wartime successes had demonstrated the critical importance of sea power. Without command of the sea, Japan could not have deployed its troops at will or supplied its expeditionary forces. Therefore, a key war objective became the long-term neutralisation of Chinese sea power.

Japanese land forces were under the command of General Ōyama Iwao (1842–1916), who had played a major role in Westernising and modernising the Imperial Japanese Army in the 1870s and 1880s and who had served as war minister from 1880 to 1891 and again from 1892 until October 1894, when he took command of the Second Army. A decade later in the Russo-Japanese War, he would serve as supreme commander of the forces in Manchuria.[26] Admiral Itō Yūkō (1843–1914), commander of the Combined Fleet, had been intimately involved in the evolution of Japanese naval tactics since the 1880s.[27] Both men had participated in the civil war that had overturned the Tokugawa Shogunate (1603–1867) and in the ensuing reforms that had Westernised Japanese political, military, economic, educational, and legal institutions during the Meiji Restoration (1868–1912).

Similarly, 71-year-old Viceroy Li Hongzhang (1823–1901), who oversaw China's land and sea forces during the Sino-Japanese War, was China's foremost moderniser. He had devoted his career to upgrading his country's armed forces and creating the necessary infrastructure and industrial base to support them. This put him at the helm of what became known as the self-strengthening movement in China. He controlled the Beiyang Fleet through his subordinate Admiral Ding Ruchang (1836–1895). Their association went back to the Taiping Rebellion (1850–64) when they had each led land forces to victory in the civil war. Both men's formative military experiences had thus been in the army and forty years in the past.[28]

The defences of Weihaiwei were concentrated in three areas: the two islands that blocked the centre of the harbour entrance, the high ground on the mainland overlooking the northwestern harbour entrance, and the high ground on the mainland overlooking the southeastern harbour entrance. These fortifications had the best artillery available and should have been extremely difficult to overrun.[29]

Li's land forces at Weihaiwei consisted of three armies, two of which had been augmented by soldiers recently trained in artillery. The Gong Army (2700 men) plus 500 new artillery specialists under Regional Commander Liu Chaopei defended the southeast, while the Beiyang's Hu Army (2000 men) under Regional Commander Zhang Wenyi was responsible for the harbour islands. The Sui Army (3100 men) had responsibility for the northwestern defenses of Weihaiwei. It was under the command of Circuit Intendant Dai Zongqian (1842–95), a long time associate of Li's who had served as one of his key military and naval strategists.[30] Dai also commanded 1500 newly trained artillery men. This disposition of forces anticipated an attack from the northwest, from the direction of Dengzhou where the Japanese would make their feint, not from the direction of Rongcheng in the southeast where the Japanese actually landed.[31]

Admiral Ding had closed the harbour with booms to prevent any unwelcome visitors, while the Japanese had laid contact mines and patrolled the entrance to prevent any possible escapes.[32] This left the Beiyang Fleet bottled up in port.[33] Once the hostilities began, Li Hongzhang's naval commander-in-chief, Ding, and his army commander-in-chief, Dai, virulently disagreed on military strategy. Ding wanted Dai to take more aggressive actions on land, but Dai had little confidence in the ability of his troops to concentrate let alone to fend off the Japanese.[34]

Japanese Joint Operations in the Battle of Weihaiwei

The Chinese defensive strategy assumed a futile Japanese naval attack on the heavily defended harbour. Unlike China, which did not coordinate the deployment of its numerous land forces, let alone unify their command, the Japanese relied on a joint strategy. Their success depended on carefully coordinated army and navy movements in order to overcome China's overwhelming superiority of numbers, resources, interior lines, and strategic depth. The Japanese had no intention of taking Weihaiwei by sea, but would take it by land as they had done in Port Arthur. The Navy would play the critical logistical role of delivering troops and supplies, and the equally important naval role of blockading the Beiyang Fleet in port to prevent its escape. Just as Captain Alfred Thayer Mahan, USN, had argued in 1890 that the British should have blockaded the French fleet in port during the American Revolution in order to neutralise the Franco-American alliance, Japan actually succeeded in blockading the Chinese fleet in port, and with the strategic effects anticipated by Mahan.[35] Japan would obliterate the Beiyang Fleet to guarantee its command of the sea.

The attack on Weihaiwei began with a diversionary bombardment of the town of Dengzhou on 18–19 January 1895 in order to permit the safe landing of Japan's land forces at Rongcheng.[36] Japanese troop carriers left Dalian between 19 and 22 January, landing a day later on the Shandong Peninsula, where the Army divided into two parts. One headed westward toward Weihaiwei by the coastal road, while the other went by a parallel road about four miles inland. While the Japanese had attacked P'yongyang on the autumn equinox, a major Chinese festival, they began their advance on Weihaiwei on New Year's Day, 26 January.[37] The annual celebrations and sacrifices for the lunar new year were meant to ensure good luck for the coming year and made it the most important holiday in China.[38] These choices for battle dates made sense both militarily and psychologically. Inauspicious defeats on the most auspicious days of the Chinese calendar cast doubt on the legitimacy of Manchu rule.[39]

On 30 January 1895, despite the bitterly cold weather and snow-covered ground, the Japanese launched a three-pronged attack and soon took the main forts to the south and east of Weihaiwei. The next day they attacked the forts in the immediate vicinity of the city.[40] This shattered the morale of the Chinese land forces so that the Japanese entered the town of Weihaiwei on 2 February to find the garrisons abandoned.[41] As the Imperial Japanese Army approached the city, even the Chinese hospital staff had fled, leaving a few foreigners to take over the medical service.[42]

Admiral Ding had only succeeded in having a few of the forts surrounding the harbour destroyed before the Japanese took them on 2 February. The conquerors soon trained the repairable guns on the remaining Chinese positions and on the fleet stuck in the harbour.[43] On the night of 3 February, the Japanese tried unsuccessfully to remove the booms blocking the harbour entrances. The following night, they were successful. Two squadrons of torpedo boats entered to face heavy Chinese fire. As two of the Japanese boats opened fire in a diversion, the others disabled the flagship. The next night, a squadron of Japanese torpedo boats repeated their attack on the Chinese men-of-war trapped in port, disabling two other warships and a transport.[44] On 7 February the Army and Navy launched a joint attack on Weihaiwei.[45] In response, the fifteen seaworthy Chinese torpedo boats mutinied and tried to run the blockade.[46] Only one escaped. The others were sunk or driven aground. Later Japan was able to salvage five of them and integrated them into the Imperial Japanese Navy.[47]

The Capitulation

The fall of Port Arthur had been the Chinese Empress Dowager's birthday present from Japan. Li Hongzhang's turn came at Weihaiwei. On February 12, three days short of his seventy-first birthday, Weihaiwei fell to the Japanese.[48] At the surrender, Admiral Itÿ wrote a letter urging his old acquaintance and former friend, Admiral Ding, to come with him to Japan. 'It is not the fault of one man that has brought China into the position she now occupies. The blame rests with the errors of the Government that has long administered her affairs.'[49] In fact, Ding had already committed suicide. Admiral Ding had a reputation for personal bravery, but was out of his element. As a cavalry officer by experience and training, he knew little about naval affairs and was not popular with the sailors.[50] Most of his officers came from the southern province of Fujian, while the admiral had been born to the north in Anhui province. Military service did not overcome such divisions in China where regional loyalties remained paramount.[51] Admiral Ding had long resisted capitulation and had sought death in action by repeatedly standing exposed during bombardments. According to a foreign adviser, 'Ting [Ding] declared at first that capitulation was impossible; but later he said he could arrange it by committing suicide, and so save the lives of many.'[52] At the very end, when mutinous sailors threatened him with knives, he went to his cabin to take a lethal dose of opium.[53] Two captains followed his example. A third, the second in command aboard the flagship, shot himself as the Japanese boarded.[54]

Admiral Ding had little choice but to take his own life. Over the summer, the Guangxu Emperor had demoted him for not keeping the Japanese Navy out of the Bohai.[55] After the fall of Port Arthur, the Emperor had degraded him again and had threatened to turn him over to the Board of Punishments. But for the personal intervention of Viceroy Li Hongzhang, this could have meant execution.[56] With the destruction of the Beiyang Fleet, China's only credible military force against Japan, death awaited him in Beijing. His family would also have suffered. By committing suicide, he chose an honourable death, which transformed him into a tragic war hero in both Japan and China.[57] The Japanese admired his final act because it was consistent with their military code of *bushido* or the Code of the Samurai, which called for ritual suicide in the event of military failure.[58] The Japanese accorded full military honours to those Chinese officers who had taken their own lives, and freed their men after confiscating their weapons.[59]

Western observers could not understand the Chinese naval strategy. They wondered why Admiral Ding had refused to leave Weihaiwei before the Japanese had blockaded him in harbour and why, once stuck there, he had not sunk his fleet rather than permitting it 'to fall bodily into the enemy's hands.'[60] His four surviving warships and six gunboats became a direct contribution to Japanese naval expansion.[61] In fact Admiral Ding had tried to scuttle his fleet at the very end, but by that time his crew had mutinied and refused to carry out the order.[62] He had also ordered the destruction of the mainland forts only to be overruled by a Chinese general, who 'declined the responsibility, declaring it might cost him his head.'[63] The general's decision reflected a system that punished officers who altered plans and sought individual scapegoats to blame for systemic problems. The Qing legal code mandated decapitation for the destruction of twenty firearms or more.[64] One can only speculate on the punishment for dynamiting an imported artillery piece, let alone for scuttling an imported capital ship. Ding did manage to dynamite the engine room of his great battleship, so that his best remaining ship did not fall to the Japanese in a repairable condition.[65]

Le Temps of Paris later provided an accounting for the Chinese Navy. During the war, 22 of its war vessels had been destroyed and 12 had been surrendered to the Japanese to become additions to the Imperial Japanese Navy. In contrast, the Japanese had lost two vessels. For Japan this became a net gain of ten ships.[66] After the Battle of Weihaiwei, the Chinese Government ordered the closing of the Admiralty Board since China no longer possessed a navy. Five million *taels* were discovered missing from the account. Allegedly these had been diverted to the Empress Dowager for her much-anticipated sixtieth birthday celebrations, which were ultimately cancelled for a lack of anything to celebrate.[67]

An article in *The Japan Weekly Mail* summed up the strategic situation after the fall of Weihaiwei:

> *In the present war between the two Eastern Powers, China has been defeated in every battle. Her frontier line has been pushed back, her cities have been captured, and her armies annihilated. The great forts of Port Arthur and Weihaiwei, that guarded the entrance to the gulf of Pechili [Bohai], have fallen, and with them China has lost defences on which she had exhausted the contents of her treasury for more than twenty years. They failed her at the critical moment. The Peiyang Squadron [Beiyang Fleet], too, formed at great cost, fared no better when put to the test. It remained concealed in harbour, and if any of its warships attempted to fight they were burnt or sunk: finally the remnant of the once famous squadron surrendered to the enemy. By the fall of Port Arthur and Weihaiwei, the gateway to the capital of China has been thrown open, and no power in China can stay Japanese entry into Pekin [Beijing].*[68]

At this juncture, the Manchus had no choice but to negotiate. Utter military incapacity meant that a continued Japanese drive to Beijing would result in their overthrow either by the Japanese or if not sooner by some enterprising Han Chinese. The Manchus correctly concluded that their only hope for survival rested with negotiations.

The War and the Naval Doctrine of Mahan

The outcome of the war seemed to vindicate the theories of Mahan, whose publications of the day were catapulting him to international fame. Japan had invested in a great navy, which it had used to take down a great continental power. Without this Navy, the Japanese victory would have been impossible: Japan could not have got its Army to the Asian mainland, much less have sustained it there. Just as Mahan had recommended, Japan had blockaded the enemy fleet in its home port, preventing it from interfering with the war effort. Then the Japanese had obliterated that fleet with the consequences predicted by Mahan: Japanese command of the sea and regional dominance. In practice, this meant the reversal of the traditional Far Eastern balance of power. Foreigners grasped this change immediately.

Four days after the fall of Weihaiwei, the Japanese allowed Vice Admiral Sir Edmund R. Fremantle and about 100 British officers to tour the fallen citadel. 'After their examination of the fortress they pronounced it to be impregnable if any real attempt had been made to defend it by the Chinese.'[69] During the siege at Weihaiwei, the editors of *Le Siècle*, who throughout the war had been very sparing in their judgments concerning the belligerents, ran a front-page article:

> *Like Port Arthur, Weihaiwei was well defended by European-style fortifications, and if the Japanese, instead of being the besiegers, had been the besieged, they could not have been subdued except by famine. But the Chinese soldier is so badly commanded, its leaders so slovenly and so incapable, that even behind the most solid walls its resistance cannot last more than a few hours.*

The article concluded: 'The facility and fullness with which the Empire of the Rising Sun has assimilated the military institutions and the diplomatic customs of our ancient Europe are truly marvellous.'[70] The official newspaper of the Russian Foreign Ministry, *Journal de St-Pétersbourg*, agreed: 'China has no military organisation whatsoever... The Japanese, in contrast, have resolutely begun a war European style.'[71]

The Japanese attack on Port Arthur, November 1894. (AUSTRALIAN NATIONAL MARITIME MUSEUM, SYDNEY)

The Times of London was appalled by the Chinese situation:

> *The actuality of ineptitude stultifies the imagination...China may henceforth be considered incapable of war either by sea or land, and incapable even of learning it... The single word rabble describes the Chinese army, untrained, unarmed, unpaid, unfed, and uncared for – so grossly ill-used, indeed, that the docility of the poor wretches who are sent in droves to the slaughter would be admirable if it carried any intelligence with it.*[72]

A front page story in *The New York Times* proclaimed, 'China is an anachronism,'[73] while *The North-China Herald* concurred, pronouncing it 'doomed to disappearance.'[74]

Such beliefs would have consequences. After the war European countries would reorient their foreign policies to take into account the reversal in the Far Eastern balance of power. For China, the cost of the war went beyond Japan, ultimately to include all of the other powers. Foreigners no longer had any faith that the Chinese could reform their decaying government, so the powers intervened much more aggressively in Chinese internal affairs in what became known as the 'scramble for concessions.' Previously the powers had contented themselves with concession areas in the major treaty ports; after the war they sought exclusive spheres of influence in whole provinces, carving up China at will.

Defeat at the hands of the Japanese, long considered inferiors by the Chinese, caused an irretrievable loss of prestige and legitimacy for the Manchu dynasty not only among the powers, but, more ominously, at home. The Boxer Rebellion (1899–1900) broke out four years later. Although the Manchus managed to deflect rebel hostilities away from themselves on to the foreign community, the latter imposed a massive indemnity on the dynasty. At this time, the Manchus finally felt compelled to follow the Japanese prescription of imposing a far-reaching Westernising reform program modelled on the Meiji reforms. These included the creation of the New Army, a unified and modernised Han Army. Within the decade, however, the New Army turned on the Manchus to topple the dynasty in 1911, showing that the Manchu fear of unified Han armies had been justified.

The fall of the Qing dynasty also seemed to vindicate Mahan who had predicted enormous outcomes for sea power. Once China was trounced at sea, a cascade of unpleasant consequences had ensued. A decade after the Sino-Japanese War, the Russo-Japanese War seemed to reconfirm Mahan's theories when the Japanese sank much of the Russian fleet in the decisive Battle of Tsushima (1905) that brought the war to a rapid close.

The two turn-of-the-century wars fixated the Imperial Japanese Navy on the single decisive fleet engagement. This unduly narrow view of sea power would hamper Japanese naval creativity during World War II, when Japan failed: to target US transports aggressively; to protect its own merchant marine; to emphasise submarines; or, to question the integrity of its codes. In the intervening years, the Imperial Japanese Navy had focused, like Mahan, on symmetrical naval warfare and on the acquisition of capital ships to fight the symmetrical fight.

Missed by Japan were lessons from the Sino-Japanese and Russo-Japanese Wars concerning the value of jointness. The budgetary wars during the lean years of the 1920s and the dire years of the Great Depression left deep animosities between the two services so that the Imperial Japanese Army and the Imperial Japanese Navy fought separate uncoordinated wars after Pearl Harbor. The Army focused on Russia and the China theatre, while the Navy focused on the United States and the

European colonies in the Pacific. Yet cutting lend lease aid to the Soviet Union was a Navy operation, while holding the vast Pacific empire was an Army operation. Both services focused on the symmetrical missions of engaging the conventional forces of the enemy. Neither service thought deeply about the non-military aspects of warfare, such as the hearts and minds of occupied populations or the economic requirements of the home economy, and both missed the problem of strategic over-extension. Had Japan not defeated two vast empires before, first China, then Russia? Luckily for the West, the Japanese also missed the key role played by joint operations in their turn-of-the-century wars. Mahan did not emphasise these either. On the eve of World War I, a British maritime theorist, Sir Julian S. Corbett, would highlight the strategic potential of asymmetric and joint operations for sea powers. Apparently the Japanese did not give much attention to Corbett.[75]

Japanese Hegemony in the Far East

If the Chinese through their military failings brought in a harvest of bewilderment in the West, they sowed a crop of utter contempt in Japan. The pre-war Japanese respect for Chinese high culture vanished without a trace. Foreign Minister Mutsu Munemitsu wrote after the war:

> *Japanese students of China and Confucianism were once wont to regard China with great reverence. They called her the 'Celestial Kingdom' or the 'Great Empire', worshipping her without caring how much they insulted their own nation. But now, we look down upon China as a bigoted and ignorant colossus of conservatism. She, in turn, mocks us as a tiny island of barbarians who have recklessly and impudently rushed forward in a mad effort to imitate the external trappings of Western civilisation.*[76]

By the end of the war intimacy had brought contempt, and the Japanese considered the Chinese and Koreans to be their utter inferiors.[77]

The war left a power vacuum in China. This meant a power vacuum extending to China's tributaries such as Korea. In the traditional Confucian order, China had maintained the peace in Asia. With China gone as a competent regional power, terrible instability ensued within China and among its tributaries. Both China and Korea rejected Japanese primacy – a position they have maintained down to the present day. But neither had a positive program to serve as a viable alternative to Japan's insistence on rapid Westernisation, a course both China and Korea also rejected. In their eyes Japan was the heretic nation that had rejected the Confucian order to Westernise and, in contravention of all rules of propriety, had turned on its benefactor and senior, China, to defeat it in war. The Chinese had assumed Confucianism to be a one-way street; others conformed, no one defected. All of these events constituted a logical impossibility under Confucianism, which put China at the pinnacle of all areas of human achievement. As a result of the Sino-Japanese War, Confucianism as an all-encompassing dynastic tributary system was done. This had far-reaching domestic implications for China. The passing of the Confucian system left not only a regional but also a domestic political vacuum. This soon devolved into a multi-lateral and decades long shifting civil war that started with warlords and ended in a Nationalist-Communist showdown in 1949.

As China imploded, Japan set out on the road to empire. In an 1894 interview, Count Ōkuma Shigenobu (1838–1922), one of the most important political leaders of the Meiji era, pointed out:

> *Those in power may think it a great achievement to have secured the independence of Korea [from China] and to have concluded new treaties with the European Powers [granting Japan full juridical equality]. But the nation...has the grander aim of laying a foundation for the dissemination of its influence throughout the world...[T]he European Powers are in a decline, and... in the course of the coming century a complete prostration will overtake them. Who are to take their place?*

He concluded that their place 'will be taken by our own descendants . . . To sum up . . . with treaties revised, and China completely humbled. . . Japan will have become a Power whose intentions and attitudes must be carefully studied by all other countries of the world.'[78]

As a result of the Sino-Japanese War, Japan acquired Taiwan and the Pescadores. The Russo-Japanese War would snag the southern half of Sakhalin Island and the extensive Russian railway concessions in southern Manchuria. In 1910 Korea would formally become a Japanese colony. Meanwhile, Japanese investments in Korea and Manchuria skyrocketed so that, by World War II, Japan had transformed Manchuria from an economic backwater of China into the most industrialised part of Asia outside the Japanese home islands.[79] The Great Depression would spur Japan to incorporate Manchuria into the Japanese Empire. Like the European colonial powers, the Japanese considered themselves to be on a civilising mission to bring modernity to the Far East. The fullest expression of this by the Japanese came during World War II, with their framework for a Far Eastern Co-prosperity Zone. Yet Japan was never able to quell concerted resistance to its bid for regional hegemony.

The Chinese were no more capable than the Japanese at constructing a stable international order to replace the Confucian system. In China Communism did not fill the philosophical void left by Confucianism because it failed to bring either political stability or economic growth. Even after the Communist reunification of China, political upheaval continued with the Great Leap Forward and Great Famine (1958–61) and the Cultural Revolution (1966–76). Rural stagnation set in until the belated introduction of market-oriented reforms in the 1980s – a de facto rejection of the Communist economic model – but political stability has been elusive. Leadership changes remain highly destabilising events.

Thus, while the Sino-Japanese War marked the death of the old order, it did not deliver a new one, like those created in Europe at the end of the Napoleonic Wars or of World War II. Indeed, the new order remains to be created in the Far East. Japan and Russia have yet to sign a peace treaty settling World War II. Instability on the Korean Peninsula has become more lethal with the spread of weapons of mass destruction. China, Japan, and Korea remain at odds. Divisive territorial issues abound, the most ominous being those affecting vital sea lanes and concerning sovereignty over off shore energy reserves.

Until World War II, Japan maintained its regional primacy militarily. After World War II it has done so economically. Recent economic growth rates – double digit for China and flat for Japan – portend great changes. China appears to be on the rise and intent upon reversing the verdict of the Sino-Japanese War that toppled it from regional primacy a century ago.

3. Maritime Aspects of the Gallipoli Operation

Robin Prior

IN THIS CHAPTER I want to investigate what support a naval attack at Gallipoli had among Britain's decision-making groups in 1915.[1] The conventional answer is that Winston Churchill and no one else wanted to undertake the naval attack. I want to challenge that conventional view.

So far as naval operations were concerned, there were two key decision making bodies – the first was the Admiralty War Group consisting of the First Lord, Winston Churchill, its First Sea Lord, Admiral Lord Fisher, the Chief of Staff, Rear Admiral H.F. Oliver, Admirals Jackson and Wilson, appointed in an advisory capacity, and Captain de Bartoleme, Churchill's naval secretary.

This group had seen many plans come and go since the beginning of the war – a scheme to capture an island off the German coast to force the German High Seas Fleet to sally forth to defend it and then be conveniently sunk by the Grand Fleet; the entry into the Baltic by a British Fleet to land a force of Russians on the North German Coast; the blockading of Holland into the war and so on. All had been rejected, essentially on the same grounds that they risked Britain's main fleet – the instrument, after all, that allowed the Army in France and Flanders to survive and grow – in waters that were sown with mines, in some cases under fire from shore batteries, and where heavy units would not have sufficient room to manoeuvre away from small torpedo craft and destroyers.

In late December 1914 and early January 1915 a series of events had drawn the attention of the Admiralty War Group towards Turkey. This situation had not, it needs to be stressed, been brought about by some obsession Churchill was said to have had against Turkey from the beginning of the war. Certainly the escape of the Imperial German Navy's *Goeben* and *Breslau* and Turkey's adherence to the Central Powers and its position athwart British Sea routes to the East had directed Churchill's attention towards the Middle East. Indeed Churchill's response was to order a bombardment of the outer forts at the Dardanelles in November 1914, but then his thoughts turned northward and the last strategic paper he had written (on 29 December) had attempted to re-focus the Government's mind on the capture of a German island.

But at the same time Churchill's mind was concentrated on northern waters, others turned towards Turkey as an object for British operations. On New Year's Eve future British Prime Minister David Lloyd George had mentioned, among other options, an attack against Turkey; on 28 December Colonel Maurice Hankey, the Secretary of the War Council, suggested a whole series of operations which would weave a web around Turkey, and which would have ended its role as an European power. Fisher, for reasons that are obscure, took up Hankey's paper with alacrity. He suggested that a multi-national army be assembled to attack Alexandretta while a Greek army landed on the Gallipoli Peninsula and a fleet forced the Dardanelles.[2]

Then, on 1 January 1915 the British ambassador in Russia had telegraphed the Foreign Office to the effect that the Grand Duke had informed him that Russian forces were being hard pressed by the Turks in the Caucasus. The Grand Duke had asked if 'it would be possible for Lord Kitchener to arrange a demonstration of some kind against the Turk elsewhere, either naval or military, [to] ease the position of Russia.'[3]

Kitchener asked Churchill for his opinion, making the point that 'we have no troops to land anywhere' and enquiring whether a naval 'demonstration' at the Dardanelles might be the most effective way of preventing more Turkish troops being sent east.[4]

It was in this climate then that the Admiralty War Group met on 3 January. It is unfortunate for historians that no minutes were kept of this meeting but we have a note from the Secretary (Grahame Greene) that the subject of Turkey was brought forward by the First Lord and thoroughly discussed. The enthusiasm for operations against Turkey was so widespread that no one in the Group challenged the telegram dispatched by Churchill to Vice Admiral Carden commanding the squadron off the Dardanelles, later the same day. It said :

> *Do you consider that forcing of the Dardanelles by ships alone a practicable operation? It is assumed older Battleships fitted with mine bumpers would be used preceded by Colliers or other merchant craft as bumpers and sweepers. Importance of results would justify severe loss. Let me know your views.*[5]

In the War Group Fisher was the key player and there is further evidence that Fisher was an enthusiast for operations against Turkey. The day after the meeting of the War Group, he wrote to Arthur Balfour, a member of the War Council, urging him to support 'Hankey's Turkey plan'. Churchill still had an open mind. On 4 January he wrote to the Commander-in-Chief of the Grand Fleet Admiral Sir John Jellicoe that 'everything convinces me we must take a German island'.[6] At the same time he reiterated the importance of this strategy to Fisher and warned him that 'we had better hear what others have to say about Turkish plans before we take a decided line. I would not grudge 100,000 men because of the great political effects in the Balkan Peninsula but Germany is the force and it is bad war to seek cheaper victories and easier antagonists.'[7]

Nevertheless, despite Churchill's enthusiasm, in the next few days not one of the War Group evinced any more enthusiasm for capturing a German island. Then on 8 January, Carden's plan for forcing the Dardanelles arrived.

In truth it was not so much a plan as a list of the order in which the Dardanelles defences would be attacked – starting with the outer forts and working towards the series of forts at the Narrows. The only new factors were requests from Carden for seaplanes to direct the fire of the ships' guns into the forts, the provision of battle cruisers to deal with *Goeben*, and the supply of the large amounts of ammunition required to complete the task.[8]

At last Churchill had some kind of operational plan. He lost no time in placing it before the War Council which met again on 13 January. He emphasised that the guns of the Fleet were more modern than those in the Turkish forts and outranged them. What the Fleet would do when it cleared the Narrows was left rather vague. All Churchill had to say on this aspect was that 'it would proceed up to Constantinople and destroy the "*Goeben*".'[9]

A brief rambling discussion (at which the War Council was expert) followed. Their conclusion was that the Admiralty should 'prepare for a naval expedition in February to bombard and take the Gallipoli Peninsula, with Constantinople as its objective'. Why did the War Council agree so readily to this scheme? This answer is not far to seek.

Most members of the War Council were born between 1850 and 1876. Their formative years had therefore been spent when British imperialism and its main instrument, the Royal Navy, was at its height. If, in their eyes, there seemed little that sea power could not accomplish, it should be no matter for wonder. Moreover, as no naval advisor had yet spoken a word against the Dardanelles operation, they had no peg on which to hang opposition to the project, even had it existed.

What did Admiralty opinion think of forcing the Dardanelles by ships alone? Captain H.W. Richmond, one of the sharpest minds on the Admiralty War Staff, initially thought it 'excellent'. By mid-February, however, he was calling for an army to be landed at Gallipoli to convert what he considered could only be a temporary naval success into something more permanent.[10] It is hard to derive from this exactly what Richmond was thinking but surely the 'local success' of which he spoke was the forcing of the straits by the Navy. In other words he thought the naval attack would succeed but that to force Turkey out of the war a military force would also be needed. It is a reasonable supposition from this evidence that he still thought the ships could get through the Narrows. This was not so much a change of heart as a change of emphasis. Anyway, the issue of troops was a military matter. What did other naval authorities and Fisher think? Oliver seemed consistent in his view. He expressed a wish to command the bombarding squadron and in his unpublished autobiography he stated that he was an enthusiast for the naval attack. In his operation order to Carden he also waxed optimistic. He considered that the mines at the Dardanelles should present no problem and that the forts could be picked off one by one at long distance by the ships without a large expenditure of ammunition. He insisted that the destruction of the outer forts would shatter the confidence of the Turks in their German advisors and might possibly result in the overthrow of the German rule at Constantinople![11]

Vice Admiral Sir Henry Jackson, who of all the naval personalities at the Admiralty studied the Dardanelles problem most closely, and who was the one who issued the final instructions to Carden, is more of an enigma. On the one hand he seemed an optimist. When Carden's plan arrived he stated that he had 'concurred generally with it'. He also thought that the forts could be dealt with at long range by ships at anchor and came up with the formula that ten rounds for guns would be needed to destroy each gun at the forts. He saw no difficulty in clearing the minefields if it could be done at night.[12]

Turkish guns in the Dardanelles were by no means ineffective. This photograph shows the aftermath of a hit on the destroyer HMS *Racoon*. (RAN)

It is worth noting, in passing, that none of these assumptions was correct. After the war the Mitchell Committee calculated that at long distance under perfect conditions a ship at anchor had between a 1 in 200 chance and a 1 in 50 chance of hitting a gun depending whether their range was 20,000 yards or 11,500 yards. Nor could the ships anchor with ease inside the Straits. The current coming down the Straits would cause them to swing at anchor. They would be forced to move away from the mobile Turkish howitzers which lined both sides of the straits. Nor was the minefield problem given proper consideration by Jackson. Of the 17 attempts by the trawlers sent to do this job not one reached the minefields. Before the major attack on 18 March only three mines out of 187 were swept.

So far this shows Jackson was definitely on the side of the optimists such as Oliver. He then went on to qualify his position considerably. At the end of this memorandum he stated: 'Only the provision of the necessary military forces would enable the fruits of the naval attack to be gathered. Transports for carrying them through the straits should be ready as soon as it is seen the forts at the Narrows were silenced'. He concluded:

> *To complete their destruction, strong military landing parties with strong covering forces will be necessary. It is considered, however, that the full advantage of the undertaking would only be obtained by the occupation of the Peninsula by a military force acting in conjunction with the naval operations, as the pressure of a strong field army of the enemy on the Peninsula would not only greatly harass the operations, but would render the passage of the Straits impractical by any but powerfully-armed vessels, even though all the permanent defences had been silenced.*
>
> *The naval bombardment is not recommended as a sound military operation, unless a strong military force is ready to assist in the operation, or, at least, follow it up immediately the forts are silenced.*[13]

This seems pretty clear. Jackson was saying that after all, troops would be needed. But in fact his message was quite ambiguous. He was not stating that the ships could not get through, in fact he was definitely stating powerfully armed vessels will get through. He was just saying that these vessels might not accomplish much of substance. Troops would have to be landed on the Peninsula to:

(a) destroy the Turkish garrisons which will harass naval operations
(b) destroy all the forts.

This would allow troop transports through the Dardanelles to compel surrender at Constantinople.

The problem with this paper, and no doubt the reason it did not carry much weight with decision makers, is that its message at the end was rather neutralised by its optimistic assumptions at the beginning and by the fact that those assumptions led Jackson inexorably to the view that heavily armed ships after all could get through. This message was too subtle in the political climate of 1915 (I would suggest in any political climate) to sway a busy group of civilian decision makers.

Fisher's opinion of the operation is an even deeper puzzle. He approved the initial telegram to Carden and had added HMS *Queen Elizabeth*, the most powerful ship in the world, to the bombardment squadron. But he was clearly uneasy. On the 18th – while not directly criticising the Dardanelles option, he urged Churchill to turn his attention to bringing in Holland – 'which would finish the war'. He then told the First Lord that Jellicoe had sunk into depression because he was losing destroyers to the Dardanelles. These were at best oblique objections. Churchill immediately rectified the destroyer problem by sending an additional squadron to Jellicoe from southern waters. As for how to bring in Holland – he had no more idea of how to do it than anyone else. There is also little doubt that Fisher was not being direct with Churchill. To others he was being much more forthright. On the 19th he wrote to Jellicoe:

> *And now the Cabinet have decided on taking the Dardanelles solely with the Navy, using 15 battleships and 32 other vessels, and keeping out there three battle cruisers and a flotilla of destroyers – all urgently required in the decisive theatre at home! There is only one way out, and that is to resign! But you say 'no' which simply means I am a consenting party to what I absolutely disapprove. I don't agree with one single step taken …*[14]

Here there is a clear sign that Fisher did not approve the naval operation to which the War Council had agreed just six days before. The conventional view assigns great importance to this. If only the War Council had known of the objections of Britain's senior sailor, this knowledge must have been fatal to the naval attack – the Dardanelles as 'ships alone' operations would never have occurred.

The fact is, however, the War Council did know of Fisher's doubts. On 20 January Hankey, who had been aware of them for some time, informed the Prime Minister. Eventually Fisher also informed Churchill directly of his reservations about the operation ('if troops were provided he would have no complaint') so he knew too. Hankey had also told Balfour, so now three members of the Council were aware of Fisher's views. Clearly the lips of these members did not remain sealed because four more members of the Council – Grey, Haldane, Crewe and Lloyd George – told the Dardanelles Committee that they also knew of Fisher's opinion and were aware of Fisher's objections.

On January 28, Fisher had a chance to inform the War Council of his reasons for opposing the operation. When the Council met he understood that the naval attack was not to be discussed that day. He then left the table. He was talked into

HMS *Queen Elizabeth*, the first of five battleships in her class. She was described as presenting the embodiment of power, speed and majesty. (RAN)

returning by Kitchener who must now also be added to those who knew. As he returned he heard Balfour summing up the advantages of the naval attack.

> *It would cut the Turkish army in two;*
> *It would put Constantinople under our control;*
> *It would give us the advantage of having Russian wheat, and enable Russia to resume exports;*
> *This would restore the Russian exchanges, which were falling owing to her inability to export, and causing great embarrassment;*
> *It would also open a passage to the Danube;*
> *It was difficult to imagine a more helpful operation.*[15]

The entire War Council was therefore of a mind to discount Fisher's views. They were, in fact, increasingly aware that the First Sea Lord was showing alarming signs of mental instability. It was known that he had recently threatened to resign over the air defence of London – an Admiralty responsibility but not one of vital importance in January 1915; that he was forced to take long afternoon naps because of the pressure of work; that his strategic views were expressed in language that was becoming increasingly violent and incoherent ('I think in oceans, shoot at sight' was one of his essentially meaningless aphorisms).

There was also another reason. The supposed ubiquity of British sea power had convinced politicians and Fisher, among others over the years, that the Fleet could accomplish almost anything – especially against second rate opponents such as the Turks were considered to be. World history was, in fact, about to enter a phase where naval forces relative to land power were about to diminish. But we can hardly blame the British War Council in 1915 for not being up with the trend and determining that (despite growing concern that troops should be sent) the Navy would indeed try to force the Dardanelles by ships alone.

My conclusion is that expert naval authorities such as Richmond and Jackson would have preferred a combined operation but were confident enough that the fleet could get through the Straits. Fisher later agreed that he had supported the operation because of the 'possible ineptitude of the Turks'. Others at the Admiralty such as Oliver and Wilson always thought the ships alone approach would work. As for the War Council, opinion at the Admiralty and especially Fisher's was not unknown to them, yet they were still enthusiasts for the naval attack. It is ironic that by the time operations started on 19 February one of the sceptics about the naval attack was Churchill – he was not only convinced that troops would be needed to help the fleet through, he considered that the fighting on the Peninsula would be difficult. This is why he tried so hard to have the last Regular British Division (the 29th) sent to the Dardanelles. Yet he too in the end submerged his doubts and let the naval attack go ahead unaided by troops. But so did the War Council and his naval advisers, none of whom resigned over the matter. The blame then for the naval fiasco at the Dardanelles rests on more shoulders than those of Winston Churchill.

4. The Royal Naval Division at Gallipoli

Chris Page

ONE OF MY MAIN INTERESTS in the past few years has been to remind anyone who will listen, and some who will not, of the vital roles played by the Royal Navy in both World Wars, in particular the huge, and largely unsung, effort made by the Royal Naval Division (RND) in World War I. The fact that this was, initially, a totally incorrect use of sailors and Royal Marines has only served to make these studies more poignant.[1] More than 40 per cent of Royal Navy casualties during World War I were suffered not at sea, but on land by the sailors and marines of the RND. More than 45,000 became casualties, of whom more than 11,000 were killed. These figures are in line with those for other elite units. To bring these losses even further into perspective, the average number of men in the RND was rarely more than 4 per cent of the Royal Navy. At Gallipoli, more than 75 per cent of Royal Navy casualties were in the RND, despite the fact that many large ships were sunk.

Most readers will be aware of how the Gallipoli adventure came about, so I will only say here that one of the reasons for the failure of the campaign was the gross underestimation of the fighting qualities of the Turkish infantry. It was this underestimation that led the British to believe that they could capture the Peninsula with only one fully trained Division, the 29th. The other land forces were comprised of the ANZACs, whose fighting power at that time was untested, whose discipline was unusual, and who were viewed without much confidence by British commanders, and the equally dodgy RND. Ironically, all these formations turned out to be among the best in the British Expeditionary Force. But it is fair to say that they were not looked upon as elite units in early 1915.

An Unusual Force

The RND was a highly unusual outfit. The majority of the Division throughout the war were sailors, with naval ranks – leading seaman rather than corporal, chief petty officer rather than sergeant, lieutenant commander rather than major. And yet, almost none of them had ever served at sea. The ratings grew beards, they referred to 'going ashore' when they took leave, they worked naval watch systems, went to sick bay when ill, called the field kitchen 'the Galley', wore naval cap badges with their khaki and their Battalions carried the names of naval heroes. When deployed to Antwerp by rail, there was even discussion as to whether units should disembark from the port and starboard side of the train! Initially the Division consisted of brigades of men from the Royal Naval Reserve, the Royal Naval Volunteer Reserve (RNVR), and the Royal Marines. Subsequently, due to a shortage of sailors, a third Brigade was formed in 1916 of soldiers from various regiments. It was never a pukka smart Army outfit, discipline was more like that found in some of the Dominion divisions.

For illustration of this, I quote the reaction of the Division to the efforts of the Army to smarten them up by the appointment of an excellent, but unsympathetic, Divisional Commander, Major General Cameron Shute. He had no time for Navy foibles and was intent on transferring the RND to the proper Army as soon as he could. During one inspection of the RND trenches, he apparently discovered some signs of excrement in improper places, and, not unreasonably, was highly critical. The result in the Division was that the famous poet and author A.P. Herbert, then serving in Hawke Battalion, wrote a scurrilous and insubordinate verse about the incident that ended:

A recruiting poster for the Royal Naval Division (D.J. PERRYMAN)

For shit may be shot in odd corners
And paper supplied there to suit
But a shit would be shot without mourners
If somebody shot that shit Shute.

Shute also attempted to get the bearded members of the RND to conform to Army regulations and shave their full sets. Sub-Lieutenant G.S. Codner of Hawke Battalion, winner of the Military Cross and three times mentioned in despatches, invoked Naval Regulations to keep his beard. This episode too was famously recorded by Herbert in 'The Battle of Codson's Beard'. The RND also showed its individuality, or 'awkwardness' to the Army, by insisting on marching off ahead of the Guards detachments when units were on courses in France.

From the start the RND attracted some interesting people, but I shall mention only three:

Major General Sir Bernard Cyril Freyburg. The Commander of Hood Battalion, a winner of the Victoria Cross (VC), and later Governor General of New Zealand. A New Zealander, he was working in Mexico when the war broke out, and immediately caught a boat to England to join up. There is an apocryphal story of how he accosted Winston Churchill in Whitehall and demanded a commission in the Naval Brigades. What is certain is that he owed his induction into the RND to Churchill's influence. He was a superb swimmer and utterly fearless.

Arthur Asquith. Nicknamed 'Oc', and the son of the British Prime Minister, he developed the very closest friendship with Freyberg, whom he relieved in command of Hood Battalion on Freyberg's promotion to brigadier in 1917. He was recommended for the VC twice, awarded three DSOs and the *Croix de Guerre*, recommended for the DSC, and seriously wounded four times. On the last occasion, in December 1917, he lost his left leg after being hit in the ankle by a sniper. I have written about him at length elsewhere.[2] Both Asquith and Freyberg ended the war as brigadiers, in Asquith's case wearing the broad stripe of a commodore on his khaki uniform.

Rupert Brooke. A brilliant Cambridge scholar and nationally respected poet who died on 23 April 1915 of septicaemia, after having been bitten on the lip by an insect, just before going into action at Gallipoli. He was buried by his friends, including Asquith and Freyburg, on Skyros Island. An Australian writer has commented that, had he lived, Brooke would 'surely have thought his poetry foolish'.

The Birth of the RND

The Division was formed at the outbreak of war by the First Lord of the Admiralty, Churchill, who realised that there would be between twenty and thirty thousand naval reservists for whom there would be no jobs at sea. Remember, this was at a time when the British Expeditionary Force stood at only six divisions. The Division was formed around cadres of officers and NCOs from the Royal Navy and Royal Marines, supported by a sprinkling of retired officers of high quality, mostly from the Brigade of Guards. A very large proportion of the reservists were miners from the North of England, or from Scotland, and Northern Ireland.

As a naval historian, I would like to say that the establishment of the RND was a masterstroke by a man who fully understood the rationale of and need for projecting sea power ashore, and that it was a logical development which grew out of the famous Naval Brigades of the Victorian era. However, many years of study of World War I have led me to understand that it was much more the opportunist creation of a First Lord of the Admiralty who seized his chance to form his own private army.

By 1 October 1914 it was decided that the Royal Marines should administer the Division. Their first action was in Belgium. It had been decided to try to delay the German advance on the channel ports by shoring up the Belgians in their fortress port. The plan was for them to hold the forts around the city while the French Army and the British 7th Division counter-attacked the German flank eventually to relieve them. Again, it would be tempting to describe this as a visionary use of naval power projection capabilities. But the simple fact was that the sailors were virtually untrained, and they went because there was no-one else to send. The battalions had no first aid kit at all, and had to borrow bandages from a local French Military hospital. Neither had the Division any cavalry, artillery, engineers, or transport. The operation was a military shambles, culminating in the loss of the major part of a brigade which was forced to march into internment in Holland.

Meanwhile, a depot for the new recruits had been set up at Crystal Palace, and it was decided to concentrate the Brigade and Divisional training at the new Blandford Camp, on Salisbury Plain, early in 1915. The Division was allocated medical teams and engineers attached, but remained dispersed, pending the move to Blandford. On return from Antwerp, the Army decided that the needs of the new Army divisions took priority, and removed the majority of the regular and ex-regular officers who had been with the RND from the start. The outcome was that the RND was thrown on to its own resources to find its officers.

Unlike the other new Army formations, Blandford Camp formed a dedicated base for the Division. Here recruits could be trained and the wounded could recover, here each element of the front line battalions was represented in microcosm, and here a genuine *esprit de corps* and friendship between officers and men was engendered, which was to last through the war. In this regard, the RND had major advantages over some of the normal Army divisions. Naval personnel in the RND could hardly serve anywhere else. Wounded men tended to return, when they recovered, not just to the Division but almost always to the same Battalion. In addition, officer casualties tended to be replaced from within the Division's own resources. From an early stage therefore, many officers were promoted from the ranks. This was done, not on the basis of having proved oneself a good administrator back at base, but as a result of demonstrated competence and courage in battle.

Gallipoli

Before the Division could be concentrated at its base at Blandford, it was on the move again. This time to Gallipoli.

Major General Sir Archibald Paris, of the Royal Marine Artillery, the divisional commander, sent 300 men from Hood and Howe Battalions under Sub-Lieutenant Dodge, an American, to act as beach parties for the landings further south. Already, Anson Battalion and the Plymouth Battalion of the Royal Marines had been transferred to assist the 29th Division, which was landing on the southern beaches. On 24 April 1915 the RND transports set sail for Bulair where they arrived at daylight on the fateful 25th. The Division was therefore split, and it was some weeks before it fought as a unit.

A demonstration at the narrow neck of the peninsula was proposed to keep Turkish forces away from the beaches. Originally it was planned that a platoon or more of the Hood Battalion would land during the night, light flares, and generally make a lot of noise. Freyberg, a company commander in Hood, pointed out that this would be suicide, as the force had already been steaming up and down in full view of the shore, and a warm welcome would be guaranteed. He volunteered to swim ashore and light enough flares to make the Turks believe that something was about to happen, and then return to the ships. Asquith, another strong swimmer, volunteered to go with him, but the commanding officer's reaction was that he did not mind risking one mad Kiwi in the venture, but he was blowed if he was going to have the son of the Prime Minister killed!

Freyberg had to swim about two miles, and lit one lot of flares after stumbling about naked on the exposed shore in low temperatures. Avoiding the Turkish fire, and that of his own diversionary barrage, he swam another quarter of a mile to light the second set of flares and, with no ship in sight, working on a compass bearing only, began the long swim back, hoping to bump into a British boat. By some chance he was picked up after a further two hours in the water, nearly dead from cold, and exhaustion. The ruse worked, and the Turks kept two of their four divisions in the north of the Peninsula. Freyberg was awarded his first DSO.

This is not the place to go into detail about the landings in the south. From the point of view of the RND, the so-called beach parties found themselves in the very thick of the action right from the start; at V beach, Sub-Lieutenant Tisdall, of Anson Battalion, won the VC for rescuing many men under heavy fire. To have been noticed that day among so many acts of selfless heroism testifies to the worth of Tisdall's achievement. He survived until the next attack on 6 May when, with Sub-Lieutenant Melland, from the family of the British Prime Minister's first wife, among many others, he was killed.

The Plymouth Battalion of the Royal Marines was landed on Y beach with an Army battalion, the 1st King's Own Scottish Borderers, and a company from the 2nd South Wales Borderers, as a flanking party. It is fair to say that this force behaved with great gallantry, but with a little more dash at the beginning, when its landing had been virtually unopposed, it could possibly have unhinged the Turkish defence. The officer in command, the commanding officer of the Royal Marines, was perhaps over cautious at the start. He arranged for the withdrawal of his force from his defensive perimeter during the following morning, having received no instructions from either the Commander-in-Chief, General Sir Ian Hamilton, or Aylmer Hunter-Weston, the Corps Commander, and after having lost about one third of the force in repulsing several Turkish night attacks. So ended the RND's involvement in the landings.

Throughout the next few weeks the big set piece battles took place, where the Allies tried to fight their way up the Peninsula from the South, and to break out from the ANZAC landing. For all this period the RND, while heavily involved, was used by the Commander-in-Chief piecemeal by battalions and brigades attached to other formations, French, British and Australian. In fact, Arthur Asquith and others received the Croix de Guerre while serving in support of the French. These actions involved all the RND battalions present.

Lance Corporal Parker of the Portsmouth Battalion of the Royal Marines won the VC, at Anzac, for extreme courage in assisting wounded and bringing up supplies. He was part of four battalions of the RND, comprising three Royal Marine battalions and Nelson Battalion, sent to relieve tired ANZACs in the line. ANZAC opinion has tended to downplay the contribution of the RND units which provided them with their well-earned respite. On first seeing the RND men, the ANZAC commander William Birdwood described them as 'children led by untrained officers'. The first part was incorrect: many of the marines and naval reservists were in fact above average age for recruits. The comment about 'untrained officers' was undoubtedly true, however, and would remain the case throughout the British Army until all units had seen some action. The ANZACs considered most British soldiers of the period as small and even 'weedy', despite the fact that many RND men were ex-miners, like Able Seaman Sayles – height 5 feet 3 inches and chest 41inches, unexpanded! Ironically, when the Americans arrived on the Western Front, they were giants even compared with the ANZACs.

The Royal Marines' performance between the 29 April and 2 May, when they held a considerable part of the ANZAC line and repulsed several major Turkish attacks, as well as Nelson Battalion's losses in a particularly difficult attack on Baby 700 on 2 May, led to a discontinuation of such critical remarks. As an example, in its two week stay at Anzac, Nelson Battalion lost 55 dead and about 150 wounded, most of them on 2/3 May, more than 20 per cent of the battalion.

Lowering the Gallipoli dead to a trawler for burial at sea. (RAN)

Achi Baba and the End at Gallipoli

At last, by 13 May, the remnants of the Division, less the Hawke, Benbow, and Collingwood Battalions, which were still on the high seas, were concentrated under General Paris on the Achi Baba Front, which now took on characteristics of trench warfare, soon to be equally familiar on the Western Front. The Division held the line on the left of the French, who were on the right of the line. Then came the 42nd Division and the remains of the 29th Division, badly cut up in the landings, on the left. During a series of night advances towards the end of May, the Division pushed its line forward nearly half a mile, so at least the rest areas were safer from shellfire. On 28 and 29 May the three new battalions joined the Division, which for the first time was now constituted as it was meant to have been, with two naval brigades and a Royal Marine brigade. Nevertheless, because of the losses, the Division was still weaker than on 25 April. Before the arrival of the three extra battalions, the Division was down to under 5000 men. The addition of these fresh, but inexperienced, troops bolstered the numbers, but not necessarily the combat power of the Division, for the next great push – the attack of 4 June 1915.

This attack took the form of the last two efforts on this front, being a frontal push all along the line in a straightforward trench to trench assault, from the French on the right to the 29th Division on the left. The basic problem was the same as earlier: any hold-up in any part of the attacking line opened the unit next in line to catastrophic enfilade fire from uncaptured Turkish positions. From the RND point of view this meant that they were utterly dependent on the success of the French on their right flank. At 1200 Howe, Hood, and Anson battalions advanced on the RND front into a hail of fire. With desperate losses, they reached and captured the Turkish front line. Only about 20 officers and 300 men had survived, and they were under severe fire. The reinforcing battalion, Collingwood, new in action and up to full strength, advanced across no man's land to support. But they were cut down, mainly from fire from the uncaptured ridge on the right, which the French, despite gallant attempts, had not taken. After further actions in the second and third Turkish lines, it was obvious that the captured positions could not be held, and the troops withdrew. Five officers and 950 men, from 70 officers and 1900 men who started the attack, returned to their front line. In view of the losses, General Paris disbanded the Collingwood and Benbow Battalions to make up the numbers in Hood, Howe, and Anson, and reduced the naval brigades to three battalions each. The disbanded battalions were never reconstituted.

And so it was back to trench warfare, until the next attack on 12 July, for which the RND was in reserve to support the newly arrived 52nd Lowland Division. Two battalions of the Division went into action, the Portsmouth Royal Marines and Nelson. Twelve officers were killed, and only one of Nelson's was unwounded.

At last, on 25 July the Division was relieved and sent back to rest. It now numbered only 129 officers and 5038 men. So ended the RND's first spell of action on the Peninsula: they took no part in the later struggles at Anzac and Suvla. This is hardly surprising: conditions were desperate, dysentery was rife. Much more influential on the future of the RND at this time, however, were two other decisions, one made by the Admiralty, and the other by the War Office: the first was an order for all stokers of the Fleet reserve to be recalled for duty with the Fleet; this affected only about 300 men, but had the effect of withdrawing the hardened core of trained and disciplined men remaining from the earliest days of the Division. The second decision was the refusal of the War Office to provide reinforcements for the Division; the Army had always looked askance at the RND, and it was not surprising that they did not bend over backwards to get the Admiralty out of a hole. This meant that, for the rest of the campaign, the RND was used only for garrison service. For a while it looked as though it was to be broken up, but they hung on and were among the last to leave in January 1916. During the Dardanelles expedition, of the 16,000 men who had passed through the Division, over 13,000 had become casualties, of whom 2628 are on the Peninsula still.

Command in the RND

The complement of ex-retired Guards officers, Royal Marine, Royal Navy and RNVR officers who had initially formed the upper echelons of the battalions was soon swept away by the battles in Gallipoli. The Army had by then suddenly found itself desperately short of officers for the Kitchener and Territorial battalions, and for the replacement of casualties in the regular units. There was no way that the General Staff were going to send officers to the rag-bag outfit run by the Admiralty in the Dardanelles. So, from the earliest times, the RND was forced back on its own devices.

When large numbers of officers were eventually recruited from the OCBs (Officer Cadet Battalions) it was decided that those who had been ratings in the RND would always return to the Division regardless of their final examination mark, rather than, as happened in the Army, being sent to the next vacant post. William Bentham, an ex-rating, records that he was recruited from OCB 18 in Bath, and that one of the advantages was that as a naval officer, one was paid more than the newly recruited Army equivalent. He got 8/6d a day rather than 7/6d. Bentham also mentions that because of the competition to get into the Naval Division, it was able to take those who passed out top in the exams. Almost invariably, RND officers would be sent back to the brigade in which they had served in the ranks, and would frequently come back to the same battalion. Since a man required a recommendation from his Commanding Officer for officer training, this proved a very wise policy. Any such recommendation was earned on the basis of performance and leadership qualities shown in action, usually under conditions of the most extreme hardship and danger; there was therefore no chance of an ex-sailor returning to his old unit as an officer and not being accepted. The other officers, the men he led, and he himself knew he was there by merit alone. Comparisons with the Dominion divisions and the Guards are obvious here.

At Gallipoli, the officer casualty rates were so high that promotion for survivors who had demonstrated professional competence was rapid. Asquith began as a temporary sub-lieutenant. By 7 June 1915 he was a lieutenant, by 5 July a lieutenant commander, and he was actually a battalion commander in command of Nelson for a few days in November of that year.

Throughout the war, the RND maintained its cohesion. In 1918, when the British Army with the help of its Allies was fighting for survival against the German offensives of March–May, and then winning the war by driving the Germans back virtually to where the war had begun, the RND was being prepared for what was believed would be the 1919 campaign. Of nearly 200 officers commissioned during this period, almost three quarters were ex-ratings from the Division and one quarter were ex-soldiers who had seen the light. In all, 80 per cent of these officers were taken from the OCBs.

On the Western Front the RND became very effective as a fighting unit. Churchill said (doubtless with an element of paternal pride) that the Division came to be regarded 'In the glorious company of the seven or eight most famous in the British Army',[3] and General Fergusson, General Officer Commanding XVII Corps, under whom the Division fought in the advances of 1918, said of the Divisional Commander: 'I asked more from his Division than any other, and neither he nor they ever failed me'. Fergusson also said that under its last divisional commander, Major General C. Blacklock, the RND in 1918 was the finest he had ever seen.

An Elite Force

So, after all this, we have not learned much that is new about command. The best fighting units have a great coherence. They have first class leadership that is maintained even after serious losses. They believe themselves special, and they train hard and learn lessons quickly. Finally, on top of all this, it is important to have a cause. There is a huge amount of evidence in World War I correspondence from ordinary soldiers, as well as officers, that the troops actually believed they were fighting in a just and noble cause. Unfortunately the verification of old truths does not always mean that the hard-won lessons of history are put into practice.

5. Naval Aviation at Gallipoli

Chris Clark

IN AUSTRALIA, THE GALLIPOLI campaign of 1915 has acquired the status of legend. But strange to say, it is chiefly remembered, in this country at least, for the land component of its story rather than for the naval operations with which it commenced, or the naval strategy which was, from first to last, its primary rationale. Many people today forget that the original purpose of the nine months of ground fighting was to make it possible for a fleet to steam through the Dardanelles and deliver a blow against Constantinople (modern Istanbul) which was expected to knock Turkey out of the War. No less odd is the lack of recognition given to the fact that the Dardanelles campaign also provided the setting for extensive and, in historical terms, ground-breaking air operations which, even now, provide an interesting case study of the uses of air power in support of military operations in a littoral environment.

What makes the Gallipoli air campaign so notable is that British planners expected so much from the Navy's air arm less than six months after the start of World War I. Attempting much with little may have been a sign of confidence and faith, but equally it could be a sign of desperation or ignorance. Quite apart from the large scale demands for air support of sea operations in the English Channel and land operations in northern France, the preparations for the Dardanelles had to proceed in tandem with other expeditionary schemes for Mesopotamia and East Africa. While the Eastern Mediterranean operations were actually underway during 1915, the Royal Naval Air Service (RNAS) had to supply men and machines for these concurrent campaigns.[1]

The fact is that the RNAS, having become an entity separate from the Royal Flying Corps (RFC) only on 1 July 1914, entered the war during the next month with just 39 aeroplanes and 52 seaplanes (not all of which were in flyable condition) and about 130 officers and 700 ratings. This was a far cry from the 3000 aircraft and 67,000 personnel on strength when the RNAS eventually merged with the RFC to form the Royal Air Force in April 1918. Although the mixture of aircraft types that the RNAS initially possessed was, as a whole, better than those in the inventory of the RFC, the reality was that most were not fit for much more than short range reconnaissance patrols in fair weather conditions. In August 1914 the service also had two aircraft 'carriers'; a week after the war began, the Admiralty requisitioned a further three cross-Channel packets for conversion to this role.[2]

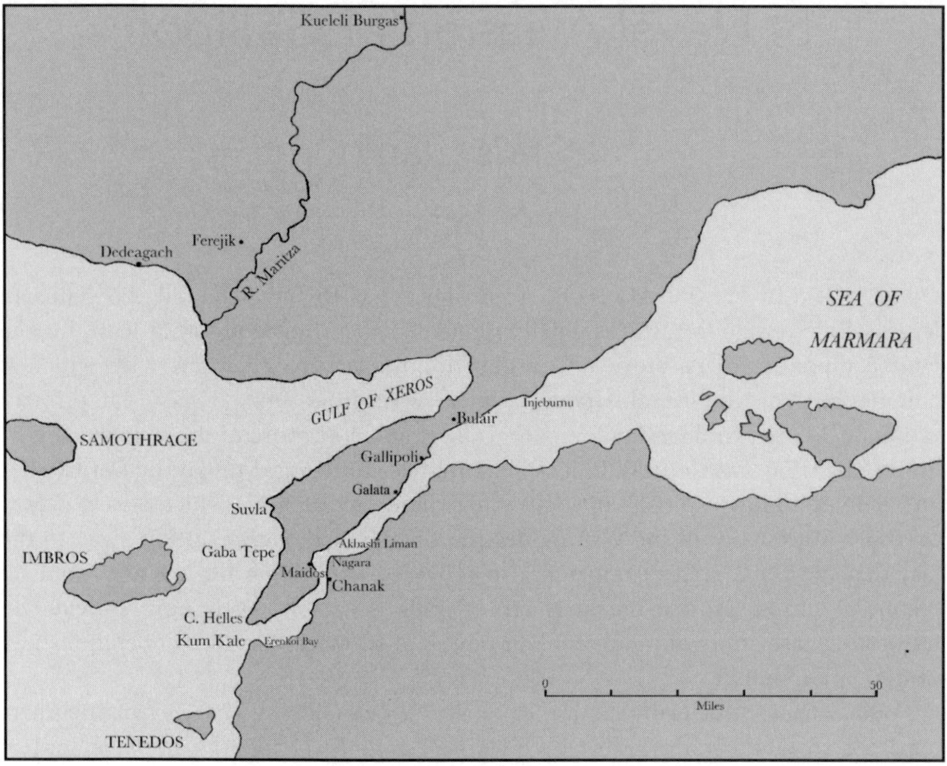

Figure 5.1. The Dardanelles Theatre.

Despite the limited amount and range of resources available, the Admiralty planned to provide the fleet assembling for the Dardanelles with a supporting force initially of eight aircraft. These were six floatplanes and two land planes, of various types, belonging to the Eastchurch Squadron under Commander Charles Samson, RN. This unit had been sent on operations to Dunkirk late in August 1914, but was withdrawn to England in December to prepare for redeployment to the Eastern Mediterranean. It sailed for the Dardanelles in the 7080-ton carrier HMS *Ark Royal* on 1 February 1915 and arrived at the Greek-held island of Tenedos sixteen days later.[3] In March it was renamed No. 3 Squadron, RNAS.

On 21 May a second unit of the RNAS was also despatched to the Dardanelles. This was No. 2 Squadron under Squadron Commander Cecil Malone, RN, which sailed in HMS *Ben-my-Chree*, a former Liverpool – Isle of Man passenger steamer that had also been converted to the carrier role. Arriving on 12 June at Mitylene on Lesbos, an island seized from Turkey in the 1912-13 Balkan Wars, the squadron had new Sopwith Schneider seaplane fighters and two Short 184 seaplanes that were configured to launch torpedoes, along with more land planes. The French also contributed a seaplane tender, *Foudre*, to the Dardanelles operation which operated from March until May, but its aircraft lacked radios and apparently achieved little.[4]

A reorganisation of the RNAS took place in June, with the two squadrons being redesignated as 'wings'.[5] On 25 August the RNAS in the Eastern Mediterranean was placed under overall command of a military officer, Colonel Frederick Sykes, whose career straddled both the RFC and RNAS. It may be remembered that it was Sykes who, as a major general in April 1918, succeeded the founding Chief of Air Staff of the RAF, Major General Hugh Trenchard, until replaced by Trenchard again in April 1919.[6] In August, Commander E.L. Gerrard took over No. 2 Wing.

Various other machines were added to the strength of the two wings in the Dardanelles, including single-seat Nieuport 10 and 11 Scouts and two seater Nieuport 12s delivered in July. By December 1915, plans were in hand at the Admiralty in London for a large expansion of the RNAS presence off Gallipoli, involving more than 50 additional aircraft and another carrier.[7] While an increase along these lines could well have had a dramatic impact on the part that air power played in overall operations, these plans came to nought when the decision was made to plan for the evacuation of the Army from the Peninsula instead.

With an idea of the scale of the air forces committed, we may turn to how the component was employed and what it achieved. First, it needs to be understood that the 'carriers' mentioned were not aircraft carriers in the form that we commonly know them today, but were actually no more than seaplane tenders. *Ark Royal*, for example, had a 130-foot long flush deck forward which resembled a launching platform (and has been described as such[8]), but in fact no provision had been made for flying off operations. The reason was that the ship's speed of just over ten knots was insufficient

The aircraft carrier HMS *Ark Royal* at Mudros. (DAVID HOBBS)

to create enough wind over the deck for successful flight operations;[9] the vessel's seaplanes were accordingly lowered and recovered over the side by crane.[10]

For the aircraft from both carriers that were fitted with wheeled undercarriages, a land base was required. Early in the campaign Samson tried using a rough landing ground within the British perimeter on Cape Helles, but as it was within sight and gun range of the Turkish positions the arrival of any aircraft invariably attracted too much enemy fire for this to be a satisfactory arrangement. At least four machines were lost on the ground due to enemy shelling.[11] By mid-April the arrangement was abandoned in favour of the safety of Tenedos, but not before Samson had left behind a dummy aircraft which, 'to the joy of the troops', caused the Turks to expend a large quantity of shells over several days to achieve its destruction.[12] In early August the main landing field was switched to Imbros, in time to provide air cover for the British landings at Suvla.

Generally, the use of either Imbros or Tenedos proved a limiting factor on the range which the land types of aircraft could achieve. Moving the seaplanes about by water before launching them gave them greater flexibility in terms of reach and targets. Even so, other restrictions applied depending on the number of crew carried and the type of weapons fitted for a particular mission. At one stage Commander Samson maintained that, although he had 22 aircraft on strength, only about five machines were considered to be of practical use. It should be remembered that in 1915 there were still great strides to be made in the area of aircraft development. Regardless of these considerations, it is amazing how many of the roles for which air power was employed were actually performed by the machines available to the RNAS contingent at the Dardanelles, although initially serviceability factors often played a significant part in determining their effectiveness.

Reconnaissance

The arrival of *Ark Royal* in theatre by mid-February ensured that aircraft were on hand to provide reconnaissance support for the opening naval operations of the campaign. The first task was undertaken the day the airmen arrived, but only one of four machines that took off was able to complete its mission.[13] The aircraft crewed by Flight Lieutenant Geoffrey Bromet, RN, and Flight Commander Hugh Williamson, RN, penetrated 'about four miles' up the Dardanelles Straits while spotting Turkish batteries and trench defences.[14]

When Vice Admiral Sackville Carden commenced the work of reducing the Turkish forts at Sedd-el-Bahr and Kum Kale, at the mouth of the Straits, from 19 February, the carrier's seaplanes were certainly sent aloft with the aim of helping to spot enemy targets and to report on the effectiveness of ships' gunfire. Unfortunately, it seems that the seaplanes proved 'so unreliable and their observers so inexperienced that, in spite of the great gallantry and resolution of the pilots, they were almost useless for reconnaissance purposes.'[15]

Much the same thing happened during later attempts to employ them in this role. On 5 March, for instance, it is claimed that the British aircraft 'failed again; of the three seaplanes used, one crashed on take-off, another was hit by rifle-fire and the observer wounded, and the third sent back only one correction.'[16] The problems were that any bad weather stopped flying, and the wireless equipment carried by the Shorts to relay Morse messages to the receiving ship was unreliable and produced only a short range signal.[17]

The aviators are also apportioned part blame for the failure of the main naval attempt to force the Straits on 18 March. The disaster which overtook the fleet, resulting in the sinking of three battleships (two British, one French), was due to the existence of a previously undiscovered minefield which the Turks had laid in Erenköi Bay ten nights before. Minesweepers had actually lifted three mines from this area in the days before the attack, but it had not been realised that these were just part of a line of twenty laid parallel to the shore. Seaplane patrols had, unfortunately, reported the area clear of mines just the day before the attack, and the Allied naval commander, Vice Admiral J.M. de Robeck – believing that aircraft could spot mines at a depth of 18 feet (5.5 metres) – accepted this advice.[18]

Adding to the picture of general aircraft ineffectiveness that emerges is the experience on the day of the first troop landings on 25 April. *Ark Royal* was assigned to support the ANZACS north of Gaba Tepe by spotting for the guns of the battleship HMS *Majestic*. Again, this effort failed, since the right flank of the Australian landing area quickly came under heavy shrapnel fire from concealed Turkish batteries which the aviators could not locate. It should be said that the balloon ship *Manica* was also allotted to this sector and its tethered balloons could not detect the enemy guns either. As the Australian official historian fairly concluded: 'The balloon ship and the seaplanes gave no solution. It was no fault of the Navy that its guns could not fire over impossible angles at undiscoverable targets.'[19]

More successful were some later efforts at reconnaissance by the Navy planes. One of the more notable was the episode on 18 May, when a seaplane flown by Lieutenant G.L. Thomson, RN, was sent from Imbros to spot for HMS *Canopus*. The battleship was attempting to shell an enemy warship which had taken up position in the Straits to fire on the ANZAC beachhead. The 'shoot' failed when it was found that *Canopus* did not have sufficient range for its guns. But as Thomson passed over the Peninsula and headed for home he noticed that several valleys east of the Australian lines were densely packed with Turkish troops. After attempting to turn the guns of *Canopus* on to this target of opportunity, without success, Thomson returned to Imbros at about 2.30 pm and reported what he had seen. A second plane was then despatched to confirm his sighting, and the pilot of this aircraft also discovered Turkish troops in the process of landing from four small steamers at Akbashi Liman, while there were also newly-arrived groups encamped with their transport on the neighbouring hills. ANZAC Headquarters duly telegraphed this news to the two divisional commanders who were warned to be on alert for a night attack. It was, of course, entirely due to this advice that the frontline trenches at Anzac were fully manned and ready when the Turks launched a major offensive early the next morning.[20]

The reconnaissance effort was naturally ongoing throughout the campaign. In October, for instance, one of the carriers was sent with some fighting ships that had been detailed to bombard the remote Thracian coastal town of Dedeagach. The seaplanes from the carrier co-operated in spotting for the ships' guns, also making a valuable reconnaissance of the Bulgarian coast and railway.[21] This effort was linked to British efforts to carry the war further afield from the air.

Aerial Photography

Apart from acting as a set of eyes over the enemy's lines, the RNAS aircraft also served to bring back invaluable visual records in the form of air photographs. Given the nature of the terrain, often these aerial views provided the only precise information which ground operations staff had regarding the layout and location of Turkish defences. Detailed maps drawn from air photos are specifically mentioned among the preparations undertaken for the Third Battle of Krithia at Cape Helles in June,[22] and likewise for the Australian attack on Lone Pine in August. In the words of one published history, it was unfortunate in the latter case that 'the science of interpreting aerial photographs was so much in its infancy that ... vital features were overlooked.'[23] These omissions included the fact that the enemy trenches to be attacked had been provided with heavy overhead protection, a situation which left the assault troops initially floundering in attempts to enter the enemy line. The timber cover of the front trenches was clearly visible in the photographs, so the failure to detect its existence was again no failure of the airmen.

The photographic apparatus originally available to the RNAS contingent was one camera, a printing frame and a dark room lamp. Notwithstanding these slender resources, the first air photos were taken in April 1915 from a Henri Farman at low altitude, and when the first operator was wounded in June, his place was taken by Lieutenant Thomson, RN, who succeeded, working on his own, in making 900 exposures and producing 3600 prints.[24] In the later words of Colonel Sykes:

> *In addition to the assistance of air photography to reconnaissance, the war gave it great impetus as the handmaid of survey and mapping. It was, in fact, the only means of mapping or correcting the maps of country held by the enemy, which in certain cases, as at Gallipoli and in Palestine, were very inaccurate.*[25]

Anti-shipping Strikes

The arrival in *Ben-my-Chree* of Short 184 seaplanes capable of carrying 14-inch torpedoes added greatly to the variety of offensive operations which the RNAS contingent could undertake. On 12 August, Flight Commander Charles Edmonds,

RN, achieved a notable first after he took off from the Gulf of Xeros and crossed the Bulair Peninsula into the Straits. There he came across a 5000 ton transport off Injeburnu on the Gallipoli coast and attacked from a height of about twenty feet and a range of 300 yards, scoring a direct hit.[26] It turned out that the ship was already beached when struck, having earlier been torpedoed and shelled by a submarine, but this detracts only marginally from Edmonds' achievement in mounting the first successful air torpedo attack in history. As his commanding officer, Squadron Commander Malone, RN, reported at the time: 'One cannot help looking on this operation as being the forerunner of a line of development which will tend to revolutionize warfare.'[27]

Five days later, Edmonds found and attacked another steamer off Akbashi Liman, this time one actually underway. He left the vessel seriously damaged and in flames, although it succeeded in reaching Constantinople.[28] Some accounts describe this vessel as an ammunition ship, but Turkish versions maintain that it was actually a hospital ship.[29] That same day, the other Short seaplane flown by Flight Commander G.B. Dacre, RN, attempted a similar attack on a large steam tug off Nagara. This pilot experienced engine trouble and was forced to land on the water. Not to be foiled, however, he taxied his machine into position before firing his torpedo and sinking the target. Thus relieved of the weight of the torpedo, the seaplane was able to take off again – although only just, after a run of some two miles.[30]

These were the only successes of the torpedo bombers in the Dardanelles campaign, although bombing attacks extended the tally against shipping to some extent. More than 70 bombing attacks were made during 1915, resulting in damage to two large steamers and a tug, while one lighter and six dhows were wrecked.[31] Another victim was the old Turkish battleship *Barbarossa*, which was found lying off Akbashi Liman on 29 May and bombed. A direct hit put the centre turret out of action, but the ship was able to make its way back to Constantinople.[32] As one writer observes, 'Although these may seem meagre results, it must be remembered that the bombs were small, and bomb-sights non-existent.'[33]

Anti-submarine Patrols

Initially it was the Allies who made the greater use of submarines at Gallipoli. The feat of penetrating the Turkish defences of the Narrows was first achieved by the Australian boat *AE2*, and then subsequently by two British submarines. Disruption of enemy shipping in the Sea of Marmara was, of course, achieved as much by Allied submarines as it was by the efforts of the aviators, but the latter played an important role in facilitating these operations.

On 12 April the commander of the submarine *E-15* was taken aloft to reconnoitre a course up the Narrows from the air. On the day that the boat made its ill-fated attempt to pass through the waterway, all available aircraft were sent up to create a diversion. Once *E-15* grounded, a similar effort was made to suppress with bombs the enemy fire directed at the hapless vessel.[34] It is also worth mentioning

The last photograph of HMS *Majestic*, sunk by *U-21* on 27 May 1915. She is seen here with destroyers in Mudros Bay. (RAN)

the experimental enterprise which the RNAS undertook in dropping parachute bombs aimed at sinking the buoys which supported the heavy wire anti-submarine net which the Turks had stretched across the Straits at Nagara.[35]

In mid-May it was learnt that the German submarine *U-21* had passed through the Straits of Gibraltar and was en route to the Eastern Mediterranean. This was the start of a serious threat to Allied shipping, fully realised on 25 May when the battleship HMS *Triumph* was torpedoed off Anzac in 'full view of both armies'.[36] The result was an immediate recall of all large warships to the safety of Imbros. *U-21* added to the drama of its arrival by sinking *Majestic* the next day, but this was the extent of the boat's success. On 30 May it was lured into attacking a transport which had been disguised to resemble a battlecruiser, whereupon it was chased by Samson's pilots until forced to enter the Dardanelles and make for the safety of Constantinople.[37]

Although enemy submarines never had quite the same impact again, several of these boats came to be active in the Eastern Mediterranean. One succeeded in torpedoing the transport *Southland* on 2 September, as it was conveying the headquarters and a brigade of the 2nd Australian Division to Lemnos en route to Gallipoli. Allied ships could never feel entirely safe, and air patrols against this menace became a regular demand imposed on the RNAS crews. Claims were later made that enemy submarines were sighted by aircraft on two occasions and attacked with bombs.[38]

Long-range Bombing

From early in the Gallipoli campaign, the RNAS machines undertook raids against a variety of Turkish targets. The enemy lines opposite Anzac or Helles were often visited at mealtimes, to catch the troops as they congregated for food. Sometimes the target was the flour mill in the Gallipoli township or some other ration target on the Peninsula. A formation of five British aircraft visited Maidos (now known by its

Turkish name of Eceabat) on 23 April and set the place alight with bombs.[39] On another famous occasion, apparently in mid-September, Commander Samson attacked a staff car he encountered travelling along a road and delivered several bombs close by the vehicle's occupants. It was allegedly discovered after the war that the car contained none other than the Turkish commander Mustapha Kemal.[40]

The bomb load carried by the available aircraft was quite small, ranging from four to eleven 20 pound Hales bombs, each equal to the explosive effect of a six-inch artillery shell. Even the larger Short seaplanes were normally fitted with just a pair of 112-pound bombs.[41] It is nonetheless remarkable that by the end of October, after Bulgaria had entered the war on the side of the Central Powers, the RNAS wings were embarking on missions involving distances of 200 miles which aimed at cutting the road and rail systems carrying German supplies to Turkey via Bulgaria.

As part of this effort, four attacks were made against the Kueleli Burgas railway bridge spanning the Maritza River, and eleven attacks against Ferejik Junction, another vulnerable point on the railway lines. A moonlight attack on the Kueleli Burgas bridge on 13 November by Flight Commander J.R.W. Smyth Pigott, RN, flying a BE2c of 2 Wing at a bombing height of only 300 feet, earned him the DSO.[42] One of the attacks on Ferejik on 19 November is chiefly notable for the fact that one of the 3 Wing pilots, Squadron Commander Richard Bell-Davies, RN, won the VC after he landed in a Nieuport 12 Scout to rescue a colleague whose Farman machine had been damaged by ground fire and forced down.[43]

Despite the bravery and daring displayed in such attacks, the actual operational impact in terms of destroying bridges or severing railway links appears to have been negligible. The most remarkable mission of this type carried out in the course of the campaign probably took place on 18 December, when Commander Samson flew from Imbros in a Henri Farman F.27 aircraft and dropped a single 500-pound bomb (the biggest of the war to that date) on a Turkish barracks in Constantinople.[44]

Enemy Air Operations

At first the Turks had no answer to Allied air power, since at the beginning of the campaign they possessed very few aircraft. A French military mission had been assisting the Turks to build up an air arm, but these advisers promptly left when Turkey entered the war on Germany's side. In August 1914, even before Turkey formally declared war, a Nieuport seaplane had been sent to Chanak (now Çanakkale) to perform reconnaissance duties.[45] When this machine became unserviceable in mid-October, another Nieuport was sent to replace it. This also became unusable after it capsized on 5 November. In the face of clear signs that the British and French were preparing for some sort of naval operation, the Ottoman General Staff ordered another aircraft sent to this area. A derelict Bleriot XI-2 was hastily rebuilt and sent forward during January 1915. A week after the first Allied naval attack on the Narrows forts on 19 January, this machine attempted to drop four grenades on to *Majestic*, but missed.[46]

Two Australians in the Royal Naval Air Service carrying out checks before take off from RNAS *Mudros*. (C. GILMOUR)

By mid-March the Turks had had a second aircraft shipped to Chanak by boat. This was a Rumpler B.I (referred to as a 'Taube') piloted by a German Army captain named Erich Serno, who had arrived the previous month to become Chief of Staff of the Ottoman air force.[47] All new machines received by the Turks were of German manufacture and were initially piloted by Germans with Turkish observers. Both the Rumpler and the Bleriot were in the air on 18 March, when the Allies launched their main assault on the Straits, and played a part in keeping the fortress commander at Chanak in touch with developments during the failed thrust by the Allied fleet. Two Albatros B.I. aircraft were received a week later, ensuring that three relatively modern machines were on hand to help the local defenders keep abreast of their opponents' moves. The Bleriot was sent back to Constantinople but was lost in a forced landing in the Sea of Marmara.

Although the French made a bombing attack on their aerodrome on 28 March, the Turkish aviators maintained their support over the battlefront, on several occasions during the first weeks of April attempting to drop small bombs on to Allied ships – generally without much success. The available aircraft played a useful role in providing an overall picture of the 25 April Allied landings and from that point onwards, reinforced by a new Albatros from 2 May, they conducted almost daily reconnaissance flights over the frontline. During May, the 1st Aircraft Company was brought into being, comprised of two flights of two aircraft each.

Despite repeated bombing attacks by Allied machines against the airfield at Chanak, the Turks also made regular flights (usually twice weekly) over the main

British bases on Tenedos and Imbros to count the ships in harbour. During these missions a load of hand-dropped bombs was generally carried and dropped as the Allied lines were crossed, along with quantities of steel darts which were thrown over the side as an anti-personnel weapon. The British, it should be mentioned, also used these flechettes,[48] although the darts which Samson and his men had to deliver reportedly 'wobbled so violently in descent that they had no effect whatever, and even bounced harmlessly off tents.'[49] Anecdotal evidence suggests that the Turkish versions were more lethal and hence more psychologically effective.[50]

On 27 May the Turks also used their aircraft to notch up a first by conducting a propaganda mission against the Allies. Pamphlets written in Urdu were dropped over Anzac and the French lines at Helles, appealing to Moslem Indians and French colonial troops not to fight fellow Moslems. The tactic was promptly copied by the Allies, who responded with a pamphlet drop over Turkish lines four days later. This return attempt to undermine enemy morale was essentially misguided, since most Turkish officers and troops were illiterate.[51]

From the end of May the tempo of air operations seemed to pick up, with bombing raids by both sides becoming more frequent. Clashes between aircraft also began to take place. On 22 June the crew of one British machine fought with the Rumpler using rifle fire for 20 minutes, eventually hitting the latter in the engine and forcing it to crash land between the opposing lines, whereupon artillery fire caused it further damage.[52] The 1st Aircraft Company was effectively wiped out a fortnight later, when on 5 July an attack on the aerodrome at Chanak by ten British and French aircraft succeeded in destroying all three Albatros aircraft in their hangar.

Fortunately for the Turks, but coincidentally, that same day a Gotha WD.1 aircraft had arrived to begin the establishment of a new seaplane unit based at Naga, near Chanak. This machine made its first reconnaissance on 6 July, overflying Lemnos and bombing Tenedos on its way back. It remained the Turks' sole answer to Allied air power until 13 July, when four new Rumplers arrived to bring the 1st Aircraft Company back up to strength. To reduce the risk of repeat destruction, the Turks moved their unit to a new airfield outside Galata, near the 5th Army Headquarters at Gallipoli.

With these new reinforcements, enemy machines began appearing over the Allied lines on an almost daily basis. Allied commanders were then in the final stage of preparation for the offensive planned for early August, so were making every effort to shield troop movements and concentrations from prying enemy eyes. General Liman von Sanders, the German commander of Turkish forces at Gallipoli, was equally looking for signs of when and where the Allies' expected attack would fall. The Australian official history records that patrolling enemy machines flew over Anzac between 29 and 31 July, and again on 2 and 4 August.[53] This despite the raids which aircraft from the French air unit in the theatre (Escadrille M.F. 98T) made on the airfield at Chanak, not knowing that this was no longer in use.[54]

During September the 1st Aircraft Company took delivery of armed Albatros C.I aircraft, while the seaplane group received additional Gothas which brought its strength up to five machines. Thus strengthened, the Turks started an air offensive of their own. Apart from more frequent reconnaissance sorties, bombing attacks were launched against the British GHQ at Imbros as well as numerous attacks on shipping. This effort came to an abrupt halt when a fierce storm on the night of 6–7 October wrecked five of the six aircraft at Galata. The remaining machine suffered an engine failure while on a reconnaissance mission south of Lesbos and was damaged making a forced landing.

Not until the end of October was the company brought back up to strength, and even then it was still significantly outnumbered by Allied aircraft. In October, however, the Germans ferried 24 aircraft on a non-stop flight from Herkulesbad (Ungara) to Adrianople (Edirne), a distance of 306 miles (492 kms) by air,[55] and numbers of these machines soon appeared on the Gallipoli front. Included were more Albatros C.Is, newer Gotha WD.2s and a Fokker Eindecker E.III fighter. On 30 November a Turkish-crewed Albatros C.I engaged in an air duel with a French aircraft and succeeded in shooting it down – the first such success of the war.

At the end of December, at the specific request of General von Sanders, a detachment of Fokker Eindeckers was transferred from the Macedonian front. These superior machines were based at a new airfield outside Chanak and quickly altered the balance of air power in the area in favour of the Turks. It was more than coincidental that in January 1916, while the Allies were in the process of withdrawing from Helles, that the British suffered four of their machines shot down in combat with the new enemy fighters, resulting in at least three crew members being killed and another made prisoner.[56] Undoubtedly, had the Dardanelles operation been prolonged into 1916, the Allies would have faced increased resistance to the way they had operated in the air throughout 1915.

As previously noted, plans had been made to bolster significantly the RNAS air strength committed to the Eastern Mediterranean, so a shift in the tactical balance during any prospective air operations during 1916 should not be assumed. What is clear, however, is that by the time the Gallipoli Peninsula was actually evacuated, Allied air strength was at a very low ebb indeed. According to Sykes, he was asked at this time 'to be careful not to do too much air work'. To this he added that 'This [request came] at a time when through stress and strain and loss we had, I think, a total of five machines left able to take to the air!'[57] Accidents and enemy ground fire had clearly exacted a toll over the course of the campaign, with Samson admitting to at least a dozen machines lost from such causes.[58]

* * *

The air operations that accompanied the Dardanelles campaign have given rise to some sharply divergent views about what they all meant and achieved. At one end of the spectrum are the opinions expressed by Sykes just a few years after the event.

In a book published in 1922 he declared that the RNAS, in carrying out all air work with the Army and Navy in the Gallipoli campaign, had 'showed how a single air force could effect really important co-operation with both services.'[59] That observation may now be unexceptional given the experience of many countries, Australia included, over the last 90 years.

Much more contentious is Sykes' assertion that:

> *when in future years the story of Helles and Anzac and Suvla is weighed, it will, I think, appear that had the necessary air service been built up from the beginning and sustained, the Army and Navy could have forced the Straits and taken Constantinople. I insistently urged the dependence of the naval and military forces upon air assistance and the necessity for carrying out a strong aerial offensive, especially by bombing, for which the local conditions governing the enemy operations on the Peninsula offered exceptional advantages.*[60]

The idea that air power – in 1915 or even 1916 – was capable of delivering an outcome to the campaign different from that which actually occurred seems a doubtful proposition. R.D. Layman, a more recent student of the Gallipoli campaign, has remarked on the rather obvious fact that despite possession of air superiority (and even supremacy) for much of the campaign, the Allies lacked aircraft with the performance which would have enabled them to inflict the sort of damage that Allied aircraft of World War II could achieve. In Layman's view, 'Despite all the contributions of aviation to the campaign, it had not yet developed to the point to which it could help solve a fundamental problem that dominated nearly all of World War I: the immense difficulty of defeating well-entrenched troops supplied with machine guns and backed by artillery.'[61]

Regardless of its failings, aviation had still been a highly significant factor in the campaign. As Layman himself later observed:

The failure of the Allies to seize the peninsula, about which so much has been

> *written, is counter-balanced by the Turkish failure, about which so little has been written, to drive the invaders off it. There can hardly be any doubt that much of the Turkish failure was caused by Allied mastery of the skies, which deprived the Turks from achieving strategic or, often, even tactical surprise … Although the Allies failed to make any real headway despite their aerial advantage, the role of aviation in helping to nullify Turkish action produced the same indecisiveness at Gallipoli as it did in France and thus was a factor prolonging the campaign. It is possible to speculate that the campaign could not have been sustained as long as it was had that factor been absent.*[62]

The precise nature and extent of aviation's contribution at Gallipoli is clearly an area for further analysis and debate.

PART II

Naval power projection during the Second World War

6. An Inferior Naval Power Ashore: German Navy Baltic, Mediterranean and Black Sea Operations

Sarandis Papadopoulos

HISTORIANS HAVE RECENTLY QUESTIONED the wisdom of the large scale German investment in naval strength in World War II. The pre-war Third Reich seemingly squandered resources on building a balanced battle fleet at the expense of its submarine force.[1] That surface force was largely ruined during the conflict by the 1940 Norwegian campaign, and further nullified by the elimination of its supply vessels as well as the major warships *Graf Spee* and *Bismarck*, sunk by the Royal Navy in the first two years of the war. Similarly, the ongoing German U-boat campaign has also been characterised as a misguided waste of resources and lives. In that light for National Socialist Germany to invest in naval power seemed misguided, even 'useless'.[2] In contrast, this chapter argues that limited German naval power proved important to the Army in fighting its defensive campaigns in the latter half of the Second World War. In doing so it measures the gain in a *joint* sense from a naval investment, showing how a strategically dysfunctional military – that of the Germans – profited operationally from possessing even a small and outnumbered naval force.[3] These operations amounted to the coastal, or littoral, employment of inferior naval power, thus furthering their relevance for today.

German Naval Offensive Operations in the Baltic

The opening of the 1941 German invasion of the Soviet Union (Operation BARBAROSSA) presented the German Navy with several serious obstacles. The Soviet Navy in the Baltic seriously outnumbered the German and Finnish naval forces committed to support the invasion. While the Axis nations employed just two coastal defence ships, 12 submarines, 34 motor torpedo boats, and 70 mine warfare craft in their maritime offensive, the Soviet Baltic Red Banner Fleet could deploy two battleships, two heavy cruisers, 21 destroyers of various vintages, 65 submarines, 39 mine warfare craft, and 48 motor torpedo boats.[4] Admittedly, the minimal seaworthiness of many of the Soviet surface warships reduced their effectiveness.[5] Still, the German Navy, the *Kriegsmarine*, seemed even to lack interest in committing larger warships to support BARBAROSSA, despite the availability of several surface vessels.[6] As a consequence early Axis naval efforts in the Baltic would involve a force inferior to that of the Soviets.

Complicating German naval planning, the focus of the service's training also lay in the Baltic Sea. Upon learning of the Army's planned attack on the USSR, the *Kriegsmarine* set forth its priorities for displacing westward all vessels in training flotillas or performing work up periods.[7] As an indication of the minimal connection between the Army and the Navy, the movement advisory noted that the U-boat training force of 90 submarines would sail west eight weeks before the start of the campaign.[8] One surprised staff officer noted concern over the projected date, for the warning had been sent just nine days before the movement was to start, a rather limited amount of time given the scale of such an operation.[9]

Plans for opening sea routes within the Baltic emerged in May 1941, before the start of the campaign. Field Marshall Wilhelm Ritter von Leeb, the commander of Army Group North, approved of an operation to take three small Latvian islands at the mouth of the Gulf of Riga. Named Ösel, Dägo and Moon, the islands range between 10 and 100 kilometres in length, and cut the opening of the Gulf into two navigable channels.[10] The larger of the two routes opens 15 kilometres wide to the west, while the eastern route separates the islands from the Latvian mainland by at least five kilometres. With von Leeb's forces due to advance into the Soviet Union on the northern third of the front line, the planning offered as justifications the need to free up access to the port of Riga, and to allow sea movement of supplies for the Army.[11] The dominant territorial objective for the Army would be the Svorbe peninsula, a feature on the southwest end of Ösel and dominating the wider channel. From the perspective of Field Marshall von Leeb, only the peninsula needed capture, in order to open up the sea route to Riga. The Navy's Baltic commander, Admiral Rolf Carls, also sought to open the sea route to Riga, but insisted on the need to clear the southern end of the eastern channel as well, to protect shipping inside the Gulf.[12]

Prior to the opening of BARBAROSSA, German officers discussed the technical requirements of the Baltic landings. The *Kriegsmarine* suggested employing troops trained the year before for the cancelled invasion of Britain, Operation SEELÖWE. The assault craft collected for the aborted 1940 landing would not be available, however, so that only the combat engineer landing boats (*Sturmbooten*) from Army Group North could be employed.[13] These boats could only carry infantry, so the Army Group planned to use motorised pontoon bridges (*Siebelfähren*) and civilian vessels from Memel to ferry heavy equipment to the islands.[14] The Army planned to employ its engineers to transport Panzer Group 4 over the Dvina River running through Riga, so the islands could only be taken after the crossing of that obstacle. To supplement the lift for the operation, named BEOWULF, the Navy provided a small experimental unit (*Erproberungsverband Ostsee*), equipped with enough craft to transport a lightened infantry regiment and artillery battery.[15]

As with many German joint operations, command arrangements proved contentious. Field Marshall von Leeb's staff argued that the overall command of the

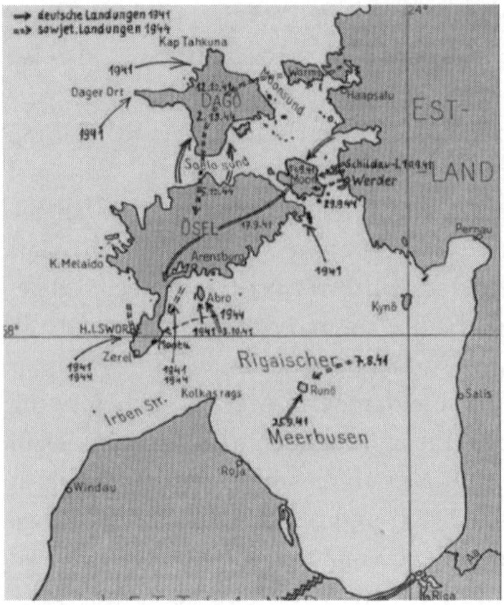

Plan of the German BEOWULF landing, 1941.
(US ARMY WAR COLLEGE, CARLISLE BARRACKS, PA)

operation needed to rest with an army officer.[16] In making its successful case, the military staff noted the isolation of the islands, the Army's predominant effort and the need to coordinate supporting fires as rationales for assuming command. As a result, the XXXXII Army Corps ran BEOWULF.[17] In point of fact the *Kriegsmarine* remained uninterested in attacking the islands, instead focusing its efforts on the defensive mining of the Gulf of Finland to preclude a Soviet Navy sortie. That effort sought to bottle the Soviets in their Leningrad base, but also allowed evacuation of sizeable Army forces from the cities of Talinn and Hangö.[18]

German planners debated how to capture the islands near Riga, with two schools of thought weighing the merits of landing on the western and eastern coasts.[19] These discussions lasted through the remaining weeks before the invasion, and in the end the planners elected to cross using the narrower eastern passage. That course of action took advantage of the engineering craft, freed up after the speedy crossing of the Dvina, just eight days after the start of the invasion.[20] In the end the BEOWULF landings would first assault the island of Moon from the east, with the troops ashore then attacking across a narrow causeway on to Ösel and completing their drive to the Svorbe peninsula. The order to the German 61st Infantry Division, the formation assigned to attack the islands, altered the rationale for the assault. From that time onward the need to curtail Soviet small craft attacks on seagoing traffic in the Baltic, as well as the threat to the *Reich* posed by aircraft based there, would be the basis for BEOWULF.[21]

In addition to the 61st Division, the German *Luftwaffe* provided three groups (*Geschwader*) totalling 250 aircraft, to support the attack on the Baltic islands. German naval participation came in the form of a single light cruiser, *Köln*, minesweeping vessels, and a mixed bag of patrol craft and fishing boats.[22] The Finnish Navy provided its two coastal defence ships, both armed with 10-inch guns, offering more muscular fire support. The Soviet defenders amounted to a mixture of infantry drawn from the 2nd Infantry Division and several odd lots of other units, totalling over 15,000 defenders, armed with over 200 pieces of coastal, anti-aircraft and field artillery, although some of the guns proved unusable.[23] The largest portion of Soviet forces remained on Ösel.

The operation opened on 13 September 1941 as the Axis ships sailed off the coast of all three islands, successfully diverting Soviet reinforcements from the real landing on Moon. These small scale operations employed flotillas of craft against the Svorbe peninsula and deployed other craft against the south coast of Ösel. During these diversions, one of the two Finnish warships struck a mine and sank, as did a German minesweeper.[24] Near dawn on 14 September, 50 commandos from the German 800th 'Brandenburger' Regiment, set to land from fishing boats and four gliders, had been assigned to attack a battery of coastal artillery stationed at Kubessarre on Ösel and threatening the landing area on the eastern side of Moon.[25] Instead the Brandenburgers mistakenly landed directly in front of the guns and, pinned down by heavy fire, had to be evacuated.[26]

By omission, the main assault on Moon proved the need for intense fire support to suppress the defences and allow an amphibious operation to succeed. The artillery and air attacks achieved little, except to dig craters in the beach landing area, providing cover for the attacking infantry.[27] While crossing the five-kilometre wide channel, one third of the battalion in the assault wave became lost and made landfall on the mainland instead, leaving just two infantry companies to secure the real beachhead.[28] Fortunately for the attackers, the confusion sown by the naval forces allowed the infantry lodged ashore, bolstered by a second wave of infantry, to remain overnight. The next day the rest of the 61st Division crossed using the assault boats and ferries, allowing the Germans to secure Moon.

The capture of much of the island of Ösel took place quickly after Moon, with German infantry crossing the causeway joining the two islands and encountering serious resistance only during the attack on the Svorbe peninsula. Beginning on 21 September over 4000 Soviet defenders, backed by the bulk of the island's artillery, defended the narrow position and gave ground slowly.[29] During 26 and 28 September the *Kriegsmarine* cruisers *Leipzig* and *Emden*, as well as three torpedo boats, provided fire support to 61st Division soldiers as they assaulted the dug-in Soviet infantry. Despite these assaults, the Soviet defenders hung on until 5 October, aided by friendly aircraft on one occasion.

The occupation of Dagö came just one week later, and with the help of supporting fires from the light cruiser *Köln* the Germans completed its capture in nine days.[30] In total the attackers took 15,000 Soviet prisoners on all three islands, capturing 212 cannon, a dozen aircraft, and 20 small vessels. Attempts by some soldiers to escape by boat from Svorbe ended in failure as German light vessels intercepted and sank them.[31] German losses totalled more than 2800 killed and wounded.[32]

How then can we assess BEOWULF? The Germans achieved their goal of opening the Gulf of Riga, winning a tactical victory, but at a cost. Apart from the human loss, the small numbers of Axis ships sunk by mines worsened the balance of forces *vis-à-vis* the Soviet Navy in the Baltic. More importantly the assignment of air support amounting to 10 per cent of *Luftwaffe* strength on the Eastern Front seems to have been high, especially as the BARBAROSSA campaign had other goals which it ultimately failed to complete, such as capturing Leningrad and Moscow.

The German cruiser *Köln* in 1940. (GRAZEBROOK COLLECTION)

Without a commensurate naval effort to support the landings, however, a heavy air commitment became necessary to ensure the success of the landings. More fundamentally, the opportunity cost of not invading the Baltic islands must be considered – the weak Soviet ground forces, with few aircraft, posed a psychological threat, but little more. With the German military running out of steam in its efforts to conquer the USSR by September 1941, the wisdom of investing so much effort in an amphibious offensive seems questionable.

German Naval Power on the Defensive: the Mediterranean

In late 1942 an unbroken series of Allied offensive operations in the Mediterranean theatre began with the 8th Army's victory at El Alamein in October, and the successful combined Anglo-American Operation TORCH the following month. These forces cleared Tunisia of German and Italian forces in May 1943. In the wake of these Axis defeats, both Allied and German leaders recognised the fragility of continued Italian participation in the war. Allied successes in the July landing on Sicily, named Operation HUSKY, and the drive to clear the island of Axis forces made more fragile the ability of Italy to contribute as an Axis power, and dictator Benito Mussolini was deposed on 25 July 1943. For a time Italian forces continued to fight, but with a diminishing will.

The balance of forces in the Sicilian campaign did not favour the Axis, especially at sea. The Italian *Regia Marina* remained strong in numbers with seven battleships, 11 cruisers, and many smaller vessels. But its lack of aerial escorts, fuel shortages, and three years of war had sapped much of its ability to fight.[33] German U-boats sent to the Mediterranean had scored initial successes, but suffered high losses since the start of 1943.[34] In contrast Allied naval forces for HUSKY amounted to six battleships, two aircraft carriers, 15 cruisers, 128 destroyers, 26 submarines, and numerous landing vessels, a comparatively much stronger and more capable force.[35] Similarly the *Luftwaffe* and the Italian Air Force, having suffered grievous losses in planes and pilots, had lost air superiority over Sicily by the time Allied troops conducted HUSKY.[36] Only on the ground did Axis forces hold any hope for a successful defence on either side of Mount Etna, where elements of four German divisions and remaining Italian forces stalled the drive of seven stronger Allied divisions on Messina in late July and early August 1943.[37]

As the Sicilian campaign wore on, the prospect of Italian collapse weighed on German leaders at the highest levels, including the *Führer*, Adolf Hitler. In late July, *Kriegsmarine* leaders suggested to him that the Italians were working a 'double game', or double cross, and outlined the need for additional German military strength in Italy.[38] German staffs created contingency plans, key among them being Operation ACHSE, to occupy Italy and take control of the *Regia Marina* in its bases of La Spezia and Taranto.[39] These orders included preparations to evacuate Sicily, Sardinia, and Corsica, and to transport eight additional German divisions to Italy.

Despite these measures, contradictions over the possible consequences resulting from the defection of their ally bedevilled the Germans. On 5 August the Armed Forces High Command (OKW) reported to Hitler that the enclave around Messina could be held for an extended time, if given replacements and ammunition, and supported by aircraft. Should Italy surrender, however, German leaders 'must reckon with the loss of the mass of currently available south Italian (including Sicilian) troops. . . . [I]f the enemy has set his foot on Calabria it is too late.'[40] A naval report of the same day counted 22 LSTs (Landing Ships Tank) and 134 LCTs (Landing Craft Tank) in Sicily, enough to put ashore a forceful Allied threat in southern Italy.[41] Clearly, the Germans confronted the prospect of a weak sea flank opening up in case of an Italian collapse.

The seeming clarity of the threat belied confusion at the top: German military service leaders appeared before Hitler to disagree over the desirability of evacuating the Sicilian bridgehead. In the first ten days of August, Chief of Armed Forces Staff Alfred Jodl, citing the logistical challenges of supplying the Sicilian forces over lines continually struck by Allied air power, and the need to concentrate strength in mainland Italy, demanded a withdrawal from the island. Jodl cited the threat from the Allies who had begun assembling landing craft in Licata, possibly to land behind the Axis positions.[42] In response, Navy leader Grand Admiral Karl Dönitz argued that the troops on the island tied down enemy forces, preventing them from conducting operations elsewhere.[43] Somewhat incredibly, as late as 11 August, the *Kriegsmarine* commander variously argued that Italy might not change sides, that the timing of such a switch remained unknown, or that the troops might escape Sicily, only to be trapped on the Calabrian mainland instead. In these discussions Dönitz appears to have been saying what he expected his *Führer* wanted to hear. These disagreements between senior commanders, though, gave Hitler sufficient reason not to order an evacuation.

The German Army formation on Sicily, XIV Panzer Corps, confronted a daunting challenge. Allied ground forces had it outnumbered and possessed ample air and naval support. In particular, US Navy and Royal Navy warships provided gunfire support throughout the Sicilian campaign. For example the cruiser USS

Philadelphia provided much needed fire support to American troops which landed to outflank German positions on the north side of the island; the battleship HMS *Warspite* did the same on the east coast.[44] German reports noted how Allied warplanes repeatedly attacked Army rear areas and shipping in the Straits of Messina, in 'undiminished' strength.[45] Despite enjoying well established defences in the hills of northeastern Sicily, four powerful formations – the Hermann Göring Panzer Division, 15th and 29th Panzer Grenadier Divisions, and 2nd Parachute Division – frequently counterattacked to restore their front line. And XIV Panzer Corps also faced serious danger. The nearest substantial German ground reinforcement lay no closer than Apulia, 300 kilometres north by air, with only Italian units occupying the Calabrian 'toe' in between. With a water obstacle preventing XIV Panzer Corps from reacting quickly to an attack on the mainland, Allied control of the sea opened up the possibility of an unopposed landing to surround the Germans.[46]

The evacuation of Axis forces from Sicily, named Operation LEHRGANG, began on 11 August 1943 at the orders of the armed forces staff (*Wehrmacht Führungsstab*).[47] Initial German reports of the evacuation laconically noted the 'carrying out as planned of the disengagement of the last bridgehead position on Sicily'.[48] The same report went on to describe the hazard posed by a new Allied landing, which 'instead of having us fix the opponents to Sicily, fixes us to be attacked there. . . We must position ourselves to meet the threatened change, namely a strike at Calabria and south Italy'. Worked out at the staff officer level, the successful evacuation of four divisions from Sicily, half the German Army in Italy, would allow XIV Panzer Corps to achieve the final goal.[49]

To permit the evacuation, German troops on Sicily staged a fighting withdrawal towards the Straits of Messina, shortening their defensive lines and disengaging combat units and their equipment in succession. Once out of their defences, German troops used one of five embarkation points in Messina and to its north, along the shoreline closest to mainland Italy.[50] There, they boarded an assortment of vessels ferrying them to the opposite side. The width of the channel, between three and four kilometres, made the crossings relatively brief. Upon arrival in Calabria, German troops drove north and to relative safety.

Early German estimates of the evacuation foresaw five days as the time required for the pullout.[51] For the evacuation of 7000 soldiers per day from Sicily, the Germans employed a variety of watercraft, and generally crossed by day.[52] In the end, German and Italian commanders used seven days to complete the withdrawal of their forces from the island. Over 39,000 German soldiers, with 47 tanks, 9600 vehicles, 94 artillery pieces, and 17,000 tons of supplies, escaped. A larger number of Italian troops, as many as 62,000, did likewise, bringing along a further 227 vehicles and 41 fieldpieces as well.[53] Key to the Axis success had been the availability of 29 of the *Siebelfähren*, and about seven 220-ton *Marine-Fähr-Prähme* (MFP) landing craft.[54] These transports, supplementing the two ferries and three steamers, with a small naval escort of coastal patrol craft, S-boats, allowed over 100,000 Axis troops to escape.[55] Total German naval losses amounted to seven transport craft sunk, and one damaged, with Italian losses somewhat higher.[56]

To the pleasant surprise of the German planners of LEHRGANG the evacuation met with little Allied resistance, despite the daylight crossings. The lack of effective opposition stemmed from several factors, key among them being a lack of planning for such an eventuality. On 3 August, HUSKY commander General Harold Alexander had told his air and naval commanders, Air Chief Marshal Arthur Tedder and Fleet Admiral Andrew Cunningham, to watch for a German evacuation and coordinate their interruption of it.[57] In fact, neither subordinate went on to plan cutting off the German forces from the mainland. Alexander himself did not realise that the evacuation had begun until the end of its fourth day, that is, 14 August.[58] His subordinates on the ground could not speed their advance either. Admiral Cunningham later suggested that the British 8th Army commander, Field Marshal Bernard L. Montgomery, failed to exploit available sea power by landing troops on the east coast of Sicily. Such an outflanking manoeuvre might have turned the German positions and sped the Anglo-Canadian advance on Messina.[59] In contrast the US 7th Army commander, Lieutenant General George S. Patton, had ordered a set of amphibious landings along the Sicilian north coast, although the degree of their success had been open to question.

Strong defences frustrated Allied air attacks, proving critical for the successful escape of XIV Panzer Corps from being trapped in Sicily. German and Italian forces committed hundreds of anti-aircraft guns, a large proportion of them the feared 88-mm flak pieces, to ward off hostile aeroplanes. As a result, Tedder did not commit his heavy bomber force, commanded by Lieutenant General James Doolittle, to attack Messina, and sent in medium bombers to attack only by night, when German vessels were in port.[60] Given the defences, Allied pilots did not apparently press their attacks as aggressively as they might have done in a clearer command climate: while one quarter of Mediterranean Air Command sorties flew to attack targets in the Straits, only 31 of them fell victim to flak in almost three weeks of strikes.[61] Allied air superiority made little difference to the withdrawal.

USAAF B-17s over Messina. (NAVAL HISTORICAL CENTER)

Land defences also frustrated Allied sea power. To protect the evacuation from sea attack a large number of Axis field and heavy artillery pieces, ranging up to 280-mm calibre, also defended the withdrawal. Here too Allied operational decisions seem to have played a large role in restraining interference, for the two naval commanders, Cunningham and US Navy Rear Admiral H. Kent Hewitt, agreed not to commit destroyers or cruisers to interfere with the evacuation.[62] Given the heavy firepower devoted to protecting the Straits, the reluctance of the Royal Navy's 32 Motor Torpedo Boats (MTBs) to press home attacks inside the channel seems understandable.[63] But, the need to preserve naval forces for the upcoming invasion of the Italian mainland also seems to have prevented British and American commanders from using greater force to stop the German retreat.[64] As Admiral Cunningham later observed, 'There was no effective way of stopping them, either by sea or air', and the Axis forces got away.[65] An otherwise outstanding senior commander, Cunningham had infamously urged his forces to trap or destroy enemy troops in Tunisia just four months earlier.[66] His avoidance of the same brand of aggressiveness in the Sicilian campaign suggests a failure to consider the possibility; and missing the start of the evacuation, when combined with a much greater degree of caution, allowed the Germans to get away.

Tactically masterful and operationally successful, the German withdrawal preserved Army troops so they could fight another day, demonstrably lengthening the Italian campaign. But viewed more broadly, Operation LEHRGANG has the character of an individual suffering from a multiple personality disorder. The operation demonstrated how confusion at the senior levels of German leadership served only to raise the risks to important combat forces. Only their good fortune, compounded by Allied command confusion, allowed XIV Panzer Corps to continue existing. Still, on 8 September 1943, the Italians capitulated and German forces initiated Operation ACHSE. Their forces successfully occupied ports and communications throughout the country, although they failed to capture the vessels of the *Regia Marina*.[67] The warships sailed into Allied hands instead. In light of the Italian surrender, LEHRGANG allowed the Germans to turn a losing hand of cards into one that let them split the pot, and no more. Perhaps most surprisingly, the opportunity to do so arose as a result of omissions and caution on the part of the first team of Allied military leadership. With little imagination and less daring, and despite a preponderance of naval and air power at their disposal, the Allied failure to stop the Sicily evacuation allowed the Germans to drag out the Italian Campaign.

German Naval Power on the Defensive: the Black Sea

The failure of National Socialist fortunes throughout Europe added to pressure upon the German Navy to support the efforts of Army forces ashore, and the Sicilian evacuation reflected the situation on a medium scale. Increasingly, *Kriegsmarine* leaders and staff officers displayed interest in joint operations on the Soviet front also, making such support a centrepiece of their thought. While laudable in terms of supporting their Army brethren, by doing so they ran the risk of politicising naval operations, attracting the attention of a *Führer* looking for any sort of military success which he could find.[68]

German Army forces entered the Caucasus region of the Soviet Union in the summer of 1942, as Army Groups A and B respectively headed southeast and east to win the oil fields and Stalingrad, more than 600 kilometres from each other.[69] Failing to use the maritime resources in the Black Sea, however, the Axis nations ignored that sea as a supply route eastward through the first 15 months of the eastern war.[70] In large measure, the lack of naval support in the central and eastern Black Sea before late summer 1942 reflected the minimal forces available until then, the absence of a multi-service vision for the region, and the dissonances of the leadership of the German military services.[71] In particular, the Axis failed to recognise the potential of the port of Novorossiysk, 150 kilometres east of the Kerch Straits, as a potential supply head for the two Caucasus army groups.

Playing catch up, in early September 1942 the German 17th Army, part of Army Group A, crossed the four to ten-kilometre wide Kerch Straits separating the Crimea from the Caucasus and captured virtually all of Novorossiysk. While a small portion of the city remained occupied by holdout Soviet troops, a more determined effort to secure and employ it could have eased Axis supply problems in the autumn of 1942. In a reflection of the short sightedness of senior German military leadership however, 17th Army proved too weak to advance farther. Subsequently, the broad and diluted efforts in the Caucasus heightened Axis vulnerabilities and led directly to the Soviet offensive successes in November 1942. These operations pocketed the German 6th Army at Stalingrad, destroyed two further Italian and Romanian armies, and opened a 160 kilometre wide gap in the front line along the Chir and Don Rivers. Only in late 1942 did Hitler order maximum use of sea traffic to Taman to supply three additional armies strung out in the Caucasus.[72] To prevent their destruction, German leaders withdrew these forces as rapidly as possible to the northwest, leaving 17th Army to hold a small bridgehead on the Kuban peninsula.[73] In this disastrous climate a naval commitment to support the Army came to the fore.

During 1943 the available German naval units in the Black Sea proved minimal under the command of *Admiral Schwarzes Meer*. Somewhat surprisingly, six small *Kriegsmarine* U-boats operated with some success in the Black Sea after 1942, having been brought overland and down the Danube for operations against the Soviets. Further R- and E-boats – patrol craft – 23 minesweepers, 26 submarine chasers, and 25 *Marine-Artillerie-Leichter* (MAL), based upon the MFP, rounded out the German Black Sea flotilla.[74] Romania provided destroyers and smaller craft to escort convoys west of Sevastopol, and an occasional submarine to operate against Soviet forces, while Italian motor torpedo boats and midget submarines reinforced the offensive flotilla.[75] In contrast, by early 1943 the Soviet Black Sea Fleet amounted to one battleship, four cruisers, eight destroyers, 31 minesweepers, 67 motor torpedo boats, and 29 submarines.[76] While the Soviet naval force worked with more diffidence than its numbers suggest, Axis sea power clearly operated at a distinct numerical disadvantage.[77]

In the air, Axis forces had no advantage either. For much of 1942, a small portion of *Luftflotte* (Air Fleet) 4 had deterred Soviet naval forces from conducting offensive operations, quite effectively supporting Axis naval elements.[78] One historian later credited this detachment with so cowing the Soviet Black Sea Fleet commander, Admiral Filipp S. Oktyabrskii, that he curtailed naval shore bombardment and major landing missions. Following the disaster at Stalingrad the Germans detached their air group to support their threatened armies. With the Germans unable to offer air support in the region the Soviet naval squadron resumed its contest for the Black Sea, reinforced by more than 250 aircraft which established air superiority, at least over the landward portions of the region.[79]

For senior German leaders the situation in the Black Sea region demanded utmost attention by the *Kriegsmarine*. On 18 February 1943 Hitler declared to Dönitz, 'the Kerch Straits transport task holds decisive meaning for the eastern campaign.' From that date, keeping 17th Army supplied assumed a much greater degree of importance to the German naval forces in the Black Sea. Four days later the *Führer* reinforced his perspective, insisting 'the Kerch task is the most important since the Norwegian operation.'[80] The *Kriegsmarine* would begin to support vulnerable Army defences on the eastern front, elevating the practice to an 'often dramatic high point', at the very least to show the German political leadership it was doing something to help, and indirectly justifying its preponderant commitment to the western theatre of the European war.[81]

The sizeable force of more than 200,000 German and Romanian soldiers remaining in the Caucasus demanded a large scale daily supply flow of between 1500 and 2500 tons.[82] To maintain the delivery rate the Germans initially built a cable railway across the ice bound straits, supplemented and eventually replaced by convoys of MFPs. Initially the landing craft sailed in groups of two or three, enlarged later to groups of five or six craft, every day.[83] As a measure of their effort, in May 1943 the total of people and supplies delivered to Kuban came to:

Personnel	85,615
Vehicles	9689
Guns	645
Mail	541 tons
Supplies	97,011 tons[84]

The small size of the vessels used does not seem to have impaired their ability to keep Axis forces supplied. In point of fact, employing light shipping helped protect the effort and ensure deliveries: on at least two occasions the shallow draft of the landing craft supplying 17th Army rendered them immune to Soviet submarine torpedoes. Over the course of five months these convoys delivered 337,000 tons of supplies, as well as over 6000 infantry replacements and 190 cannon.[85]

The collapse of the southern wing of the German armies in winter 1943 did not allow the traffic to flow solely to the east, however. The need for troops mandated the transfer of some 17th Army formations to the north coast of the Black Sea. As a result, 105,000 soldiers, 45,000 horses, 7000 motor vehicles, and 12,000 horse-drawn vehicles moved westward from the Kuban bridgehead, over the Kerch Straits, and through the Crimea to reinforce the main front.[86] Given the thin margins by which the Germans stabilised the southern end of their line, the redeployment must have proven important. During the movement of the ground force, a mixture of minefields and German MAL gunboats kept at bay attacking Soviet light craft, preventing them from striking at the evacuation shipping.[87]

After the spring thaw and the failure of the German July offensive at Kursk, the situation on the southern end of the Axis front line deteriorated further. On Hitler's personal orders, in September 1943 a small force of MFPs, MAL gunboats, and other patrol craft offered fire support for Army troops retreating from Taganrog on the north shore of the Sea of Azov.[88] The same month the need for troops farther north had also increased, and the exposure of the Kuban position meant that the bridgehead would never become a springboard for another Axis offensive.[89] Indeed without a withdrawal to defend the Crimea, the survival of 17th Army could not be assured. These considerations mandated the departure of the 12 remaining divisions from the Caucasus, in Operation BRUNHILD.[90]

During the Kuban withdrawal the four flotillas of 100 MFP available to *Admiral Schwarzes Meer* proved instrumental in ferrying Army troops across the Kerch Strait, beginning early on 16 September 1943.[91] The extraction began by taking troops first from Novorossiysk, the most forward position, as demolition charges were used to destroy port facilities. One day after the evacuation started, *Kriegsmarine* staff reckoned the Soviets could respond with amphibious landings on the Crimean peninsula in an effort to cut off the withdrawing troops crossing to the Kerch side of the straits.[92] Such expectations proved pessimistic, as only Soviet aircraft and mines hindered the evacuation for its first 12 days, allowing the Germans to withdraw 'as planned' and with minimal loss. Soviet naval forces did land troops on 21 September at the abandoned port of Anapa, however, at the southern end of the former German defensive position, and followed up with a further landing on the morning of 25 September.[93] But the first half of the evacuation proceeded unhindered, as two picket lines of MAL gunboats, reinforced by MFPs, kept the Soviets away from both the north and south coasts of Kuban.[94]

Beginning on the night of 28 September Soviet small craft tangled with the lines of German pickets shielding the evacuation routes.[95] In all cases the gunboat line drove off these attackers. Facing such resistance by an aggressive German naval force, the small Azov Flotilla commanded by Rear Admiral Sergei Gorshkov could do no more. In open Black Sea waters the Soviets did worse. On 6 October two of their destroyers attempted to break through the gunboats, while a third shelled the port at Yalta. Engaged by the S-boats in the defensive line, the pair of destroyers withdrew and awaited the return of the larger vessel operating farther west.[96] Shortly before daybreak the submarine *U-9* attempted to torpedo the destroyers and missed. At first light, however, a squadron of German Junkers 87 'Stuka' dive bombers struck at the trio, damaging the largest. As one undamaged destroyer attempted to tow the damaged ship to safety, the German aircraft returned twice, damaging and then sinking all three of the warships.[97] Soviet forces did not lack aggressiveness in this case, but did show a marked lack of coordination between naval and air units. Against such a one dimensional threat, small Axis naval and air forces proved able to protect the evacuation.

Operation BRUNHILD successfully removed 17th Army from the Kuban bridgehead. In the period between 7 September and 10 October the Germans withdrew 202,447 Axis personnel, 54,664 horses, 15,237 motor vehicles, 20,600 wagons, 1196 artillery pieces, and 94,937 tons of materiel.[98] Such accomplishments came at an inexpensive price. Naval losses amounted to just four MFPs and one army landing craft sunk, most of them by air attack, and an additional four MFPs and five MALs stranded and destroyed by their crews.[99] Against relatively minor opposition the Germans had executed an impressive extraction of their forces.

The withdrawal completed, the Soviet military made good use of its own amphibious capability. Less than three weeks after the pullout of the German 17th Army from the Caucasus, a small Soviet Army force established a bridgehead on the Crimean side of the Kerch Straits, and their forces expanded the bridgehead during the autumn. Worse for the Germans, the success of their evacuation may have led them to underrate the Soviet threat. The subsequent retention of portions of the Crimean Peninsula, supplied by the *Kriegsmarine* for another six months, at Hitler's insistence for 'political reasons', did little for the German war effort. Indeed much of the profit realised from the BRUNHILD operation would be frittered away by chaining a German Army Corps to the defence of Sevastopol, with the approval of the senior leadership of the Navy.[100] Once again, operational cleverness did not mandate strategic talent.

Conclusion

More broadly speaking, what do these operations say about projecting naval power ashore in World War II? Put briefly, offensive amphibious operations, executed by an adequate force, took a great deal of preparation to be successful. One need only mention the German attempt to land troops by sea on Crete in 1941, the Japanese landing on Wake Island the same year, the British attempt to exploit the Italian surrender by occupying the Dodecanese islands in 1943, or the Anglo-American landing at Anzio in 1944 to realise the price of underestimating the needs of an amphibious attack. Successes took adequate strength and good planning. The fragility of the 1942 US lodgement at Guadalcanal shows how a navy enjoying even a rough parity of power with respect to its opponent ran stiff risks by attempting to seize the initiative and conduct offensive operations against a hostile coastline.

The defensive employment of naval power ashore proved a different matter. Given space to retreat, a modicum of planning, and an unsuspecting, unwilling, or unprepared foe, evacuation operations could save large formations of troops from capture or destruction. Exercising sea denial against such operations proved difficult, even when a navy enjoyed a preponderance of power. Examining other cases, including the British evacuations from Norway, Dunkirk, Greece, and Crete in 1940–41, and the Japanese withdrawal from Guadalcanal in 1943, as well as the German evacuations from Sardinia in 1943, from the Crimean peninsula in 1944, and from eastern Germany in the winter of 1945, would highlight the defensive importance of naval power in the conflict. That such operations were successful when confronting a stronger, and in some cases, well led enemy, suggests that creating a naval shield required a relatively low investment to yield a serious profit. In that sense, for those nations conducting defensive operations during the Second World War, a minimal naval capability could go a long way.

7. Sea Power Ashore: North Africa to Normandy

Stephen Prince

> *Since men live upon the land and not upon the sea, great issues between nations at war have always been decided – except in the rarest of cases – either by what your army can do against your enemy's territory and national life, or else by the fear of what the fleet makes it possible for your army to do.*
>
> Sir Julian Corbett, *Some Principles of Maritime Strategy*, 1911

Corbett's statement indicates the fundamental connection that should always exist between maritime and land operations. Sometimes this connection is obvious and intimate, while at other times it will be more subtle and structural. Allied operations from North Africa in 1940–43, through to the Normandy campaign of 1944 illustrate both the full scope of these linkages and their relevance to contemporary defence issues. Both the North African and Normandy campaigns have been subject to significant and original study in recent years, but usually this has focused on the land or air land battle and has neglected areas of fundamental maritime contribution.[1]

Before World War II much of the application of sea power ashore was obvious and direct. Following Trafalgar in 1805 the Royal Navy exploited the dominance it had secured at sea to support the British Army in Spain. Throughout the 19th century naval brigades were frequent elements of Britain's imperial expansion. At Jutland in 1916, Jellicoe and Beatty were both fighting in their first major naval battle, but both had also had first hand experience of fighting ashore in 'brown water' imperial operations.[2]

Generally these operations ashore were either against opponents with significantly inferior technology or organisation, or they were able to exploit maritime flexibility to avoid initial opposition in the littoral. Neither of these advantages applied at Gallipoli, leading to the first modern example of the most challenging form of applying seapower ashore – an opposed amphibious landing on an unprecedented scale. The impact of the Gallipoli experience is controversial. The doubtful strategic basis, heavy losses and ultimate failure of the campaign certainly caused serious discrediting of amphibious warfare and were seen by many as a confirmation of altered circumstances, whereby land power with the aid of technology had now countered many previous maritime advantages. It is important to remember, however, that the landings of April and August 1915 were both successful in establishing bridgeheads and were frustrated by an immobility of land warfare that was also seen on the Western Front.[3] As troops were withdrawn from Gallipoli, serious consideration was given to a further landing against Alexandretta and, in 1917, there were extensive plans for a 'Great Landing' in Flanders.[4]

Overall though, Gallipoli cast a long shadow and in British inter-war defence planning this was extended by the lack of amphibious requirements in defence planning assumptions about likely conflict scenarios. Amphibious warfare was studied jointly by the separate staff colleges, but it remained an area of almost everyone's interest but no-one's responsibility. Although reasonable *Manuals of Combined Operations* were produced in 1931 and 1938 it was only in the latter year that a small Inter-Services Training and Development Centre (ISTDC) was established at Eastney.[5] This reflected the assumption that, although a cadre of expertise and prototype equipment would be useful, any large-scale amphibious activity was an unlikely or distant prospect.[6]

The summer of 1940 witnessed a strategic earthquake which devastated Britain's and indeed much of the rest of the world's defence assumptions. German domination of Europe in alliance with Italy, coupled with the loss of all French resources, had never been considered by Britain. The situation was a crisis but maritime factors were significant in ensuring that this crisis did not slip into despair. Germany's limited naval power, particularly after its losses in Norway, and its complete lack of preparation for any form of amphibious warfare, meant that Hitler could not translate his formidable military power into a feasible invasion of the British Isles. This would have remained true even if the Royal Air Force's (RAF) Fighter Command had suffered heavier losses in the Battle of Britain, as the Royal Navy would have continued to operate from bases outside German fighter range and would have been particularly effective in disrupting shipping in the English Channel at night.[7]

Also, although the resources controlled by Germany had expanded dramatically, its economy still suffered fundamentally from lack of access to the diversity of global maritime trade. Recent scholarship has depreciated this effect but failed to appreciate the structural distortions which the maritime blockade of Europe imposed.[8] While the impact of shortages of key alloys was important, the most significant constraining issue remained the paucity of oil and associated lubricants and rubber. Germany never had access to more than 3 percent of the world's oil while the Allies were able to draw on 90 percent of it from across the globe.[9] Some substitution was possible, but it was both costly and marginal. Germany's considerable efforts at blockade running were likewise insufficient to change the picture.[10]

The results of the oil shortage manifested themselves across the Axis war effort, with the search for fuel resources being a major contribution to Germany's decision to invade the Soviet Union. Due to the lack of oil, the Army that launched the invasion was largely horse-drawn, limiting its mobility and flexibility in comparison with increasingly motorised Allied forces. Oil deficiencies, both in terms of volume and quality, also progressively degraded the vital air cover on which German ground forces depended. As the *Luftwaffe* declined in effectiveness German troops were increasingly forced to move only by night, something that further reduced their mobility, as horses are not naturally nocturnal.[11]

While many of the maritime strictures on Germany's continental forces would only emerge slowly, and as part of a wider mosaic of Allied pressures, one achievement of 1940–41 would underpin much of the British Empire's landward activity for the rest of the war. This was the series of evacuations from Norway, France, Greece and Crete. In these operations over 626,000 personnel were withdrawn, over 200,000 of them being from Britain's allies.[12] While equipment losses were heavy, the personnel saved would form the basis for enduring and expanded armies, and their survival was vital to maintaining Allied cohesion. This was especially true for Australia and New Zealand, where the proportionate losses could have been devastating. As the commander of the British Mediterranean fleet, Vice Admiral Andrew B. Cunningham, put it during the evacuation of Crete: 'It takes two years to build a battleship; it would take two hundred to rebuild the Commonwealth.'[13]

Indeed, it was the Commonwealth that provided 10 out of the 14 divisions which fought in North Africa from June 1940 to June 1942.[14] In late 1940 and early 1941, these forces prosecuted a devastating campaign against the Italian Army in North Africa, capturing over 130,000 prisoners and over 800 guns. While this success was largely due to the quality of the Commonwealth forces and their commander, Lieutenant-General Richard O'Connor, vital facilitation came from their maritime flank. The Commonwealth navies could operate freely along the coast because of the physical and psychological damage they had already inflicted on the Italian Navy, culminating in the attack on the enemy's major naval base at Taranto. The benefit of this Allied maritime activity was summarised by General Wavell, Commander-in-Chief Middle East:

> *The maintenance problem in this quick-moving operation over a distance of 500 miles would have been insurmountable without the navy's assistance in keeping open the sea supply lines and opening up of Salum, Bardia and Tobruk, thereby shortening the lines of communication and releasing motor transport for the vital task of stocking up successive field supply depots.*[15]

The limits of sea power's effect ashore were subsequently illustrated by its inability to prevent the advance of the German *Afrika Korps* under General Erwin Rommel. But its continuing relevance was also demonstrated by the naval sustainment of the besieged garrison in Tobruk between April and December 1941. Continued Allied possession of Tobruk both denied Rommel a major supply port and forced him to address a second front, factors which limited his advance. In

1942 Rommel's greater penetration into Egypt owed much to the rapid fall of Tobruk in June, allowing for improvement in German logistical sustainment together with access to the stocks captured there.[16]

The fall of Tobruk resulted in a significant Allied reinforcement of the Middle East theatre, now using not just Commonwealth but also American resources. Between January and August 1942 over 220,000 personnel were delivered to the Middle East, virtually without loss, a small part of the ten million administrative movements of Allied personnel by sea during the war, a global transport achievement which involved fewer than 8000 casualties.[17] In the run up to the Second Battle of El Alamein this was also supplemented by vast quantities of battle winning equipment, such as the 950 tanks delivered without loss between July and September 1942.[18]

A major part of achieving safe delivery of these resources was achieved by the extensive use of evasive routing, in modern terms a 'soft' loss of the optimum use of shipping space. This was principally achieved by sending shipping around the Cape of Good Hope rather than through the Mediterranean. The contrast with the very short maritime supply line for the Axis forces, however, could not have been stronger. The latter was subject to intensive joint attack by Royal Navy ships and submarines and RAF, Fleet Air Arm (FAA), US Army Air Force (USAAF) and Commonwealth aircraft, some of which operated from Malta. Using excellent signals intelligence, and exploiting the weakness of Axis maritime forces, great damage was done. Between August 1942 and May 1943 a third of all Axis supplies crossing the Mediterranean were lost, 64 percent of the casualties being caused by the Royal Navy, the great majority by Malta based submarines.[19] This 'hard' effect was also compounded by the further 'soft' effect of the unpredictability which such losses imposed on the Axis supply system. This uncertainty led to inefficient loading in an attempt to prevent complete losses in certain categories of equipment. While Allied ground forces in North Africa still had to endure stubborn and bloody fighting to achieve victory, the context of their campaign was significantly influenced by these Axis losses. Rommel described the loss of 40 percent of his fuel supplies at sea, just before the start of Second El Alamein, as 'shattering'.[20] By comparison, the heaviest loss rate Axis forces managed to inflict on Allied shipping crossing the Atlantic was 1.4 percent.[21]

A further factor in sustaining the success of Allied forces after El Alamein was the opening of a new African front in November 1942 – the first large scale amphibious assault since Gallipoli. The landing – Operation TORCH – marked a shift in the role of seapower in the Euro-Mediterranean theatre. Evolving from its shaping and supporting role, sea power was now connected immediately and intimately with the land battle. The first attempt in the theatre to apply sea power ashore directly had ended in disaster. The Dieppe raid of August 1942 had resulted in 4000 casualties, with the vast majority of the landing force never making it off the beaches.[22] The raid demonstrated the obstacles to developing a sufficient, professional amphibious capability from the tiny pre-war cadre of personnel and equipment. Also, both the Royal Navy and the RAF had shied away from providing the overwhelming firepower required to support the landing. The First Sea Lord allegedly told the Chief of Combined Operations, Admiral Mountbatten: 'Battleships by daylight off the French coast? You must be mad Dickie!'[23] Furthermore, innovations already in development, such as specialist command and control ships, were not yet available.

The 'under-insurance' of Dieppe would not be repeated in the later, sustained Allied invasions. But the oft claimed linkage between the lessons of Dieppe and subsequent success is not supported by the evidence.[24] The improvement was achieved by analysing a much broader range of experience, from a widespread range of operations and exercises, coupled with the growth of Anglo-American resources and inter-service cooperation.[25] The invasion of French North Africa was deemed viable only by taking the risk, which proved to be justified, that only short-term combat operations against Vichy French forces would have to be sustained.

The invasion of Sicily in July 1943 – Operation HUSKY – was also launched on the basis of limited opposition, but saw important developments in the sustainment of major formations by over-the-beach supply, such as the use of the DUKW amphibious truck. HUSKY was initially mounted with the aim of opening the Mediterranean for Allied shipping, but its success actually had much broader consequences ashore. The opening of a second front persuaded Hitler to terminate his already frustrated offensive on the Soviet front at Kursk. It also resulted in the deposition of Mussolini, although Italy's losses in Russia as well as North Africa had already set the context for this event. When Italy's armistice with the Allies came into force in September 1943, Germany was forced to deploy the equivalent of

Testing Grumman F4F-4 machine guns aboard USS *Ranger* en route to North Africa, November 1942. (NHC 80-G-30362)

twenty divisions to the Balkans, to fill the vacuum left by Italian troops, as well as a similar number to Italy.[26]

The Italian armistice was dependent on the next Allied invasion, that of the Italian mainland. The landings at Calabria were virtually unopposed, but those at Salerno encountered much greater opposition and extensive counter-attacks. The survival of the Salerno bridgehead was determined by a mixture of factors, and its destruction was never as likely as is often represented.[27] The advance of Commonwealth forces from Calabria was probably the single most important factor and extensive land-based bombing was also significant. Carrier-based fighters were important, however, in providing the dawn and dusk air cover which land-based aircraft could not, and naval gunnery limited the effectiveness of German combat power in the littoral. Heavy guns gave reliable fire support and broke up many German counter-attacks before they could engage the limited ground forces.[28] In the words of Kesselring's Chief of Staff, 'The greatest distress suffered by the troops was caused by the fire of ships, guns of heavy calibre, from which they could find no protection in the rocky soil.'[29]

The same effect would be achieved, on a more sustained basis, during the next amphibious assault – Operation SHINGLE – the Allied landing at Anzio in January 1944. The operation failed to achieve its primary aim of unhinging the German defences south of Rome through to a combination of factors. They included the failure of the linked Allied offensives in the vicinity of Monte Cassino as well as the small scale of the landing, because of the priority accorded the Allied build-up for the Normandy invasion. Overall, the landing relied on achieving a psychological advantage over the German command that was not actually accomplished. Once the Germans counter-attacked the psychological factor was reversed, as destruction of the bridgehead would have been a serious blow to Allied morale just before the Normandy landing. This determined the heavy scale of German counter-attacks. The survival of the bridgehead was primarily determined by the sacrifice of the Allied troops but naval gunfire again played a significant part, reducing the scale and cohesion of German counter-attacks as they approached the coast. This was particularly important as air power's role was again limited by range, given the vulnerability of any air basing within the small perimeter of the bridgehead.[30]

Sea power's influence in the land campaign in Italy is generally perceived to have declined after Anzio, given the lack of any further major landings and the concentration of maritime resources in North West Europe. The well established threat of amphibious assault still influenced the German command in Italy, even without activity and at a time when amphibious lift in the Mediterranean had been reduced to enough for only a single division. When General Alexander launched his Army Group offensive in May 1944 – Operation DIADEM – the deception plan was based on the threat of a further amphibious landing north of Rome. The overall commander of German forces in the Mediterranean theatre, Field Marshal Albert Kesselring, took the prospect of such a landing – and its potential to destroy his

Army – so seriously that he held his mobile reserves along the coast for over a week after the start of the ground offensive. By the time they were committed, his front had already ruptured and his forces were defeated in detail and pushed north of Rome, which fell only two days before D-Day at Normandy. By 6 June, six German divisions had been diverted to shore up the 20 or so reduced divisions in Italy.[31]

The diversion of these German forces had particular relevance for the Normandy landings because Germany's interior lines, and the extensive communications system of western Europe, meant that relatively little could be done to prevent German reinforcements arriving in Normandy. While their advance could be impeded, by sabotage and air attack, the most effective measure was to force, or persuade, the Germans to deploy their forces elsewhere. The contrasting mission for the Allied forces, deploying almost entirely from the sea, was summarised by Admiral Ramsay in his Special Order of the Day of 31 May 1944:

> *Our task in conjunction with the Merchant Navies of the UN and supported by the Allied Expeditionary Forces is to carry the Allied Expeditionary Force … to establish it in a secure bridgehead and to build it up and maintain it at a rate that will outmatch that of the enemy.*[32]

The growth of the Allied forces ashore was dependent upon the capacity of the landing ships and craft available, a crucial restriction which did not apply to Germany's land forces. The operational methods of the Allied armies were also based on the use of extensive fire support, particularly artillery and air support, and on complete mechanisation. Both these characteristics required extensive and reliable logistics for their effective application.[33]

Much of the Allied logistic chain reached back to North America and so trans-Atlantic security was a pre-requisite for a continental land campaign. This became even more urgent as large numbers of troops also had to cross the Atlantic. For this reason the defeat of the U-boats had been confirmed as a task of the first priority at the Casablanca Conference of January 1943. By the spring of 1944, 1,750,000 servicemen had been delivered safely to the UK, 500,000 of them Canadians. In the narrow waters of the English Channel, the problem of maintaining the sea control required to deploy the land forces, involved a more complex mosaic of threats. In addition to U-Boats, there were the threats of mines, air attack, torpedo boats and fast attack craft (FAC) – known as E-Boats to the Allies. The potential of the last to inflict mass casualties on landing forces was illustrated on the night of 27-28 April 1944 when nine E-Boats attacked the forces taking part in Exercise TIGER off the coast of Devon. A series of mishaps, and underlying confidence in Allied command of the sea, resulted in there being an escort of only one corvette which was unable to provide local sea control. Two US Landing Ships Tank (LST) were sunk and over 700 US personnel killed.[34]

The local defeat did not, however, indicate a general lack of preparation against this threat. Exploiting the confined Channel waters, there was an increasing program from the New Year onwards to establish Allied dominance. This important preparation is often neglected by coverage of Normandy that begins with the landings or focuses purely on the development of landing technology and plans. Operations TUNNEL and MAPLE concentrated on reducing German activity in the Channel by mining enemy units into their harbours and by night sweeps to destroy them directly. Many of the small German coastal units were only preserved by non-deployment until the invasion was launched, and this allowed Allied minesweeping almost up the German-occupied coast on the eve of D-Day. Consequently, the Germans concentrated their previously dispersed forces in Le Havre, close to the invasion beaches, providing RAF Bomber Command with a viable target on the night of 14-15 June. The raid destroyed three torpedo boats and eleven E-Boats, as well as killing many key staff officers. The fact that the FAC threat never recovered from this blow, indicates the joint coordination capabilities of the Allied campaign.[35]

The German U-Boat campaign took longer to develop, since the surprise which the Allies achieved, in terms of both timing and location of the landings, meant that it took time for the U-Boat force to deploy and come to readiness. It would be the end of June before a U-Boat sank an Allied landing ship, and throughout the campaign the 'exchange rate' between U-boats and Allied logistic shipping would remain at more than one submarine per ship. Much of this success was an endorsement of the Admiralty's long-term policy of aggressively hunting down U-Boats, even after convoys had been secured from attack. The policy had effectively 'de-skilled' the U-Boat crews so that, by the summer of 1944, 47 percent of German submarine commanders had less than three months experience and 84 percent had less than twelve months.[36]

Frustrated in their conventional response, the Germans also launched an extensive asymmetric assault with midget submarines, explosive boats and pressure mines. By early August, the bridgehead was subject to up to 80 attacks a night, largely against the eastern flank because of the basing that was available to the Germans in Le Havre and the short range of their units. This assault was countered by the 'Trout Line', a picket force of armed landing craft, 200-300 metres apart, extending out from shore. This force used close-in weapons systems against both minor surface and sub-surface craft and mine-laying aircraft, as well as marking mine drops. Many Allied sailors and marines served up to 80 nights on this mission, comprehensively frustrating the German assault. The secure maritime communications on which the land battle was based were preserved.[37]

The security of Allied maritime communications should also be contrasted with another virtually unknown element of the campaign: the frustrated desire of the German forces to use coastal shipping. This was a natural element of economic activity along the French coast and was extensively used by German forces earlier in the war, with the first action in the Dieppe raid being a clash between the raiding force and a German coastal convoy. The removal of this option forced the German logistic system back on to a fixed land infrastructure, much more limited in its capacity to support major land battles and vulnerable to air disruption. This was demonstrated when the German land-based logistics began to fail in mid-July, a significant factor in the subsequent Allied breakout.[38]

Beyond the advantages of sea control, it is worth examining how that control was exploited to achieve advantage ashore. The most obvious means was through naval gunfire, sometimes using platforms of the Gallipoli era. On D-Day the Allied fleets opened fire at 0550 hours and continued to bombard the shore until between 0630 and 0745 hours, depending on H-Hour at the beaches they were supporting. They would fire approximately 20,000 shells on D-Day, ranging from four to fifteen inches in calibre. These were supplemented by 20,000 sixty pound rockets fired from specialist landing craft, as well as artillery that fired from landing craft as they approached the shore.[39] This naval fire support had the advantage of continuing until just before H-Hour, in contrast to the less graduated abilities of air bombardment, which was also more extensively disrupted by marginal weather conditions. The effectiveness of naval gunfire was, however, still intimately linked to Allied air superiority. Much of its accuracy derived from firing while ships were at anchor, something only possible when free from the threat of air attack. The impact of naval fire support was also increased by the reporting of the Air Spotting Pool, a combined RAF/FAA and USN group which was also effective due to its general freedom from self-defence tasks.[40]

The effect of naval bombardment has to be carefully assessed. Britain's official history of the war at sea considered the achievement fairly modest, principally on the grounds of the low – about 5 per cent – rate of targets destroyed. This seems too narrow a judgement, however, primarily reflecting a naval perception of effects. By contrast, land artillery often seeks to achieve a suppressive or neutralising effect, of great value to advancing troops even with very limited destruction.[41] Inspection of many of the remaining German defences in Normandy helps to illustrate this. Most gun emplacements feature a limited traverse and thick walls to the seaward side. This helped to preserve their guns from naval fire but also severely limited the period when they could engage Allied landing craft. Given that the assault troops were at their most vulnerable when packed in these vessels, when single missiles could cause mass casualties, this was a 'soft' effect of great significance. Added to this is the physical and psychological disruption of the defenders, even when their weapons remained intact. The overall effect was assessed by British Army Operational Research immediately after the landing. They estimated that, given the defences available, Allied infantry should have suffered an average of a 41 percent casualty rate crossing the invasion beaches; if those weapons were seriously disrupted they estimated the rate would be 15 percent. The actual rate was 14 per cent.[42] In some areas, such as sections of Omaha Beach, naval fire was also essential to regaining freedom of manoeuvre for assault forces fixed to the beach.[43]

There is even less discussion of naval bombardment post-D-Day, but another 40,000 shells were fired in the first month of the land campaign, proving

Officers from the battleship USS *Texas* examine a damaged German pillbox at Pointe du Hoc on 6 June 1944. Earlier in the day *Texas* had bombarded the point in support of the Omaha Beach landings. (NHC 80-G-235595)

particularly valuable in the period before the weight of Allied land artillery could be established ashore. The assault by lightly equipped US airborne forces on Carentan, an operation which linked the beaches into a single front, relied heavily on gunfire from US cruisers, and from D-Day onwards naval bombardment was again used to fragment German counter-attacks.[44] These attacks suffered physical damage, and psychological intimidation which often weakened the resolve of German commanders. On 10 June the designated German counter-attack headquarters, Panzer Group West, lost 17 key staff officers to intelligence-led attacks by RAF Typhoons. Naval gunfire completed the victory by destroying the headquarters radio regiment. Panzer Group West was non-operational for a month.[45] Naval gunfire had its last significant use during Operation CHARNWOOD, a British offensive against Caen on 7-8 July, which was also the first occasion when heavy bombers were used in direct support of Allied troops. This firepower compensation, as the Allied troops moved out of the littoral, would be repeated for all subsequent major offensives in Normandy, a development long foreseen by the senior Australian officer at the War Office.[46]

Beyond naval bombardment there was much maritime activity to achieve the rate of build-up and supply on which the Allied ground campaign was based. The effects of the Italian offensive, the Soviet offensive from 22 June onwards, the air attacks and exceptionally successful Allied deception measures all inhibited the German build-up, but could not prevent it. Many of the Allied vehicles deploying to France were waterproofed and brought over the beaches but this system imposed great limits on the vehicles, and Britain's stocks of the materials required for the process – tinplate and Bostick glue – were virtually exhausted early in the campaign.[47] From the evening of D-Day onwards block-ships were being sunk off all the invasion beaches to provide the sheltered water required for efficient unloading.

Two of these 'Gooseberries' were expanded to create the famous 'Mulberry' harbours, at Arromanches and off Omaha Beach, by D+10. Recent historiography has tended to depreciate the contribution of these harbours, seeing them as ingenious rather than significant.[48] But the system was in place just in time for the 'Great Storm' in the Channel from 19 to 22 June. While it is a myth that the storm was exceptionally strong, it was exceptionally destructive because it came from an unusual direction. The flow of Allied supplies was reduced from 20,000 tons a day to 5000, almost all of it through the artificial harbours, which also provided the shelter which allowed the small craft, vital to any unloading, to survive. Because of this and the security provided by air superiority the Allied logistic system was able to regenerate rapidly. By D+24, the Allies had landed 861,000 troops, 157,000 vehicles and 501,000 tons of stores. More than 97 percent of the troops and over 99 percent of the equipment had been delivered by sea. General Eisenhower's Chief Administrative Officer stated that, 'Commanders in the field had complete freedom of action as far as administrative arrangements were concerned.'[49]

This build-up came to fruition with the breakout achieved by the US offensive, Operation COBRA, at St Lô from 25 July. This success was achieved through a variety of factors. The German logistic system was breaking down through high levels of consumption and the vulnerability of its fixed land infrastructure, in contrast with the Allied system described above. The US Army had also skilfully adapted to the Normandy environment, and its air-ground element was exceptional, in terms of both scale and cooperation. But a strategic factor – almost never remarked upon – was also present. The Allied offensive exploited the safe sea flank by breaking through the German lines and then trapping an enemy corps against the coast. In contrast with the Allied forces at Tobruk or Anzio, this German corps was completely isolated from supply and could be defeated in detail, leaving the wide gap in the German lines that facilitated a genuine breakout.

This breakthrough was then followed up by a Canadian offensive in the east – Operation TOTALISE. This is of interest from the maritime point of view because confidence in the security of communications facilitated an important operational innovation. Self-propelled 105 mm guns were being given up by the Canadian Army, as they converted back to British pattern 25-pounders. The supply situation permitted the exceptional measure of decommissioning artillery pieces in the midst of a campaign, and converting them into the first armoured personnel carriers, dramatically reducing Allied infantry casualties.[50]

By late August, the Allied armies had broken out of Normandy. Their rapid advance towards Germany would eventually be arrested as much by the inability to transport supplies from the maritime beachheads, as by enemy resistance. The necessity of basing expeditionary land operations on firm support from the littoral would again be illustrated that autumn with extensive joint operations to open the Belgian port of Antwerp, a factor also recognised by Hitler when his Ardennes offensive targeted the same port that winter.[51]

This factor endures today as was demonstrated by the Coalition assault on the Al Faw peninsula during the Iraq War in 2003. This air and amphibious assault, with support from Anglo-Australian naval gunfire, was sufficiently flexible to accommodate last minute changes in the Coalition timetable. Its success allowed minehunters to sanitise the channel leading to Iraq's only major port, Um Qasr, and therefore met the strategic requirement of early and reliable delivery of large-scale humanitarian aid. The assault also secured the bulk oil exporting infrastructure which the Iraqi economy requires for recovery. The Royal Navy's current and planned future investments in people and capabilities to achieve the effective projection of sea power ashore have thus been recently vindicated, but the effect is entirely consistent with that which was demonstrated over 60 years ago.[53]

HMAS *Anzac* fires her five-inch gun at targets on the Al Faw peninsula, 21 March 2003.

8. Together Again for the First Time: The Army, the RAN and Amphibious Warfare 1942-1945

Ross Mallett

WHEN HE SURVEYED THE war situation in early 1942, Major General S. F. Rowell, the Army's Deputy Chief of the General Staff, conceded that a plan for the ultimate defeat of Japan was unlikely to materialise for some time, but suggested that combined operations would certainly be required. He therefore recommended that a start be made on the acquisition of the special equipment needed, such as landing craft and the establishment of a school of combined operations.[1]

The Naval Staff had heard that there was such a school in the United Kingdom and an approach was made to the head of the British Combined Operations Command, Commodore Lord Louis Mountbatten, RN, for assistance. He offered advisers, including Commodore G.L. Warren, RN, the Royal Navy's Senior Officer Assault Ships and Craft, and Commander F.N. Cook, RAN, who was then in command of HMS *Tormentor*, a Combined Operations Centre. Commander Cook was relieved at *Tormentor* by Commander A.E. Buchanan, RAN, who received six months intensive training in combined operations before himself returning to Australia.[2]

General Douglas MacArthur, the recently appointed Supreme Commander, South West Pacific Area (SWPA), agreed with Prime Minister Curtin that this was a good idea and concurred that the advisors should be authorised to communicate directly with the British authorities on technical matters.[3] The Australian chiefs accepted the offer, with an exception in the case of Commodore Warren, as they felt that he was too senior and might provoke an unfavourable reaction from the Americans.[4] Commander Cook arrived in Australia by air on 3 June 1942 and a couple of days later attended a conference on combined operations at MacArthur's general headquarters (GHQ). The attendees were informed of GHQ's intention to control combined operations.

A survey of prospective training centre sites along the coasts of northern New South Wales and south eastern Queensland was carried out, resulting in a June 1942 report, co-authored by Commander Cook, that recommended the establishment of a Combined Training Centre at Port Stephens in NSW, as a central establishment for the training staffs, beach parties and small boat crews, and a Combined Training

School at Bribie Island in Queensland for Army units.[5] A major change in amphibious warfare since the Great War had been the development of landing craft. Several types had appeared by mid-1942, when the Americans and British agreed on a standard nomenclature, including the 36-foot Landing Craft, Vehicle, Personnel (LCVP), the 50-foot Landing Craft, Mechanised (LCM) and the 120-foot Landing Craft, Tank (LCT).[6] Provision of landing craft was a problem, for the RAN had none. GHQ calculated that the three divisions required for an assault on Rabaul would need 324 LCVPs and 216 LCMs for operations, while training would require an additional 102 LCVPs and 42 LCMs.

Since one of the three divisions involved in MacArthur's proposed attack on Rabaul would be Australian, responsibility for manning the landing craft was split between the USN and RAN on a two to one basis, with the RAN's share being 50 officers and 660 ratings. MacArthur asked Washington to supply landing craft to the RAN on a Lend-Lease basis, reasoning that they could be obtained faster from the United States than they could be built locally, but was advised that his needs could not be met for some time became the supply and production of landing craft was inadequate, even for the minimum needs of the European theatre. Nor could the UK provide any.[7] The cancellation of operations in Europe in 1942 freed up landing craft for other theatres and 284 LCVPs and 172 LCMs were allocated to the SWPA. They still had to be shipped as deck cargo, however, limiting the rate at which they could be delivered. At most, a Liberty ship could carry twelve in this way, but had to compete with aircraft and other bulky items that could not be carried in the cargo holds. By 4 January 1943 only 66 LCVPs, 10 LCMs and 3 LCTs had actually been shipped.[8]

An alternative was local production. In February 1942, the Australian Naval Attaché in Washington was approached by the USN with a proposal for the construction of 1200 36-foot, 600 50-foot and 100 120-foot landing craft in Australia. The US would supply propellers, transmissions, shafts, and technical expertise. The plan foundered in June 1942 due to shortages of these very items arising from an augmented American program.[9]

Australian Army engineers designed and built their own landing craft, taking as their starting point the design of the British LCA (Landing Craft, Assault) supplied by the UK. Motor vehicle manufacturers were used because what few shipyards there were in Australia were fully occupied with ship repairs. Not only was there war damage, but routine maintenance had to be carried out in Australia because enemy occupation had closed ports in Europe and the Far East.[10] Designs were therefore kept very simple. The result was a series of Australian landing craft, manufactured by Ford at its plants in Brisbane and Geelong and powered by Ford V8 Mercury petrol engines imported from the United States under Lend-Lease 'for training, lighterage, ferrying and if approved by the Navy, as assault craft'.[11]

In late 1942, the amphibian engineers of the US Army's 2nd Engineer Amphibian Brigade arrived. Getting their 600 boats to Australia was a challenge.

They came up with the idea of sending them disassembled in cargo holds and having a special unit reassemble them in Australia. In such form, as many as 1000 LCVPs could theoretically be carried in a single ship. Of course, so many valuable craft would not be risked in just one ship. Cairns was chosen as the site for their boat assembly plant. [12]

In April 1943, the US Army concluded an agreement with the Commander-in-Chief, US Fleet (COMINCH),[13] Admiral E.J. King, USN, under which the US Navy would assume responsibility for amphibious training and operations, but MacArthur's three amphibian engineer brigades were exempted.[14] The result was that the majority of landing craft in the theatre were operated by the US and Australian armies.

A ship-to-shore operation required transports to carry the troops and boats, preferably fast ones that could outrun submarines and spend the minimum amount of time in areas subject to air attack. In seeking to make its agreed one-third contribution, the Naval Board considered how it might provide such ships. The RAN had three Armed Merchant Cruisers, liners that had been converted into naval auxiliaries. This expedient had not been entirely successful and the Royal Navy either converted its 50 to troop transports or returned them to trade.[15] The Naval Board decided to convert its three into amphibious transports, known as Landing Ships Infantry (LSI) by the British and Assault Transports (APA) by the Americans.[16] The first, HMAS *Manoora*, began its conversion at Garden Island in October 1942 and was recommissioned as an LSI on 2 February 1943.[17] HMAS *Kanimbla* followed her in December 1942 and HMAS *Westralia* in April 1943. Initially, *Manoora* could carry only 850 troops due to British style messing arrangements. When the others were converted, a cafeteria system was substituted so they could carry up to 1250 troops and *Manoora* was subsequently altered to bring her into line.[18] They were initially equipped with a mix of early model American and Australian landing craft, but by December 1943 production of LCVPs at Cairns reached the point where surplus craft became available to replace the petrol driven Australian landing craft, thus simplifying fuel stowage and reducing the fire hazard involved in storing large quantities of petrol.[19] Ultimately, *Kanimbla* and *Manoora* each carried 22 LCVPs and 3 LCMs, while *Westralia* carried 15 LCVPs and 2 LCMs.[20]

The Army was approached to provide experienced hands to work the ships' gear, handle landing craft and tactically load assault equipment. These became the 1st, 2nd and 3rd Landing Ship Detachments. Each consisted of a captain, a lieutenant and 65 other ranks. When troops were embarked, the captain handled matters pertaining to their billeting and messing while the lieutenant was responsible for tactical loading.[21]

The three LSIs were the largest and most important ships in the RAN. They were essential to the conduct of major amphibious operations but their impact was even greater. Because Australia was contributing the largest and most important element of the amphibious force, at least for the time being, its wishes with respect to amphibious training and doctrine had to be considered and respected.[22]

The Australian Army's Commander-in-Chief, General Sir T.A. Blamey, urged General MacArthur to press ahead with Combined Operations facilities, 'notwithstanding the difficulties in connection with equipment' and on 9 August 1942, MacArthur issued an order establishing a Joint Overseas Operational Training School (JOOTS) at Port Stephens. *Westralia* was allotted as an accommodation ship until the shore establishment was completed, with HMAS *Ping Wo* as a tender. The War Cabinet voted £73,800 to build the base, which contained both JOOTS and an RAN Training School, at an annual maintenance cost of £49,700.[23] The subsequent addition of extra berthing, repair and accommodation facilities caused cost increases and nearly £154,000 was spent by the end of 1944.[24]

Commander Cook commissioned the new naval establishment at Port Stephens as HMAS *Assault* on 1 September 1942, and training of officers and ratings as boat crews, beach parties and signal sections began.[25] The biggest obstacle was the lack of landing craft, with only two reaching Port Stephens before December 1942. Training was carried out using Army folding boats towed by twelve motor launches, much like the tows used for the landings in New Guinea and Gallipoli in the Great War. During December enough American landing craft arrived for full-scale exercises to begin. On being commissioned as LSIs, *Manoora* and *Kanimbla* were stationed at Port Stephens and *Assault* provided them with boat crews and beach parties.[26]

Meanwhile, a Combined Operations Training School had opened at Toorbul Point in Queensland on 1 July 1942, with the RAN represented by Lieutenant Commander J.M. Band, RANR. Construction was taken in hand by Army units, which gradually expanded the facilities over time, adding a slipway, loading jetty and workshops. A 12-man RAN party arrived with five motor launches on 5 August, just in time for the training of the 7th Division. As at Port Stephens, training was conducted by means of the launches towing folding boats. An old lighter was substituted for an LCM and in September, HMAS *Koopa*, a 416-ton requisitioned pleasure steamer, began masquerading as an assault transport. An American detachment of two officers and 40 ratings arrived in December. American craft soon arrived in reasonable numbers and by mid-March there were 20 LCVPs, 20 LCMs and three LCTs, which allowed training to be conducted with tanks and artillery and full-scale exercises.

To advise General Blamey and the Chiefs of Staff on training for and mounting of combined operations, a Combined Operations Section was formed under Commander Buchanan in January 1943.[27]

The North African and Guadalcanal operations of 1942 were conducted with amphibious transports similar to *Kanimbla*, *Manoora* and *Westralia*, but in 1943 another development in amphibious warfare occurred with the introduction of landing ships. These were larger than landing craft and capable of ocean voyages. The Landing Craft, Infantry (LCI) had no ramp as it was designed to carry infantry only, disembarking them by means of gangways on either side of the bow. The Landing Ship, Tank (LST) was a 1490-ton vessel with a bow ramp, its design ingeniously reconciling sufficient draught for seaworthiness with shallow draught for landings by using ballast tanks like those of submarines. If the beach grade was just right it could beach and discharge tanks or trucks into shoal water.[28] Their priority was so high that the keel of a US aircraft carrier was removed to enable LSTs to be built in her place.[29]

The cancellation of more operations in Europe provided another windfall for SWPA in the form of 72 LSTs, 36 LCIs and 72 LCTs. In December 1942, Admiral King created Amphibious Force Southwest Pacific and appointed the head of his Amphibious Warfare Section, Rear Admiral D.E. Barbey, USN, to command it. Arriving in Australia in January 1943, Barbey found that his new command existed mainly on paper and as late as April 1943 it consisted only of *Manoora*, the barely serviceable American APA, USS *Henry T. Allen*, and five LCTs. The rest arrived during 1943 as they were completed and made the trip across the Pacific.[30] This was quite a feat for small ships with very inexperienced crews.

GHQ charged Amphibious Force Southeast Pacific with responsibility for the coordination of all amphibious training except that related to the shore-to-shore operations of the amphibian engineers. Doctrine was based upon the US Navy's *FTP 167*, the US Army's *FM 31-5* and, the British War Office's *Combined Operations* 'with such changes as may be necessary for operations with United States naval and air force'.[31] This left a question mark over the role of the Combined Operations Section. General Blamey proposed that it remain responsible for the training of Australian forces.[32] This was not acceptable to MacArthur:

> *Combined operations to be successful must be accompanied by the closest cooperation between Army, Navy and Air Forces participating therein. By the agreement underlying the establishment of the South West Pacific Area, there is only one agency which has all three forces at its command. This agency is General Headquarters ... it is inadmissible that combat elements be placed under the jurisdiction of any Operations Section operating under the Australian Chiefs of Staff and not actually under my command. While the training of combat units will continue to be along national lines to the maximum extent possible, coordination must be complete and the units, while assigned to the South West Pacific area, must remain at all times under my operational command.*

He went on to suggest that Australians be assigned to Barbey's staff.[33] After discussing the matter with the admiral, Blamey decided to accept the offer, appointing Brigadier R.N.L. Hopkins, and the Combined Operations Section was discontinued.[34]

Because the same ships and resources were required for operations as well as training, the Australian Chief of Naval Staff, Admiral Sir Guy Royle, RN, at the request of the Commander, Allied Naval Forces, Vice Admiral A. S. Carpender,

USN, assigned *Assault* and *Manoora* and the entire naval side of Toorbul to the Amphibious Force Southwest Pacific on 26 February 1943. *Kanimbla* and *Westralia* were assigned when completed.[35] *Henry T. Allen* and *Manoora* sailed down to Port Phillip Bay where training was conducted with the US 1st Marine Division. On 1 October 1943, Admiral Barbey directed that the Amphibious Training Center at Port Stephens close and that all personnel, equipment and facilities be relocated to Toorbul.[36] Since opening in September 1942, it had trained 100 officers, 100 coxswains, 120 beach commandos, 453 boat crews, 250 stokers and 40 signalmen. The majority of the beach commandos were appointed to beach parties carried in the LSIs. The RAN School continued to train boat crews and beach parties until May 1944, when the Naval Board decided this was no longer needed. *Assault* was reduced to a small caretaker establishment in August 1944 and transferred to the Royal Navy in April 1945.[37]

General Blamey's 1943 campaign plan involved a double envelopment of the enemy position at Lae. Originally it was intended that the amphibious phase would be conducted by the amphibian engineers alone, but the size of the landing necessitated the involvement of VII Amphibious Force, as Amphibious Force Southwest Pacific was renamed on 15 August 1943.[38] The Army learned many important lessons from this operation. One was the need for a naval beach party to take soundings, mark the beaches and channels and handle communications between ship and shore. This task had been assigned to the amphibian engineers but they were not trained, equipped, or organised to deal with the Navy. The lack of a proper naval beach party was keenly felt, with confusion arising over beach lights and signs and the control of vessels.[39] USN doctrine held that these parties should be drawn from the assault transports but none were involved in the Lae operation.

For the next operation, at Finschhafen, Admiral Barbey proposed cutting out the amphibian engineers and replacing them with an Australian beach organisation. He felt that a more homogenous force would give better control and that an RAN beach party could work with the USN.[40] The Australian commander, Lieutenant General Sir E.F. Herring, toyed with the possibility, but it was shelved as there was insufficient time to organise a proper Australian organisation.[41] However, an eight-man RAN Beach Party was organised under Lieutenant Commander J.M. Band, RANR.[42]

The beach party was divided into two groups of four. Band's group landed from the destroyer transport (APD) USS *Brooks* at 0450 hours on 20 September 1943. They found themselves, along with most of the first wave, in Siki Cove, south of the intended landing beach, and under machine gun fire from ten log pillboxes. They advanced to the beach at the northern end of the cove, where they were forced to ground by the crossfire. Band stood up, shouted 'Come On!' and was hit by a burst of machinegun fire. He was evacuated to Buna on an LCI but died of his wounds and was buried at the Soputa War Cemetery. He was posthumously awarded the US Navy Cross.

The second group, under Sub-Lieutenant R.R. Barnes, RANR, arrived at Siki Cove on an LCI with the second wave at 0505. As the landing beach, Scarlet Beach, had not yet been marked, Barnes and his party flashed lights to seaward indicating the beach at Siki Cove and three LCIs beached there. His party moved up to Scarlet Beach, meeting up with the first party at around 0600 hours. By this time, LSTs had started to arrive off the beach, waiting for signs to be erected to mark what was believed to be foul beach at the south end of Scarlet Beach. Barnes travelled out to the LSTs in a landing craft to pilot them on to the beach. He met the commander of the amphibian engineers, who was unaware of the beach party and its role but, after consulting with the LST group commander, agreed to assign men and equipment, including an LCVP, to Barnes. They took soundings off the beach and discovered that what had appeared to be a shallow sand bar in aerial photographs was actually a white shingle bottom and that the beach was suitable for all types of landing craft.

That evening, with the help of the amphibian engineers, Barnes marked Scarlet Beach for the second wave of LSTs. He ascertained the locations of Army dumps from the Military Landing Officer, Major J.R. Broadbent, and arranged for the LSTs to beach in front of them, thereby hastening unloading. Unfortunately, due to a misunderstanding, the beach lights were not turned on until 2230 hours, the time at which the three LSTs were supposed to arrive on the beach, instead of a half-hour earlier when they arrived offshore.[43] The result was that they beached half an hour late and, although they stayed an hour later than scheduled, they ended up retracting with some bulk stores still on board.[44]

In October 1943, General Blamey approached Admiral Royle with a request for the release of beach parties from *Manoora* and *Westralia* for training with the 6th Division around Cairns.[45] Rightly or wrongly, the Army preferred British amphibious doctrine and felt that American training methods were not thorough enough. As the ships were assigned to VII Amphibious Force, Brigadier Hopkins obtained Admiral Barbey's agreement. Barbey mentioned that he intended to overcome the lack of properly trained beach parties by forming two independent USN beach parties and Hopkins presumed that they would replace the ships' beach parties in upcoming operations.

A few days later, Barbey reversed his decision. Hopkins tried to broker a deal, suggesting that the USN beach parties should work with the Australian ships, allowing the RAN parties to remain with the 6th Division. Barbey objected to this on the grounds that USN doctrine was that beach parties were an intrinsic part of their ships' complements. RAN doctrine was different, however, 'Beach parties for landing ships should not be regarded as appropriated to those ships. They belong to *Assault* and are allocated from time to time to the landing ships as necessary.'[46] Hopkins also felt that the use of USN beach parties would be 'repugnant to the RAN'.[47] Ironically, in most subsequent operations, the LSIs would operate with USN beach parties.

The alternative was for the RAN to raise a special beach party along USN lines. General Herring was particularly keen on RAN rather than USN beach parties, but

Admiral Royle was cooler in view of the RAN's tight manpower position.[48] A few days later, he went over Barbey's head to Carpender, who ordered *Manoora* and *Kanimbla* be made available for training with the 6th Division, along with a boat company of amphibian engineers.[49]

After a detailed examination of its requirements, the Army issued orders for the formation of the 1st Beach Group on 10 December 1943. This contained the logistic units that the Army had found to be required in amphibious operations and the RAN agreed to raise a beach commando for service with it. It was felt that Barbey could offer no objection so long as it was additional to the LSIs' beach parties.[50] The beach commando was organised along the lines of the RN Beach Commandos, which had been involved in every British amphibious operation since Dieppe in 1942.[51] It consisted of a headquarters of two officers and two ratings, three beach parties, each of 3 officers and 24 ratings, and a boat repair and recovery section of 32 ratings. Commander R.S. Pearson, RAN, who had been working on the beach group concept with the Army in Cairns, was appointed Principal Beachmaster. To round out the naval complement of the 1st Beach Group, a Beach Signal Section was formed consisting of an officer and 28 ratings. Ratings for the new units were drawn from *Assault* and the Flinders Naval Depot and all personnel were carried as permanent supernumeraries of *Assault*.[52]

In March 1944, the Army began forming a 2nd Beach Group and requested a second beach commando.[53] In line with British practice, commandos were lettered rather than numbered, the new unit becoming Beach Commando B. By this time the LSIs had carried out several major operations but their beach parties had not been utilised and their boats and crews had been put ashore to make way for those of the amphibian engineers. Now the beach parties were reduced to nuclei to provide personnel for the new commando.[54]

The Army then asked for another two beach commandos, two more beach signal sections, two naval demolition units and three lieutenants to be attached to Army landing craft companies.[55] By withdrawing excess officers from the LSIs and trimming boat crews, beach parties and boat repair sections, Beach Commandos C and D were raised. These were smaller than A and B, with only two beach parties and no boat repair sections.[56] A request for information about naval demolition parties was sent to the UK, eliciting a reply that the Navy was responsible for clearing underwater obstacles, a task which was carried out by Landing Craft Obstruction Clearance Units (LCOCUs). As the RAN had no personnel trained in demolitions, the Army agreed to train Army sappers to carry out the task.[57] The RAN agreed to an Army request that Commander Pearson be assigned to the 1st Combined Operations Section, a planning and training body. In reality, he was already performing this function. In November 1944, all four beach commandos were grouped together as the RAN Beach Unit (RANBU) under Pearson, whose title became Senior Naval Officer Beach Unit (SNOBU).[58]

When Colonel C.J. Cummings, commander of the 1st Beach Group, asked for clarification of the legal status of the RAN Beach Commandos, he was informed that:

(a) RAN [Beach] Commando A is under 1st Beach Group for rations and quarters only. For all adm[inistration] purposes the unit is under NOIC [Naval Officer in Charge] Cairns through Commander Pearson, RAN.
(b) Under no circumstances are naval personnel subject to military law. No Army officer can give a lawful command to a naval officer, petty officer or rating.

When Able Seaman A.E. Jones of Beach Commando A went on leave he discovered one of the drawbacks of his job:

I was dressed in my Army jungle greens with Army issue webbing belt, American cutdown gaiters and black military boots and my round Navy cap on my head which we had been told to wear whilst on leave and not our slouch hat. The provosts came up to me and said, 'Where did you get the sailors' cap soldier? I replied that I was not a soldier but a sailor; they asked me for identification which we did not carry. No sailor does as the Navy uniform is the thing that identifies a sailor. They promptly hauled me off to the Police Office in Fremantle and handed me over to the Naval Patrol, who said they had never heard of any Naval Commandos and put me in a cell. After about an hour and a half a high ranking naval officer appeared with an equally high ranking Army officer ... They proceeded to apologise for what had happened and said that they had not been notified that men of our unit would be on leave in WA.[59]

The beach commandos finally went into action on 30 April 1945, on Borneo. A seven-man beach party from Beach Commando B under Sub-Lieutenant J. Dighton, RANR, landed on Sadau Island. There was no opposition but the beach was backed by a razorback cliff and difficulty was experienced getting men and equipment ashore. USS *LSM 171* and USS *LCT 1331* were beached at the southern end of the island and 25-pounder guns were unloaded and soon in action, providing cover for the sappers of the 2/13th Field Company, who were attempting to clear the beach obstacles on Tarakan.[60]

They needed all the help they could get. Working in deep mud under small arms and mortar fire, the sappers blew gaps in a barbed wire obstacle with Bangalore torpedoes and removed wooden logs with gelignite charges. These were not heavy enough for the most menacing row of obstacles consisting of steel rails and wrought iron pipes, almost all of which had to be blown twice. Nonetheless, by H-Hour, the sappers had accomplished their mission and twelve 20-metre gaps had been cleared through the obstacles. Not since Tarawa had beach conditions been so formidable. The tidal range was about three metres with the low water mark about 100 metres from the dune line on the beach. To top it all off, there was a neap tide on P-Day, 1 May 1945, resulting in the tide being 30 minutes late and the first wave having to land on a 4 knot flood. The beach itself was mud which would not

LSTs high and dry at low tide following the Headquarters 26th Infantry Brigade amphibious landings at Tarakan, 1 May 1945. (AWM 090854)

support a man's weight when wet. Beyond was a wall of earth and logs, a water-filled ditch and soft clay before one finally reached the road. There were numerous bomb and shell craters. In selecting such a difficult beach, the Australians were relying on the Japanese adhering to their doctrine of not opposing landings on the beach.[61]

Supporting the landing on Tarakan was the 2nd Beach Group with RAN Beach Commando B, under Lieutenant Commander B.J.B. Morris, RANVR. Landing beaches were designated Red, Yellow and Green, from north to south, each with its own beach party: Beach Party 1, under Sub Lieutenant A.M. Aitchison, RANR, for Red Beach; Beach Party 3, under Lieutenant F.E. Dadd, RANR, for Yellow Beach; and Beach Party 2, under Lieutenant G.M. Maynes, RANR, for Green Beach. USN Beach Party 10 was also subordinated to Beach Commando B.[62]

The assault landings themselves were made on Red and Green, with the assault waves on Red in LCVPs and those on Green in amphibian tractors (LVTs).[63] The main beach was Red, which came under fire from a Japanese gun on the first night. Telegraphist W.F. Ryan was killed, Telegraphist J.R. Brady fatally wounded and Signalman C.A. Ikin wounded and evacuated. They were replaced by telegraphists from Naval Beach Signal Section D. A fourth beach commando, Sub-Lieutenant S.G. Lands, RANVR, was accidentally wounded in the leg on 3 May.

There were two piers in the assault area. The South Pier on Green Beach was 500 metres long with a 120-metre T-head in 10 metres of water at low tide. A 50-metre section of the pier had been destroyed, with two other sections damaged and a sniper had the pier in his sights, causing ten casualties on P-Day and P+1, but infantry could move along it over the 6-inch oil pipes. Lieutenant Maynes ordered personnel from *Westralia* and *Manoora* to disembark on the causeway section of the pier instead of Green Beach as planned. The pier was repaired and opened to motor vehicle traffic on P+3, by which time 400 tons of stores had been dumped on the 'T'. This was cleared away by the next day, allowing the attack cargo ship USS *Titania* to go alongside. Once *Titania*'s No. 5 hold was cleared, Commander Morris ordered the ship to use its gear to unload a US Army supply ship while still discharging its own cargo from its other holds. *Titania* completed unloading and sailed on P+8.

North Pier, on White Beach, immediately north of Red, was very narrow and could not handle traffic larger than a jeep and trailer. It too had been damaged, but the beach commandos opened it to LCIs carrying personnel and LCMs and LCVPs carrying stores from *Westralia* and *Manoora*. The 2nd Beach Group made a company of the 2/2nd Pioneer Battalion available to work the pier, which was repaired on P+1. Beach commandos developed the Sebenkok River as an LCM and LCVP unloading point. An LCM was beached with a bulldozer to perform this task, allowing LCMs to unload there from the evening of P-Day. Unfortunately, it was useable for only two to three hours around high tide.

Seven LSTs, each carrying a pontoon causeway, beached. But unloading was delayed for three hours until earth ramps were created to connect the pontoons to the shore. That evening, five more LSTs were beached and all twelve were unloaded during the night. In two cases, the exits from the beach were unavailable or obstructed and vehicles and stores were unloaded on to the pontoon causeways. The LSTs attempted to retract on the morning high tide, but were unable to do so. On P+2 they were lightened by pumping out their fuel and water tanks and five managed to retract. The remainder had to await the return of the spring tide, retracting on P+11, P+12 and P+13.

Commander Morris attended a conference with the Attack Group commander, Rear Admiral F.B. Royal, USN, on board his flagship, USS *Rocky Mount* on P+1. With twelve LSTs stuck on the beach, it was decided to unload the remaining eight in the stream. Top priority was given to unloading *Westralia* and *Manoora* and all available craft were deployed to help, allowing the two ships to depart that evening. To discharge the LSTs, Seabees of the USN 111th Naval Construction Battalion put eight 170-foot 'strings' of pontoon causeway together to form a pier, with two additional strings making a 'T' at one end, with two cranes. Five LSTs were unloaded in the stream while another 19 were unloaded on to the causeway, using LCMs as tugs. The remaining two strings of pontoons were used to construct an unloading point on Yellow Beach for LSMs.[64]

Despite the difficult nature of the Tarakan operation, the well-trained 2nd Beach Group and Beach Commando B worked in harmony and the commanders of the LSTs and Landing Craft Control Group gave Commander Morris high marks.[65]

The next operation, at Labuan in British Borneo, was carried out by the 1st Beach Group with RAN Beach Commando A, under Lieutenant Commander R. McKauge, RANVR, landing on Labuan Island. The beaches were reconnoitred by USN Underwater Demolition Team (UDT) 11 on the two days leading up to Z-Day, 10 June 1945, but no obstacles were found.[66] The assault beach parties travelled, somewhat uncomfortably, in LSTs with their stores preloaded in LCVPs. The rest of the beach commandos followed in *Manoora* and LCIs. Beach commandos of Beach Party 1 under Sub Lieutenant M.G. Parkinson, RANR, and Beach Party 2 under Lieutenant P.V. Merewether, RANR, landed with the second and fourth waves on Brown Beaches 1 and 2 near Victoria Town, followed by Beach Party 5, while USN Beach Party 10 landed with the third wave.

Members of the first wave of troops of 2/17 Infantry Battalion scrambling down nets slung over the side of HMAS *Kanimbla* into their LCVPs for the landings at Labuan 10 June 1945. (AWM 108937)

Meanwhile, RAN Beach Commando C, under Lieutenant E.H. Richards, RANR, handled the landings on the southern end of Brunei Bay, with Sub-Lieutenant Aitchison in charge of Beach Party 3 on Green Beach, Lieutenant M.P. Willsallen, RANR, of part of Beach Party 3 on White and Red Beaches, and Beach Party 4 under Lieutenant W.J. Rigg, RANVR, on Yellow Beach. Beaches were of soft sand and LVTs were used instead of LCVPs on all beaches except Green, where high surf was expected, and Red, where intricate manoeuvring was required.

This time, the beaching of LCIs, LSMs and LSTs was the responsibility of the LST Beaching Officer. The division of responsibility resulted in LSTs beaching without regard to priorities ashore. Some beached before the ones carrying vehicles required for their unloading and others before ARC mesh had been placed to enable their vehicles to cross the sand. Due to lack of Beach Group personnel and equipment, unloading came to a standstill during the night although this gave the sappers time to get the roads and beach exits readied and unloading proceeded smoothly over the following days.

The Army had trouble with its inter-beach communications and these were routed through the Beach Signal Section, whose SCR610 radio sets proved excellent for seaward communications. The 1st Beach Group erected a signal tower to facilitate visual signalling and the Beach Signal Station operated around the clock for 20 days.[67]

General MacArthur, observing the operation at first hand, pronounced it 'flawless',[68] and the task force commander, Rear Admiral F.B. Royal, USN, reckoned 'that it was the neatest job he had ever seen as far as the control and clearance of the beaches and stacking at dumps was concerned'.[69]

The final operation, at Balikpapan in Dutch Borneo, utilised Beach Commandos B and D, under Commander Morris. USN UDTs 11 and 18 arrived off Balikpapan on 24 June 1945. They found no mines but plenty of obstacles in which gaps had to be blown. This they did under enemy fire. 'The job at Balikpapan', recorded the US Naval Historian, 'was one of the bravest and best by the 'frogmen' during the entire war' and UDT 11 was awarded a US Presidential Unit Citation.[70] On F-Day, 1 July 1945, Beach Party 1, under Sub-Lieutenant Aitchison landed on Red Beach, Beach Party 3, under Sub-Lieutenant R.C. Eales, RANR, on Yellow Beach and Beach Party 2, under Lieutenant Dadd, on Green Beach. USN Beach Party 12 was also under Morris's command but the experiment of dividing command with the USN was not repeated.

Red Beach was soon closed, as it was unsuitable for craft larger than LCVPs. Yellow Beach took up some of the load but only Green Beach was capable of handling LCTs. Even so, at high tide LCTs had a 3 to 5-metre water gap with 60 to 90 centimetres of water at the ramp, which was sometimes too much for vehicles. DUKWs (amphibious trucks) had to be used for unloading, which was consequently slowed. LSMs and LSTs could not beach and had to be unloaded into LCMs and LCTs using labour supplied by the 2nd Beach Group.

Commander Morris ordered construction of a pontoon dock near the junction of Green and Yellow Beaches, using twelve of the fourteen available pontoon strings. The remaining two were used for a causeway on Green Beach for LSMs. The first LSM unloaded over the causeway. Seabees of the USN 111th Naval Construction Battalion worked on a U-shaped dock capable of handling four LSTs at a time. Morris chose the site on the advice of UDT 11 but Sub-Lieutenant W.A. Jaffray, RANR, of the Hydrographic Survey Group discovered a sand and coral bar off the end of the dock and Morris was forced to relocate it to the centre of Green Beach. Opened on F+2, it never handled more than two LSTs simultaneously due to the action of wind and surf on the horizontal part of the 'U', which was removed entirely on F+6. Still, it was never vacant, and greatly sped up unloading of LSTs.

As the road to the airstrip at Sepinggang was not capable of handling the RAAF's heavy construction equipment, the Beach Group commander requested a

A LCVP landing troops of the 7th Australian Division at Balikpapan, 1 July 1945. (AWM 128283)

hydrographic survey of the beach at Sepinggang which revealed that a 4-string pontoon causeway was required. Amphibious Group 8 provided this. On 6 July Sub-Lieutenant Aitchison was sent to open the new beach, called Orange Beach. RAAF equipment was unloaded on to the dock at Green Beach and then transhipped to Orange Beach in LCTs. At first it was thought that the hard sand could be used for a beach lateral road but a trailer immediately bogged in a soft patch. The Army sent an LCT with ARC mesh to cover it, which was laid by the RAAF and Beach Commandos. Unfortunately, that evening it was washed away by a storm which also created a sand bar off the end of the pontoon. The next day, Army sappers laid the lateral and the Seabees added a fifth string to the pontoon causeway.

On 4 July, Pearson and Morris selected a new landing site for LSTs on the south side of Balikpapan Harbour. The approaches were swept by USN minesweepers and ramps constructed by the 2/11th Field Company. The new beach, named Brown Beach, was opened on 10 July and was capable of handling eight LSTs at a time, unloading them in an average of ten hours. Pontoon strings were towed from Green Beach and laid side by side. Each was capable of handling two cranes and two LCTs. Cargo ships were then unloaded in the harbour by LCTs, using the LCT docks.[71]

Once again the Beach Commandos received high praise for their work and Commander Morris was awarded the US Bronze Star. After the war, both the Army and RAN agreed on the need for Beach Commandos and the preservation of amphibious warfare skills. In practice, however, the hard-earned experience was not maintained at any effective level.[72] The Beach Commandos, for example, found that their service records contained no mention of their Beach Unit service and, not surprisingly, felt that their accomplishments, along with the techniques of amphibious warfare, were being forgotten. It is to be hoped that in what seems to be an era of expeditionary warfare today, this will no longer be the case.

9. Tethered to Okinawa: The US Navy's Role in the Okinawa Operation

Jeffrey G. Barlow

THE BATTLE FOR OKINAWA proved to be the last major battle in the Pacific, and it was highly fortunate for the United States and its Allies that this was so.[1] During the fighting on the island, in the air above it, and in the nearby waters surrounding it, during the months of April, May, and June 1945 American soldiers and Marines, and Allied ships and aircraft suffered unexpectedly high casualties to Japanese defenders determined to fight to the death in order to stave off defeat. For the US Navy, forced to remain off Okinawa continuously during these many weeks in order to provide gunfire and logistic support to the troops ashore who were fighting desperately against the entrenched defenders, it proved to be a time of grim sacrifice.

In early October 1944, following the realisation that there were insufficient forces available to Admiral Chester W. Nimitz's Pacific Ocean Areas (POA) Theatre for carrying out an invasion of Formosa, the US Joint Chiefs of Staff directed that Nimitz plan to seize one or more positions in the Ryukyu Islands beginning on 1 March 1945. Two days later, the admiral notified his command that forces of the Pacific Ocean Areas would assault the island of Iwo Jima on 20 January 1945 and positions in the Ryukyus on 1 March.[2]

The Ryukyu Islands (called the Nansei Shoto by the Japanese) are a series of some 140 volcanic islands that stretch in a shallow arc southwest from the southernmost Japanese home island of Kyushu toward Formosa (Taiwan). The islands are divided into three major groupings: the Amami Gunto in the north, the Okinawa Gunto in the centre, and the Sakishima Gunto in the south. Together the islands of the Ryukyu archipelago form the boundary between the East China Sea and the Pacific Ocean.[3] Okinawa, the largest island in the Nansei Shoto chain and located only 360 miles south of Kyushu, was soon picked to be the principal target.

Preliminary planning for the invasion, designated Operation ICEBERG, followed the issuance of a CINCPOA joint staff study on 25 October 1944. The staff study directed the accomplishment of two principal tasks. That first was to '[c]apture, occupy, defend, and develop OKINAWA island and establish control of the sea and air in NANSEI SHOTO area'. The second was to extend control over the archipelago by capturing and developing other islands in the chain. In addition to providing a staging base for the eventual invasion of Japan's Home Islands, ICEBERG would serve to sever communications between Japan and its conquered Southeast Asian territories and secure Allied sea and air communications through the East Asian Sea to the coast of China and the Yangtze Valley.[4]

The major naval commanders selected for the operation were Admiral Raymond A. Spruance, Commander Fifth Fleet, Vice Admiral Richmond Kelly Turner, Commander Amphibious Forces Pacific Fleet and Commander Task Force (TF) 51 (Joint Expeditionary Force), and Vice Admiral Marc A. Mitscher, Commander Fast Carrier Task Force (TF 58). In Operation Plan 1-45, issued by Admiral Spruance on 3 January 1945, the date for the invasion of Okinawa (L-Day) was established as 1 April. On 9 February, Admiral Turner distributed his detailed plan covering the amphibious aspects of ICEBERG.

The assault on Okinawa and its nearby islands was to begin with a naval bombardment carried out by the Amphibious Support Force (TF 52), under the command of Rear Admiral William H.P. Blandy, eight days prior to the landings. Two days later (L-6), TF 52 was to seize a circular group of small islands called Kerama Retto, located some twenty miles southwest of Okinawa, in order to furnish the Allied forces with a close-in logistics anchorage and a seaplane base. The amphibious landings on Okinawa on L-Day were to take place across a five-mile expanse designated the 'Hagushi beaches' located on the southwest coast of the island. The expeditionary troops of Army Lieutenant General Simon Bolivar Buckner's Tenth Army would be landed by two Attack Forces. The Northern Attack Force (TF 53) under Rear Admiral Lawrence F. Reifsnider would put ashore two Marine divisions north of the town of Hagushi, while the Southern Attack Force (TF 55) commanded by Rear Admiral John L. Hall landed two Army infantry divisions to the south. Once ashore, the Marines were to push east and north, while the Army troops attacked to the east and south. In addition, on L-Day and L+1, a Demonstration Group (Task Group 51.2) under Rear Admiral Jerauld R. Wright, with a Marine division embarked, would stage simulated landings on the southeast coast of Okinawa as diversions.[5]

By early 1945, Japanese planners were fully aware that in order for the United States to mount an effective invasion of the Home Islands it would first need to seize one or more of several important off-shore island positions, including Formosa, the Nansei Shoto, and the Bonins. On 20 January 1945, Japan's Imperial General Headquarters issued the 'Outline of Army and Navy Operations' that called for waging a campaign of attrition in defence of these citadels. By this stage of the war the air forces of the Imperial Japanese Army (IJA) and Navy (IJN) constituted the country's most powerful offensive arms. But since the effectiveness of their conventional bombing capabilities had decreased drastically since mid-1943, both because of the low state of training of the IJA and IJN's existing pilots and because of sharp increases in the lethality of Allied air defences, the Japanese military had turned largely to the use of *kamikaze* (Divine Wind) special attack forces, that relied on suicide attacks against enemy shipping.[6] In early February 1945, the IJA and IJN decided to concentrate their air forces in the Home Islands at bases around the periphery of the East China Sea and to further emphasise *kamikaze* training. It was at this time that planning commenced for the *Ten-Go* Operation – employing massed air assaults to defeat invasions launched against the key island targets.[7]

On Okinawa itself, the 24th and 62nd Divisions and the 44th Independent Mixed Brigade, comprising Lieutenant General Mitsuru Ushijima's 32nd Army, had spent months expanding the natural caves that festooned the island and digging tunnels that would allow the Japanese troops to be protected from the expected air and naval bombardment that would accompany an American invasion. As historian Thomas Huber noted, 'The 32nd Army intended to, and did, move its entire force underground. The caves were made large enough to hold all personnel, weapons, ammunition, provisions, and 'all other material'.[8]

On 2 March, Admiral Spruance directed the Fast Carrier Task Force to strike Kyushu as the first action of the ICEBERG Operation. His hope was that by doing so, Mitscher's aircraft carriers would drastically reduce the Japanese planes available to attack the Allied forces off Okinawa in the initial days of the invasion. As his message to Mitscher explained, 'Enemy's present strategy seems to be to conserve forces by dispersal on ground and await favorable opportunity all out blow. My principal concern is to achieve surprise in order to prevent withdrawal to fields to north thus permitting destruction high percentage of grounded planes.'[9] Accordingly, TF 58 sortied from the lagoon at Ulithi at dawn on 14 March, headed for Kyushu.

Four days later, at 0545 hours, when at a position some 100 miles off the southern tip of Kyushu, Mitscher's carriers launched fighter sweeps against targets on the island. Because the Japanese had been alerted to TF 58's approach by aerial reconnaissance, the fighters on morning strikes found few enemy planes to hit on airfields close to the coast. However, afternoon sweeps ranging farther inland caught substantial numbers of Japanese aircraft on the ground. Carrier strikes on the second day (19 March), directed against warships in the Inland Sea on Honshu as well as aircraft and airfields there, proved fruitful as well.[10] In a preliminary report on the results of the air strikes on 18 and 19 March, Spruance stated that Mitscher's force had destroyed 200 Japanese aircraft in the air and 275 on the ground and had probably damaged or destroyed several hundred more.[11] In an amplifying report sent following the four days of action (the two days of strikes against Empire targets and fighting on the two subsequent days), Fifth Fleet reported destroying a total of 556 planes and damaging or probably destroying 175 more.[12] In addition to the hundreds of Japanese aircraft, the TF 58 fighter sweeps had also demolished a large number of hangars, shops, and other installations at many airfields on the two Home Islands.

Mitscher's carriers did not emerge unscathed from this action, however. While initial Japanese air action against the task force proved slight, three of the carriers were slightly damaged on 18 March. Things took a decided turn for the worse on the 19th, though, when Task Group (TG) 58.2, located some twenty miles north of the rest of the force, was attacked by Japanese planes. The carrier USS *Franklin* was halfway through launching a strike when a bomb dropped from a horizontal bomber struck her. The explosion started heavy fires that touched off her own munitions, and the ship was wracked by six major explosions. USS *Wasp*, in the same task group, was also hit by an enemy bomb, which exploded between her second and third decks. Although she was able to put out the fire, she had suffered serious damage. The following day, the carrier USS *Enterprise* was badly damaged when struck by two 5-inch anti-aircraft rounds fired by another American warship.[13] Because of the serious damage to these aircraft carriers, Admiral Mitscher was forced to reorganise his force. *Franklin*, *Wasp*, and *Enterprise* (organised as TG 58.2) were directed to proceed to Ulithi for repairs, accompanied by Destroyer Squadron 52. On 22 March, Mitscher headed towards Okinawa with TF 58, now reduced to three groups of carriers.[14]

Even as Mitscher's force was steaming south towards Okinawa, a new addition to Spruance's Fifth Fleet was entering the lists. TF 113 of the British Pacific Fleet had reported to Admiral Nimitz for duty on 15 March.[15] The British force, equivalent in size to an American carrier task group, was commanded by Vice Admiral H. Bernard Rawlings, RN. It consisted of the 1st Battle Squadron – the battleships HM Ships *King George V* and *Howe*; the 1st Carrier Squadron – the aircraft carriers HM Ships *Indomitable*, *Victorious*, *Illustrious*, and *Indefatigable*; five cruisers of the 4th Cruiser Squadron, including the New Zealand warship HMNZS *Gambia*; and 15 destroyers, including HMA Ships *Quickmatch* and *Quiberon*.[16] At Ulithi from 20 through 22 March, Rawlings' and Spruance's staffs worked out plans for the force's participation in ICEBERG. Redesignated TF 57 and operating separately from the rest of the Fifth Fleet, it was to be assigned the mission of neutralising the airfields located on the islands of Sakishima Gunto – the island group situated between Okinawa and Formosa.[17]

On 23 March aircraft from the American carriers struck targets in the Kerama Retto and on Okinawa, Amami, and Minami. Despite being hampered by bad weather, the planes bombed, rocketed, and strafed towns, airfields, and installations on the islands.[18] Aircraft from the Fast Carrier Task Force also took numerous photographs of Okinawa. When developed and analysed, these photos revealed a sobering situation for the American troops scheduled to invade the island in just nine days. As Admiral Mitscher informed Admiral Turner, the commander of the Joint Amphibious Force, on 25 March, 'From photos whole island of OKINAWA appears to be honeycombed with caves, tunnels and gun positions particularly along raods [*sic*]. Tanks and armored cars seen entering caves. It will probably be tough.'[19]

The first step in the ICEBERG Operation kicked off successfully on 26 March when Admiral Blandy's TF 52 landed troops from the 77th Division on five islands in Kerama Retto. Facing little or no opposition, the soldiers had secured the islands of Hokaji and Geruma by noon on the same day.[20] That night and early the following morning, however, Rear Admiral Morton L. Deyo's Gunfire & Covering Force (TF 54), which had been providing gunfire support for the landings on the 26th, was attacked by about 15 Japanese aircraft. Although six of these were destroyed by the combat air patrol or anti-aircraft fire, another six successfully crashed into TF 54 ships. Yet even though most of these ships remained operable, the battleship USS *Nevada* had her number three turret knocked out and her observation planes destroyed, while the destroyer USS *O'Brien* suffered extensive damage to her superstructure. In all, the ships lost 43 men killed or missing in action and another 98 wounded in these *kamikaze* crashes.[21] Unfortunately, this action provided only a first glimpse of what the naval forces arrayed off Okinawa were to suffer from Japanese suicide attacks during the following several months.

As the initial landings were taking place at Kerama Retto on 26 and 27 March, the carriers of Admiral Rawlings' TF 57 were carrying out their first combat operations against two islands in the Sakishima Gunto. During two days of strikes, British aircraft conducted some 574 sorties against airfields, installations, and barracks on Miyako and Ishigaki. They shot up more than twenty aircraft on the ground and destroyed the one Japanese plane that was airborne. But this first blooding was not without cost. The carriers lost six aircraft to enemy flak and another 11 to operational causes.[22] This was a not insignificant cost to a force whose complement of planes numbered just over 200.[23]

Because of the minimal number of Japanese troops present at Kerama Retto, the 77th Division completed its seizure of the individual islands quickly. The seaplane base in Kerama Retto was established on 28 March, and by 1800 hours the following day all of the islands were reported to have been secured.[24] On 31 March the Army occupied the small, undefended islets lying some eight miles off the southwest coast of Okinawa that made up Keise Shima and emplaced 155-mm guns in battery positions to provide supporting fire for the Okinawa landings.[25] Everything was in readiness for the following day's invasion.

When the US Army and Marine troops landed on Okinawa on Easter Sunday, 1 April, the Japanese, following their plan, held back from artillery fire against the beachhead and from responding to the initial American probes beyond the beaches. Indeed, the 32nd Army's main units remained in their interlocking underground positions situated for the defence of Shuri.[26] And because General Ushijima's troops chose not to oppose the landings on the Hagushi beaches, the initial stages of the US invasion went off without a hitch. As the CINCPAC/CINCPOA summary of the landings explained:

> *At 0830I [Item – local time] 1 April landings were made on schedule on the western beaches of OKINAWA. All beaches were rapidly secured and the advance inland commenced against light opposition. By 1100I our troops had seized the airfields YONTAN and KADENA and considerable numbers of tanks and*

artillery together with regimental reserves were being landed. Opposition throughout the day was light and by 1800I we had penetrated to a depth of approximately 5,000 yards from the initial landing points. By 1800 the beachhead was secured from small arms fire and 50,000 troops had been landed.[27]

The Marine 6th and 1st Divisions that landed north of the Bishi River were directed to push east across the island and north and the resistance offered by the Kunigami Detachment, although increasingly stubborn after the first few days, was gradually overcome. For the Army's 7th and 96th Divisions, landed south of the river and tasked with pushing east and south directly into what turned out to be the principal Japanese defences, it was a vastly different matter.[28] The CINCPAC/CINCPOA daily summaries of the ground fighting for 7 and 8 April – just a week after the invasion – tell the tale. The summary for the 7th noted:

In the north the 6th MarDiv continued to advance against little opposition ... In the south the XXIV Corps advanced over rough and clifflike terrain finding the enemy in deliberate defensive positions consisting of blockhouses, pillboxes, and in extensive trench systems protected by double apron barbed wire.[29]

Similarly, the summary for 8 April explained:

In the north, light opposition is being encountered; however, to the south limited gains were made only after exceptionally difficult fighting. Mine-fields were encountered together with fire from all types and calibers of weapons.[30]

For the Fifth Fleet ships lying off Okinawa this indication of increasingly difficult ground combat meant that for many weeks to come they could expect to be needed to provide the air and gunfire support and combat resupply operations desperately required by Buckner's Tenth Army troops.

For Spruance and the other commanders, one issue that concerned them during the initial days following the invasion was the timing of a major attack by Japanese *kamikaze* aircraft on the assembled ships. For more than a week prior to the 1 April landings, they had been receiving intelligence based on the decryption of Japanese messages that large-scale suicide attacks would be launched against the invading forces. Yet L-Day had passed with relatively little evidence of the *kamikazes*.[31] In fact, the Japanese effort to concentrate aircraft for Operation *Ten-Go* had been seriously disrupted by TF 58's mid-March attacks on Kyushu. During the first week in April, however, the IJA and IJN managed to complete a partial concentration of force to allow *Ten-Go* to proceed.[32]

Fortunately, Admiral Turner had made detailed preparations for defending the transport areas against Japanese suicide attacks, both from small boats and, more importantly, from incoming aircraft, by establishing anti-small craft and radar picket screens. The anti-small craft (or 'flycatcher') screen, designed to protect the transport area, particularly at night, consisted of station patrols by landing craft, infantry (LCI) vessels (backed up by destroyers), covering specific areas of the coastline.[33] The far more elaborate radar picket screen consisted of a series of up to 16 stations situated at sea from 21 to almost 96 miles from a point located just north of the Hagushi beaches – a cape (Zampa Misaki) designated Point Bolo. The line of picket stations looked something like a horseshoe, with the overwhelming number of stations located to the north and west of the island, the directions from which the *kamikazes* were expected to come. As it turned out, at the height of the battle only nine of the radar picket stations were manned, and after 16 May, when radar stations were in place ashore at the northern tip of Okinawa (Hedo Saki) and on the island of Ie Shima, the number of picket stations was reduced to five.[34] Each manned picket station was patrolled initially by a destroyer or by a high-speed minesweeper (DMS) and two landing craft, support (LCS) vessels. The destroyers were equipped with air search radars and had embarked fighter-director teams to furnish the fleet with early warning and vectoring information on incoming raids. During daylight hours, each station was to have four to six combat air patrol fighters operating overhead for protection.[35]

The first *Kikusui* (floating chrysanthemums) massed *kamikaze* attack against the US ships at Okinawa, which had been ordered by the Combined Fleet on 3 April, was finally launched on 6-7 April. The attacking force, consisting of some 699 aircraft, including 355 *kamikazes*, managed to sink two destroyers, a fast minesweeper, a landing ship tank (LST), and two ammunition ships, and caused major damage to ten other ships. It was a frightful loss of ships and crewmen. For the aircraft carriers of Admiral Mitscher's TF 58, it had fortunately been a much less horrendous day. Although the light carrier USS *Cabot* was just missed by a *kamikaze* that skimmed over her flight deck, only the light carrier USS *San Jacinto* was hit, only suffering slight damage.[36] An overwhelming number of the attacking planes had been shot down by combat air patrols and ship anti-aircraft fire in the course of the engagement, but enough suicide aircraft had got through to crash against their targets.

The second massed suicide attack occurred just five days later, on 12 April. That day, in all, the IJN and IJA sent some 380 planes against the US naval forces operating in the waters surrounding Okinawa. One hundred and eighty-five of the aircraft were *kamikazes*, and they included a number of *Oka* (cherry blossom) rocket-powered piloted

USS *Aaron Ward* after she had been damaged by suicide planes, May 1945. (NHC, NH62571)

bombs hung under Japanese 'Betty' bombers. This weapon, which the US Navy designated the *Baka* (foolish) bomb, was primarily a glider armed with a 2645-lb. warhead. It was designed to be released from its launching aircraft at a maximum altitude of 27,000 feet, after which it would glide at a speed of 230 miles per hour to its target area. Upon the pilot's ignition of the rocket engine during the final three miles of its approach, it would obtain a maximum speed of some 535 mph (in level flight) before striking its target.[37]

Heavy attacks by groups of suicide aircraft began coming in from the north just after 1300 hours and continued into the evening. The destroyer USS *Mannert L. Abele* was steaming in patrol area RP 14, north-northwest of Point Bolo, when she was attacked at about 1445 hours. She was struck on her starboard side just below the main deck and immediately behind her after engine room by a *kamikaze* carrying a large bomb. Just seconds later, a *Baka* bomb hit her on the same side at the waterline, behind the forward fireroom. The combined effect of the two nearly simultaneous explosions knocked out her power and broke her keel. She split in two and sank three minutes later.[38] This attack proved to be the first successful use by the Japanese of this new suicide weapon. The fast minesweeper USS *Jeffers* was patrolling in another area that afternoon when a second *Baka* bomb just missed her. As she later reported, '[O]n afternoon of 12 April while under attack by several planes quote twin engine bomber released apparently large bomb at 14,000 yards. Bomb had head on appearance of Tojo, traveled at very high accelerating speed, splashed 50 yards port beam. Believe bomb was pilot controlled though no pilot observed in bubble canopy.'[39]

The ships providing gunfire support off Okinawa's coast were also visited by the *kamikazes*, despite the protection provided by combat air patrol fighters from the carriers. One officer who recalled the attacks that day was Lieutenant Robert F. Wallace, USN, the assistant anti-aircraft officer on board the battleship USS *Idaho*. The ship was attacked by two Japanese 'Kate' suicide planes that were operating in tandem. Although the first aircraft was obliterated by a lucky direct hit from one of the battleship's 5-inch, 38-calibre guns, the second *kamikaze*, armed with a bomb, sneaked in close to the ship through the smoke of the first plane's explosion. Although the aircraft finally came under fire from *Idaho*'s 20-mm Oerlikon cannons, it didn't explode until just before striking the hull of the battleship. As Wallace remembered:

> *A bomb exploding so close to you does not make a booming sound, as heard in the movies. It makes a crackling sound. Even in our high-up position in the forward of the ship, we were literally showered with debris. A piece of engine block landed in Sky One, the 5-inch director above us, and I picked up from our deck an aluminum rivet with a copper wire attached.*
>
> *The official reports gave different estimates of the plane's distance from the ship when it exploded, ranging from 25 feet to 50 yards . . . I watched the plane*

to the moment of impact, and I would estimate it was more like 15 or 20 feet away, at most. A more distant explosion was not likely to do all the damage that this one caused. Eight blisters were flooded on Idaho's *port side, and the ship suffered extensive [superficial] damage from flying shrapnel and the shock of the explosion.*[40]

Luckily, however, sailors in the ship suffered only minor shrapnel wounds.

This second *Kikusui* attack proved costly for Spruance's forces. In all that day, despite the fact that TF 58's fighters shot down 153 Japanese aircraft and Admiral Turner's TF 51 ships accounted for another 128, enough suicide planes got through to damage 23 ships in addition to sinking *Abele* and the support landing craft *LCS 33*.[41]

During the continued fighting for Okinawa in the days following 12 April, another eight *Kikusui* mass attacks were flown against the Fifth Fleet. These attacks occurred on 15-16 April, 27-28 April, 3-4 May, 10-11 May, 24-25 May, 27-28 May, 3-7 June, and 21-22 June. As the attacks continued, their overall level of success diminished, both because the number of aircraft available decreased and because increasing numbers of Japanese pilots carrying out suicide missions lacked sufficient flying experience to evade the American fighter pilots flying combat air patrols. Nonetheless, US losses in ships and men steadily mounted.[42] The destroyers on the radar picket stations suffered particularly from the *kamikazes*. In fact, by the second week of April 1945, the losses to TF 51's escort ships had become so serious that on the 13th, Admiral Turner informed Admiral Spruance:

Since current operation commenced 6 escort vessels have been sunk due to enemy action. 22 have been so severely damaged that their return to duty in the near future is highly problematical. 16 have been damaged to the extent that they require overhaul in forward area . . . My [message] 020135 of March requested additional escort vessels over and above those already assigned TF 51 as it was believed that an insufficient number had been assigned. The losses incurred to date have made this situation even more critical.[43]

Spruance passed this alarming message on to Chester Nimitz at Guam.

Nimitz, who was already concerned that the prolonged fighting for Okinawa was leaving his naval forces open to increasingly heavy casualties from the *kamikazes*, decided to fly to the island to see for himself how the operation was progressing. He arrived at Okinawa on 22 April, and the following morning he and Spruance visited the frontline commanders. In a meeting with General Buckner, Nimitz declared that the Tenth Army would need to speed up its offensive in order to free the Fifth Fleet from its role of supporting the fighting ashore. When Buckner responded by telling the admiral that this was a ground operation, an angry but composed Nimitz retorted, 'Yes, but ground [operation] though it may be, I'm losing a ship and a half a day. So if this line isn't moving within five days, we'll get someone here to move it so we can all get out from under these stupid air attacks.'[44]

Yet despite the best efforts of the American Marine and Army divisions on Okinawa, the fighting slogged on into May and then into June. The longer the American aircraft carriers of TF 58 operated off Okinawa in support of the fighting ashore, the greater their chances of being hit and seriously damaged by the *kamikazes*. The British carriers of TF 57 striking targets in the Sakishima Gunto, south southwest of Okinawa, were exposed to significantly fewer numbers of Japanese suicide aircraft on a day-to-day basis, but, in addition, also possessed an advantage over their American counterparts. The British carriers had armoured flight decks. This advantage paid significant benefits in reducing their time out of action following *kamikaze* damage. For example, on 9 May 1945, TF 57 bombed the islands of Miyako and Ishigaki. That day the force was attacked by several *kamikazes*. Although British fighters splashed one, four others were able to dive on the warships. As Admiral Rawlings' subsequent report recounted:

> *Enemy next seen diving in succession. 1st [aircraft] steep dived [carrier] VICTORIOUS [and] hit flight deck forward. [Plane] 2 shallow dived VICTORIOUS glanced deck and fell in sea. Both enemy hard hit by gunfire. 3rd [plane] shallow dived [battleship] HOWE but [was] splashed close by gunfire. 4th [aircraft] struck [carrier] FORMIDABLE flight deck setting [her] after flight deck alight. Damage: VICTORIOUS hole in flight deck. Forward lift, accelerator and 10% armament out of action. Can operate few aircraft at reduced handling speed. . . . FORMIDABLE: two pompoms [antiaircraft guns] out of action [but] ship remains operational.*[45]

Although TF 57 withdrew that day to make repairs and replenish consumables, its commander expected it to be able to resume strikes against Sakishima Gunto within four days. It is no wonder that the report written by Captain Edward C. Ewen, USN, the American Special Naval Observer serving with the British Pacific Fleet, noted 'During the period covered by this report the effective resistance of the armored flight deck to suicide planes has again been demonstrated.'[46]

The American aircraft carriers, lacking such protection, suffered the consequences. Admiral Marc Mitscher, the Fast Carrier Task Force commander, experienced the situation first hand on 11 May, when the carrier he was riding – USS *Bunker Hill* – was hit. Captain William A. 'Gus' Read, USN, a member of Mitscher's staff who was the duty officer on the flag bridge that morning, recalled the attack:

> *Within a few seconds the first plane hit the ship. He dropped a bomb through the flight deck, then dove into the fighters warming up on the stern, demolishing three, and then going overboard himself. Since these planes were fueled to capacity, this immediately started fires aft.*
>
> *A second plane flew in on his wing, swung past the port side with a roar, climbed steeply, did a wingover and dove into the ship, crashing it into the flight deck at the side of the island and dropping a bomb through it into the hanger*

deck, prior to impact. The plane itself tore across the gallery deck and exploded in one of the pilots' ready rooms there, killing all hands. This, of course, started fires amidships and, unfortunately, wrecked the damage control station on the hanger deck . . .[47]

With thick smoke from the fires creating impossible working conditions on the flag bridge, Admiral Mitscher was forced to transfer his flag to the carrier *Enterprise* that afternoon. Ironically, three days later *Enterprise* was hit by a suicide plane, and Mitscher and his remaining staff were transferred for a second and final time to the carrier USS *Randolph*. It was in this ship that the weary and ailing admiral rode out the last days of his participation in the Okinawa campaign.[48]

USS *Bunker Hill* afire off Okinawa, after *kamikaze* hits, 11 May 1945. (NHC, K-5274)

Admiral Nimitz was convinced that the Fifth Fleet command team needed a rest after four months in combat. Accordingly, Spruance turned over command of the fleet to Admiral William F. Halsey on 27 May, and Marc Mitscher was relieved by John McCain two days later. On 10 June the Fast Carrier Task Force (now designated TF 38) was finally released from duty off Okinawa, after eighty-nine days at sea. Ashore, the Tenth Army units penetrated the last Japanese line of resistance on 14 June, and on the 21st of the month the island was declared secure.[49] It had proven to be a terrible, bloody campaign for everyone involved.

For the US Navy in particular it had been the most expensive naval campaign of the war. During 12 weeks of continuous combat, the Navy had lost 36 ships and craft sunk and a further 243 damaged – most of these losses from *kamikaze* attacks. The Navy's picket destroyers had been particularly hard hit, accounting for 31 percent of the ships struck by suicide aircraft, even though they made up only 11 percent of the forces involved. In all, the Navy had suffered more than 4900 men killed or missing in action, with another 4824 wounded, during the campaign. For their part, the IJN and IJA had lost more than 1800 pilots and crew in their aircraft attacks.[50] The US Navy could only be overwhelmingly grateful that Japan's surrender in September 1945 meant that it did not have to face the even greater test that would have been posed by the invasion of Kyushu scheduled for later that fall.

PART III

Naval power projection in the Cold War era

10. The US Navy and Marines in the Civil War in China 1945-47

Ronald Spector

SHORTLY AFTER THE FIRST atomic bomb was dropped on Hiroshima, General Albert C. Wedemeyer, Commander of US Forces in China, sent a message to the Pentagon: 'I view Asia as an enormous pot seething and boiling', Wedemeyer declared.[1] This was a somewhat surprising statement from a general whose country was about to win World War II. But Wedemeyer was in China where the Japanese had never really been defeated and where the US had never really been able to complete all of its elaborate plans and programs. Wedemeyer was painfully aware that the American backed forces of Chiang Kai-shek were poorly positioned to take advantage of an early end to the Pacific war.

By 1945 Chiang's government was war weary, ineffective and riddled with corruption and nepotism. Chiang had never been able to completely destroy his old rivals, the Chinese Communists, now led by Mao Zedong, and during the war had formed 'The United Front' with the Communists against the Japanese invaders. When Japan unexpectedly collapsed in August, Mao's guerilla forces were in effective control of large areas of northeast China and they demanded that the Japanese armies in the region surrender to them. The Allies directed the Japanese to surrender only to Chiang's forces, but the Communists were on the scene and Chiang's Nationalist troops were hundreds of miles away in southern and south eastern China.

Wedemeyer wanted seven American divisions sent to China as soon as the war was over, and he requested as an absolute minimum that two American divisions be sent to ports near Beijing, one to Shanghai and a fourth to Canton (now Guangzhou). But having just passed through the fires of World War II, Americans in Washington felt little desire to jump into Wedemeyer's boiling pot. General George C. Marshall, the Army Chief of Staff, promptly advised Wedemeyer that, 'Your proposal that we give China first priority over Japan and Korea will not be acceptable'. He told Wedemeyer the most he could expect would be two US divisions whose arrival would be dependent on the availability of shipping.

Despite this unpromising situation, the strength of American sea power in mid-1945, its amphibious capabilities and its sea lift capacity, all of which were a residue of the great fleet that had been assembled for the invasion of Japan, was more than sufficient to largely reverse the military situation in Northern China. The United States organised a gigantic air and sea lift operation to transport thousands of

Chiang's best troops to Northern China, at the same time, thousands of US Marines backed by an amphibious taskforce secured control of the principle port cities, the railroads and of Beijing.

The US Marine 3rd Amphibious Corps, commanded by General Keller E. Rockey, was all that General Wedemeyer received in response to his request for seven American divisions. To be sure, this was no small force; it consisted of two Marine divisions, veterans of Iwo Jima and Okinawa with their airwings and various logistical and support units. Once ashore, one division of the Marines would secure the important commercial hub of Tsientsin along the Tsientsin-Beijing railway line, as well as the key deep water port of Chinwangtao, the entry point for much of North China's coal supplies.

A second Marine division, the 6th, would land at Tsingtao on the Shandong Peninsula, which had the best port north of Shanghai. Meanwhile, Chiang's forces would be airlifted into Nanjing and Shanghai, and finally the ships that had brought the Marines would transport Nationalist troops to the north where they would arrive some time in October or November.

The Marines' mission was to assist the Nationalists in reoccupying key areas in Northern China and to supervise the disarmament and evacuation of the Japanese Army and civilians in China. While the presence of the Marines would have the practical effect of denying the cities and railroads in the North to the Communists, Wedemeyer's instructions from Washington insisted that American forces avoid involvement in China's internal war.

On 26 September landing craft carrying the first battalions of the 1st Marine Division crossed the bar at the mouth of the Hun Ho River and headed for the town of Tangku where the Marines boarded trains for Tientsin. Disembarking from their trains they were greeted by cheering crowds waving small American flags. 'It seemed that each one of Tientsin's two million people turned out for the parade', wrote an American correspondent:

> *Bottles of vodka, champagne, Chinese lanterns, flags and stray pieces of women's clothing were handed up to the men on the tanks. Watching the tumultuous welcome as American landing craft sailed up the river, Sergeant Richard Earle of the 5th Marines wondered whether the Chinese had done the same thing with all their many conquerors.*

A few days after these landings a battalion of the 7th Marines came ashore at Chinwangtao, and on 11 October the 6th Marine Division began unloading from transports in the harbour of Tsingtao. In the city, buildings lining the main street were covered with posters reading 'Welcome the US Navy' and 'The cooperation of the US and China is earnest'. Nobody among the Americans knew exactly what that meant, but they took it as friendly. Reporters also observed that Chinese restaurants which once served shark's fin soup or eight precious grains were now offering steak, with or without eggs.

Tsingtao was packed with thousands of refugees from the Soviet advance in the North, as well as with renegade puppet soldiers – that is Chinese who had fought for the puppet regimes of the Japanese – some of them still in uniform. Rioting and looting had broken out in the Japanese and German sections of the city. So within a few days of their arrival, heavily armed squads of Marines were patrolling the streets of Tsingtao trying to bring an end to the widespread unrest in the city. While the Marines were suppressing rioting, the city's mayor was concerned with more pressing problems, namely the need for cabarets, dancing girls and prostitutes for the American forces. Cabarets were available but there were no dancing girls in the city. The mayor tactfully suggested to his American liaison officer that the Americans import a boatload from Shanghai. Meanwhile, to meet the emergency, he had ordered an inventory of White Russian girls in Tsingtao.

In moving into their occupation areas, the Marines encountered opposition only on the road from Tientsin to Beijing. While attempting to remove a series of roadblocks at a narrow point in the road, a party of engineers was fired on by 40-50 Chinese Communists. The engineers withdrew, and returned the next day with a platoon of tanks and an entire company of Marines. As Navy fighters circled overhead, the Marines dismantled the roadblocks and proceeded to Beijing without incident.

Despite the incident on the Tientsin-Beijing road, the Chinese Communists had no real plans initially to try to oppose the Marines by force, although they believed that the American presence in Northern China could only hurt their struggle for control of the region. The American landings had come as a surprise to Mao Zedong, because he knew the Americans were withdrawing Army and Air Force units from China, and he did not believe that the US would risk a rupture with the Soviets by sending troops to the doorstep of Manchuria.

Mao's confusion was understandable. The US was demobilising its forces worldwide at a pace which Under Secretary of State Dean Acheson said amounted almost to disintegration. In China the Pentagon was rapidly disbanding Wedemeyer's Chinese Combat Command, which was the primary organisation for providing American military advice and training to Chiang's officers. Yet the US was far from disengaging in China. General suspicion and fear of the Soviets had been growing in Washington during the spring and summer of 1945, primarily in reaction to Russian actions in Eastern Europe, and this had spilt over into misgivings by some over Russian actions in China and Manchuria. Whatever Washington's motives, the Chinese Communist leaders knew that their forces were in no way up to a military confrontation with the Americans.

The Central Committee instructed party leaders in Shandong, Tientsin and Beijing to try to be on good terms with US military personnel and not to appear hostile: 'If the Americans open fire on us, commit illegal acts, or assist the Kuomintang in suppressing the people, make reports so that we may alert international public opinion'. Party members were assured that while US

Ambassador Patrick J. Hurley and General Wedemeyer were very reactionary, 'US political circles and public opinion sympathise with us to a high degree'. The single exception to the Communist policy of temporary passivity came at the port of Yantai, which was then called Chefoo, on the northern coast of Shandong across the gulf from Dairen in Manchuria. At Chefoo, unlike other cities in China, the Japanese garrison simply withdrew from the town as soon as the surrender had taken place and the puppet troops in the town quickly capitulated to the Communist forces.

Chefoo was very important to the Communists. From there they could ferry their troops across the gulf into Manchuria. The local communist commander was instructed to resist an American landing at all costs: 'only when not able to hinder or beat back their advance should you retreat'. When the American convoy carrying the 29th Marines and escorted by cruisers and destroyers appeared on 4 October 1945, however, the local party leaders decided that might not be a good idea, opting to negotiate and inviting the senior Americans to come ashore for talks. The American commanders, Admiral Daniel E. Barbey and General Keller E. Rockey, were also in a tight spot despite the menacing guns of their warships and the presence of a heavily armed regiment. They had no authorisation to force a landing, and were under instructions not to become involved in a Chinese civil war. The communist mayor of Chefoo told Barbey that the Communists held Americans in high esteem, and that they desired to cooperate with the United States. But he could see no reason for American forces to land there. There were no Japanese in the city, and Chefoo was in the hands of Chinese troops. These troops were supported by the people who had fought the enemy for many years. Chiang Kai-shek's troops would be resisted to the last man.

Vice Admiral Daniel E. Barbey (left) is relieved by Admiral Charles M. Cooke Jr. as Commander Seventh Fleet onboard USS *Estes* (AGC-12) off Shanghai, China. (NHC, 80-G-358247)

Two days of negotiations convinced Admiral Barbey that any landing of the Marines – with or without Chinese Nationalist troops – would be opposed. Barbey was ready to land, but he believed that this situation should be appreciated by the higher command. The admiral also expressed concern about the fate of Allied prisoners and internees still in Chinese Communist held areas. General Rockey recommended to Wedemeyer's headquarters that the landing at Chefoo be cancelled, and that the 29th Marines be landed with the rest of the 6th Marine Division at Tsingtao. General George E. Stratemeyer, Wedemeyer's deputy, was the acting theatre commander at the time and took the recommendation to Chiang Kai-shek for his opinion. Chiang understood perfectly well the strategic importance of Chefoo and was privately furious about the recommendation, but nevertheless saw no way to disagree.

Communist Party historians refer to this episode as 'the victory of Yen-Tai', and in important ways it was, because the communists kept the Americans out of this strategic port. But had there been a landing things would probably have been much worse. As General Rockey observed, 'a communistic opposition would not have been very serious but there would have been serious repercussions'. The US would have been involved in combat against the Chinese Communists at the very timewhen it was encouraging peace talks between Chiang Kai-shek and Mao in Chongqing. The Central Committee had in fact hoped that heroic fighting at Chefoo would cause international public opinion to become 'greatly disturbed', as they put it. For the US, however, the Chefoo incident was an early illustration of the mutually contradictory aspects of American policy in China: to help Chiang while remaining neutral in the civil war; to thwart the Communists but not to fight them; to confront the Soviets militarily while carrying out a worldwide demobilisation.

The Marines now held Tsingtao and Beijing as well as their connecting railroads, but vast areas of the surrounding countryside still belonged to the Communists. During mid-October transports of Admiral Barbey's 7th Amphibious Force began unloading troops of the 13th and 52nd Chinese armies at Kowloon and Haiphong in former French Indochina. They were to be transported to Northern China supposedly to relieve the Marines. When US Navy representatives had met with Chinese officers early in the month to plan the movement, it had appeared likely that there would be problems. Each ship, for example, would require American liaison officers who spoke Chinese, yet the China theatre's Chinese Combat Command – Wedemeyer's training and advisory organisation – was being disbanded, creating an acute shortage of qualified interpreters. This lack of communication may have accounted for some of the unwelcome surprises which occurred as the loading began.

The Chinese troops had brought with them large quantities of fresh meat and vegetables to feed themselves. But American doctors refused to allow any of these provisions to be stored in their ships, because the animals were suspected of carrying parasites and the vegetables had been grown using night soil. In addition to their edible animals, each Chinese division wanted to transport 2500 horses which the American ships, mostly assault transports, could not accommodate. There were hurried negotiations, and it was agreed that the Chinese would swap their animals for three dozen American trucks and that they would accept American C rations in place of their fresh produce.

As the Chinese, many of whom had never seen a ship before, began to come aboard medical officers gave them only a cursory inspection, and in effect locked them up in one end of the ship for the entire voyage. By the end of October all of

Barbey's ships, crammed with Chinese troops, were at sea bound for the port of Dairen at the tip of the Liaodong Peninsula in southern Manchuria. Dairen was still in the hands of the Red Army, and the Soviets informed the Chinese government that they would not permit Chiang's troops to disembark there since under the Sino-Soviet Treaty Dairen was a commercial port where the landing of troops was not allowed. Chongqing disputed this interpretation of the treaty vigorously but the Soviets still said no.

Desperate to get his forces into Manchuria quickly, Chiang accepted a Soviet suggestion that the Nationalist troops land at the smaller port of Huludao. General Stratemeyer agreed, and Barbey arrived off Huludao in his flagship on 27 October, only to be informed by the commander of the Chinese Communist New Fourth Army that the Communists already held the town and would not let the Nationalists land unless the Americans were ready to fight. Once again Barbey did not have authorisation to begin a war with the Communists, and once again he asked for instructions from China Theatre. China Theatre, anxious to free up the 7th Fleet's shipping, had already suggested to Chiang Kai-shek that the Nationalist forces land at Chinwangtao, an important port but not actually in Manchuria. Chinwangtao was already held by the Marines. Chiang Kai-shek very reluctantly agreed, and the 13th and 32nd Chinese Armies arrived there at the end of October. The Chinese Nationalist Forces would now be obliged to fight their way overland into Manchuria through mountain passes and by way of the long and insecure railway lines between Tientsin and Mukden.

Undiscouraged by all this, Chiang Kai-shek went ahead and attacked Manchuria anyway. He opened an offensive at the strategic pass at Shanhaiguan where the Great Wall meets the sea at the very edge of Manchuria. By 20 November 1945 he had captured both the pass and the town, and two days later Chiang's armies entered Huludao from the landward side, so this setback was not a critical one. But the repercussions of these battles in Manchuria threatened to involve the Marines and the Chinese 8th Route Army forces. The Marines were charged with keeping the railway lines open and ensuring the vital flow of coal into Beijing and Tientsin. Yet the same lines which carried the coal also carried troops and supplies for Chiang Kai-shek, and were critical in the battle for the entrance to Manchuria.

Nationalist troops were arriving by sea and air in Northern China. Yet Chiang Kai-shek still showed no inclination to use them to relieve the Marines. He did not even really want to send the Japanese troops home, as he had managed to enlist many of them in his army. 'The Generalissimo is determined to retain in their present areas the Marines in North China', reported Wedemeyer:

> *He visualises using the Marines as a base of manoeuvre. The Generalissimo would like to concentrate on plans based on conducting a campaign against the Chinese Communists instead of repatriating the Japanese. Such a campaign may require months or years.*

The Chinese Communists also saw the Marines as an important ally of the Nationalists. During October and November, incidents involving communist saboteurs and raiding parties increased. But despite the mutual frustration and suspicion, both the Americans and the Chinese Communists made extraordinary efforts to avoid open hostilities. The Communists continued to assist and return American flyers forced down by accident in the countryside, and the Central Committee directed that 'when we cannot dissuade US military personnel from assisting the Kuomintang our troops shall act only in self-defence, take up defensive positions and under no circumstances fire first'.

The Marines also received very restrictive orders. Wedemeyer instructed General Rockey that if Marines received fire from a village they were to inform the military leader or responsible authority in that village in writing that such fire was endangering American lives and must stop. Only then, if a satisfactory reply were not forthcoming, were they allowed to take action 'after ensuring that your warning to said military leader has been received and understood, and no action should be taken that jeopardises innocent lives'. Marine aviators were at first not issued live ammunition for patrols. Later they were permitted to shoot back if fired upon and the source was unmistakable, the target was in the open, the fire was in some volume and innocent people would not be endangered – otherwise they could go ahead and fire at will.

On the same day that Marines were working to protect the railway line and Chiang's armies were fighting in Manchuria, General Stratemeyer reported to Washington that now that Chiang's troops had reached Northern China it might be time to begin planning for the withdrawal of the Marines. Wedemeyer had already argued in messages to the Pentagon that if Washington really wished to avoid involvement in fraternal strife in China then the US must cease acting as the Generalissimo's transportation corps. The US must also withdraw American forces before they found themselves in combat with the Chinese Communists. If, on the other hand, the US really wanted to continue actively supporting Chiang, then General Wedemeyer would require a new directive.

In other words, Wedemeyer wanted Washington to make up its mind. He was concerned about Chiang Kai-shek, whom he felt was out of control. He thought that Chiang ought to concentrate on reforming his government and consolidating his control in the south. Instead Chiang seemed obsessed with regaining control of the whole country immediately and winning back Manchuria.

Among the handful of people in Washington privy to the secret cables from China, there was a lack of consensus about the mission of the Marines and its relationship to larger policy issues, about which there was also disagreement. There were those who accepted the public explanation that the Marines were there primarily to manage the surrender and repatriate the Japanese. There were those like Wedemeyer who saw American forces as indispensable in shoring up the Generalissimo against the communists and preserving order in China. And there were those like Secretary of War, John J. McCloy, and Navy Secretary, James

Forrestal, who saw the Marines as a way of keeping the Russians from becoming too ambitious in Manchuria. In the State Department the Assistant Secretary of State for East Asia, John Carter Vincent, saw the employment of the Marines as a likely path leading the US deep into the tar pit of China's internal wars. Vincent, a professional diplomat who had spent most of his career in China, had concluded that the Kuomintang had become a sterile bureaucracy dependent on moneyed interests and the military for support.

Wedemeyer's messages therefore touched off a flurry of meetings and memos in Washington. At a meeting of the Secretaries of State, War and the Navy on 6 November, McCloy called the group's attention to Wedemeyer's recommendations that the US cease transporting Nationalists troops to the North and make plans to withdraw the Marines. According to McCloy, the US faced a basic decision about how far to back Chiang Kai-shek. Chiang's prestige would suffer if the Marines were withdrawn. The debate continued over the next three weeks. The service Secretaries and the Joint Chiefs of Staff wished to continue supporting Chiang Kai-shek against the Communists by moving additional Chinese troops to the North, keeping the Marines in place. The State Department argued for withdrawing the Marines as soon as the Japanese had been sent home, and for pressuring Chiang Kai-shek to negotiate an agreement to end the war. Secretary of State Burns proposed a characteristic compromise – the Marines should stay but no additional Chinese troops should be moved to the North.

Meanwhile, American public uneasiness was increasing over the presence of the Marines in China. Congressman Mike Mansfield of Montana, recently returned from a visit to China, delivered a speech on the floor of the House warning of the US being drawn inextricably into China's internal affairs. Mansfield warned the State Department privately that the Russians might use the Marines as an excuse to stall in evacuating Manchuria. He reminded Truman that if the US became involved in a civil war it would violate our fundamental policy of non-interference in the internal affairs of other nations; it would also cause trouble among American troops who were already unhappy. Truman, of course, was well aware of the widespread discontent among American troops overseas with the apparently slow pace of post-war demobilisation. The Marines, however, were not so much discontented with living in China, which was a traditional job for Marines, as with the exceptionally ambiguous and restrictive aspects of their mission. The Administration's public explanation of the Marine deployment, that they were there to assist in disarming and repatriating the Japanese, was becoming less and less credible as time wore on and the Japanese were repatriated to Japan. The debate about the Marine's mission and the purpose of their presence in China sputtered on for the next three years.

By 1947 Chiang was losing his war with the Communists and incidents between Marines and Communist forces had become more serious. There were altogether at least eighteen Communist ambushes of Marines, resulting in casualties for both sides, between 1945 and 1948. Secretary of the Navy Forrestal continued to argue that the Marines were the basis of order in China, but by late 1947 they were simply irrelevant, and by 1948 they had been withdrawn to Tsingtao, which also served as the China base for the US 7th Fleet.

On 1 October 1949 Mao Zedong stood atop the Gate of Heavenly Peace and declared that 'the Chinese people have stood up … Ours will no longer be a nation subject to insult and humiliation'. The last Marines had already left Tsingtao six months before. In the new China there would be no room for foreign troops – friendly or otherwise, Western or Socialist – and the only warships based in Chinese ports and rivers would be Chinese.

The American occupation of North China demonstrates the capability of a dominant sea power to impact significantly upon a military and political contest ashore. Yet it also demonstrates the limitations of such intervention from the sea when there is neither a firm political basis for such action nor a settled policy to guide it.

11. The Shanghai Blockade and British Diplomatic Recognition of the PRC

Bruce Elleman

DURING THE NATIONALIST RETREAT from the mainland in 1949, the Republic of China (ROC) on Taiwan instituted a naval blockade of the People's Republic of China's (PRC) southern coastal regions, including a blockade of Shanghai. British trade, especially between Hong Kong and Shanghai, was seriously disrupted. The Nationalist blockade of the PRC was destined to last, in one form or another, from 1949 through to 1958.

Although porous, the blockade generally succeeded in interdicting China's coastal and international trade. It created significant monetary losses for British merchants, especially in Shanghai and Hong Kong. Immediately after the Nationalist blockade was instituted, the UK declared it illegal. Under the direct protection of British warships the North-South trade continued, albeit at reduced levels. Instead of strengthening Anglo-Nationalist relations, the Shanghai blockade arguably helped to alienate British public opinion, especially in Asia, which soon turned against the Nationalist government. In January 1950, the British government recognised the PRC, making it the first major Western nation to acknowledge the existence of Communist China.

This chapter will examine the Shanghai blockade's effect on Sino-British diplomatic relations. In particular, by recognising the newly formed PRC, the British government appeared to be supporting the Communists even while its Cold War ally, the United States, continued to support the Nationalists on Taiwan. This chapter will also ask whether British recognition of the PRC was primarily intended to break the blockade and reopen trade with China. Or was it perhaps part of a larger plan orchestrated by the British and Americans to break China away from its Communist mentor, the USSR?

The Chinese Nationalists Adopt a Blockade Strategy

In June 1949, the Nationalist government announced that all Chinese ports not under their control would be closed as of 18 June, and beginning at midnight on 25 June all of China's territorial waters would be closed to foreign vessels from a point just north of Taiwan all the way to Beijing (Peking). Major ports to be closed included Qinhuangdao, Tianjin, Shanghai, Ningbo, and Wenzhou. If foreign ships chose to ignore this prohibition, then the Nationalist government 'will accept no responsibility for the consequences'.[1] Because Shanghai represented a major trading port for British Far Eastern shipping, a debate immediately erupted within British circles over whether to support the Nationalist blockade, and thereby undermine the Communists, or oppose the blockade in an effort to keep Shanghai's port open as the final capitalist bastion within a Communist-ruled China.

The Nationalist blockade became a reality on 21 June 1949, when the Blue Funnel liner SS *Anchises* was bombed by Nationalist fighter bombers during a passage from Woosung to Shanghai on the Wangpu River. Her engine room quickly flooded and the ship sank into shallow water. *Anchises* was later refloated and towed up river where she discharged her cargo; on the way, she was bombed again but without any damage. Towed to Kobe, Japan, she was repaired before resuming commercial service.

The immediate British response to these events was embodied in a 22 June 1949 confidential report from the British embassy in Guangzhou (Canton) to its representative in Nanjing. This report concluded that the Nationalist blockade was illegal for two reasons. First because the Nationalist government had not accorded 'belligerent rights to the Chinese Communists', and second because the blockade was not effective, since 'Intermittent aircraft attacks on shipping cannot be held to constitute an effective blockade'. In addition, the recent Chinese attack on *Anchises* was 'flagrantly illegal', since to 'attack a ship without warning is not a legal method of enforcing a blockade'.[2]

Based on this analysis, the British government protested directly to the Nationalist government on 1 July 1949 that their action was not 'a valid proclamation of blockade in International Law' since war had never been officially declared and the British government had received 'no indication from the Chinese Nationalist government that they recognize, or are about to recognize, the Chinese Communists as having belligerent status.' Because the Nationalist Navy was not actively enforcing the blockade, 'a mere decree of a lawful Government purporting to close ports occupied by insurgents, without the maintenance of a real and effective blockade, cannot be regarded as valid'.[3] Furthermore, the Nationalist government was attempting to impose a blockade to close Chinese ports to all foreign trade, including 'even those ports which they do **not** control'.[4] This included, in particular, the port of Shanghai, whose loss could potentially cripple British trade throughout the region.

However, the Royal Navy was more optimistic about the Nationalists' ability to impose a true blockade. A 9 July 1949 report by the Admiralty's Intelligence Division concluded that retaining the Miao Islands would allow the Nationalists to blockade most of the Bo Hai and provide them with a '50% cover' of Qingdao. From the naval base at Dinghai, they could mount an 'effective' blockade of the Yangzi River in league with troops on the Miao Islands, but only a 'restricted' blockade if forced to act independently. Finally, if the Nationalist forces lost the Miao Islands and the entire Zhoushan area including Dinghai, the blockade would have no effect north of the Yangzi River, only a 'limited and sporadic' effect on trade entering and leaving the Yangzi, and the 'Nationalist blockade could only be effective south of Fuzhou'.[5]

On 18 July 1949, the Commander-in-Chief (CinC) of the Far East Station, Admiral Sir Patrick Brind, concluded that breaking the blockade by force by escorting 'our Merchant Ships to their destination', would 'certainly seem an intervention on [sic] the Communist' side. Even though some were arguing that the best way to defeat the Communists was to depend 'on Outposts of Western Civilisation like Shanghai at work behind the curtain', and so the blockade had to be terminated before it destroyed the foreign position in Shanghai, the CinC concluded that the most promising plan was to use diplomatic channels to 'persuade Nationalists on humanitarian grounds to allow an agreed number of ships to enter Shanghai and Tientsin with food and essential[s]'.[6]

Furthermore, the CinC argued that 'the fact remains that blockade is effective and is delaying Communists', and that when joined with the massive political problems in China 'may hold up Communist advance and bring about accommodation between last elements in the country. This presumably accords with our aim.' Therefore, he recommended: 'I am strongly averse in present circumstances to breaking the blockade which Nationalists appear to be operating very reasonably'.[7]

The British response to the Nationalist blockade was initially mixed, with the Foreign Office condemning the blockade and publicly advocating the continuation of normal trade with China, even if it fell to Communist rule. Meanwhile, the British military and naval authorities were generally more supportive of the Nationalist blockade against the Communists, and the CinC advocated that nothing be done to interfere with its success. Such divergent views meant that the British reaction to the Nationalist blockade was bound to be incoherent and uncoordinated.

Early British Negotiations with the PRC to Evade the Blockade

The Nationalist blockade of Shanghai, rather than solidifying British support for the Nationalist government in exile on Taiwan, created a division between those who supported the blockade and those who denounced it. Eventually, those who opposed the blockade gained the upper hand, and argued that the British government must consider opening diplomatic relations with the PRC. This policy shift did not happen at once, but came in gradual stages. Each stage merely emphasised how a naval operation – in this case the Nationalist blockade of Shanghai – had a direct bearing on the ongoing political discussion in London leading to the British recognition of the PRC.

There were many reasons why the British government might have wanted to break the blockade, including financial, humanitarian, and political concerns. According to one estimate, the British stake in China was estimated to be about

£200,000,000, with 'upwards of two thirds of this . . . concentrated in, or controlled from, Shanghai'.[8] According to another estimate from late July 1949, British firms in Shanghai were being forced to remit £350,000 sterling per month just to stay in business, and companies such as Sassoon and the British-American Tobacco Company were on the point of closing down their China operations. Meanwhile, thousands of foreigners, including many British citizens, were trapped in China. Food and medical supplies were running short. Relief ships, assuming that they could get through to Shanghai, 'would materially ease the[ir] position and firms might thereupon be ready to carry on at any rate for some time longer'.[9]

Finally, there were important political considerations. The British political goals included keeping 'a foot in the door in China' so as to have something to bargain with once the Communists consolidated power throughout the country. Although radically different from the US government's continued support for the Nationalists, this approach did not necessarily conflict with Washington's more hardline treatment of the Communists, and 'the United States agreed with this policy'.[10]

The first stage in the negotiating process took place rather quickly, on 19 July 1949, when a secret meeting was arranged between British representatives and a newspaperman named Han Ming, who appeared to have intimate contacts among the Communist leaders. Since the Communist forces did not have any naval assets that could challenge the Nationalist ships, it would be up to the British warships to escort merchant ships as far as Woosung, at which point 'Han said they would have to take their chance between Woosung and Shanghai but the [Communist] authorities would provide anti-aircraft artillery.' According to Han, the 'Communists were favourably disposed to turning a blind eye on the presence of escorting [British] warships below Woosung', and 'Han promised that a meeting could be arranged with the military in order to obtain any necessary assurance that the warships would not be fired on'.[11] Ten days later it was reported that Chou En-lai himself had agreed that British ships could penetrate as far as Woosung without any fear of being fired upon, that the Chinese Communists would provide anti-aircraft protection, and that 'They guarantee a speedy turn around'.[12]

These negotiations were rudely interrupted by the 30-31 July 1949 'Amethyst' incident, during which a British frigate that had been trapped for ten weeks upriver on the Yangzi River, HMS *Amethyst*, broke through the Chinese Communist barrier at the Woosung fort. During this incident, *Amethyst* was fired on by the Communists, and she returned that fire with her 4-inch gun. The British ship sustained 54 casualties of whom 22 were killed. It was a considerable embarrassment to the Communists when *Amethyst* escaped and rejoined the British fleet without any major damage or further casualties. But this event also made it more difficult for the Royal Navy to resort to force against the Nationalist blockade. As one study was to opine six months later: 'Public opinion would find it hard to reconcile action such as that previously taken by the 'Amethyst' in firing on the Communists with a state of affairs in which HM ships now had to fire at

Nationalists'.[13] But in late August 1949, Admiral Brind provided a top secret operational plan for escorting blockade-breakers that purported to do just this.

Included in his plan was the full expectation that should the Nationalist naval forces take 'hostile action at sea' then the British ships 'shall treat the operation as one of war [and] shall have full freedom of offensive and defensive action against any Nationalist military forces which might affect the attainment of our aim'. If conflict broke out between the UK and Taiwan, then Brind warned that 'In peace time we can hardly do less than allow for the most serious circumstances because a failure would be disastrous and casualty risks cannot be accepted as in War'.[14]

The scene on board HMS *Amethyst* when she came alongside at Hong Kong, showing the damage sustained to the bridge and superstructure. She came under heavy gunfire on two occasions during her dash down river, but suffered no further casualties. (A.L. PERRYMAN)

However, in a separate top secret memorandum logged in London six minutes later and focusing on 'wider military and political aspects', the CinC warned that escorting convoys to Shanghai would pull much-needed forces away from defending Hong Kong, and that 'It is hard to see how the possible gain to our hard pressed community at Shanghai will compensate for the risk we must consequently accept at Hong Kong'. Furthermore, any benefits that Britain would receive from Communist China by breaking the blockade would be illusory, and further he warned: 'I am convinced they will not … alter their hostile attitude towards us, provide any long term security for our trade or relax any intention they may have against Hong Kong, unless it is clearly in their own interests, or unless they are made to do so by force or circumstances'.[15]

During August 1949, the British government requested that the Nationalists provide 'safe-conduct for relief ships' bound for Shanghai, but their response was 'totally unsatisfactory'. In addition, the Legal Adviser, Vallat, warned that any British attempt to escort ships past the Nationalist blockade would involve trespassing into China's territorial waters, which could not be construed as 'innocent passage' and so the 'Nationalist Government might try to take us to the International Court where, if Mr. Vallat's doubts are justified, the findings of the Court might be against us'.[16] On 7 September 1949, the Foreign Office reported that on 29 August the Cabinet had decided to ask its own Law Officers whether a British escort through Chinese territorial waters would be legal, and the Law Officers agreed with Vallat that it would not.[17]

Clearly, so long as the British government refused to recognise the Communist government in China, then using force to break the Nationalist blockade not only risked conflict with the Nationalist government on Taiwan; Britain might also be brought before the International Court. One way out would be to recognise the

Communist government instead, since: 'It is conceivable that, by the time the matter can be taken to the Court, there will no longer be a National Government of China, or at any rate one which we recognize'.[18] This naval quandary would soon become one element underlying discussions in Britain over recognising the PRC.

Changes in the British Attitude toward Breaking the Shanghai Blockade

On 1 October 1949, Mao Zedong declared the formation of the People's Republic of China. The establishment of a central Communist government had been eagerly awaited for a long time, in particular because it was difficult to know with which Communist official to deal. As British recognition of the PRC became more likely, the British government once again became more concerned about upsetting the Chinese – this time the Chinese Communist government, not the Nationalists. This interval, lasting approximately 2-3 months, was the transitional period between disavowing the Nationalists and recognising the Communist government in their place.

On 9 September 1949, Admiral Brind recommended a policy to the Admiralty intended to 'prevent the arrest of British ships', rather than act as 'an excuse to break the closure' of Shanghai. This involved telling the Nationalist government that 'ships attempting to break the closure contravene no law, but it is up to the Nationalists to make the closure effective by intercepting and turning ships back'. The British government did 'not recognise the right to visit, search or arrest ships in these circumstances either outside or inside territorial waters', and British 'warships are only engaged in ensuring compliance with International Law and are not in any way helping to run the closure'.[19] But the Prime Minister's instructions expressly precluded any British warships from entering 'Chinese territorial waters save for humanitarian purposes'.[20]

On 18 September 1949, however, several foreign ships were able to break through the Nationalist blockade, largely because the crew of the Nationalist destroyer on station mutinied and then defected to the Communists. This incident even led to speculation that the Nationalist blockade was over.[21] By late September, a reported 13,000 tons of shipping had left Hong Kong, bound for Shanghai.[22] The number of British and American blockade-runners increased from late September through to the middle of December.

British relations with the Nationalists changed following the 1 October founding of the PRC. On 5 October 1949, Admiral Brind sent a more strictly worded message to Admiral Kwei, the Chief of the Nationalist Navy. Although the British government recognised 'the right of Nationalist warships to control merchant vessels within Chinese territorial waters', it did not 'recognize any right to interfere with British merchant vessels outside territorial waters or to hold them under arrest when they leave territorial waters.' In addition, the British interpretation was that

China's territorial waters extended 'strictly three miles from the low water mark on mainland and islands', which meant that both the 'Tungsha Band and Kiutoan Spit Buoys are outside territorial waters'. Finally, 'HM ships will stay outside territorial waters unless British merchant ships are endangered'.[23]

The British government's policy toward China and the blockade was most clearly set out in the document CP(49) 214, approved on 24 October 1949. This policy envisaged keeping a British foothold in China, and so 'advice and encouragement was given to British interests to stay put'. Not only was this policy intended to retain Britain's enormous investments in Shanghai, but 'in the absence of all trade and contact with the West, the field in China would be left exclusively open to the Soviet Union and China's subservience to the Soviet Union might well become complete and irrevocable'.[24]

With the fall of Canton to Communist forces in October, pressure increased on the British government to end the blockade. According to *The China Weekly Review*, 'The question now arises as to how long the foreign powers, especially America and Britain, will continue to grant de facto recognition to a blockade which day by day comes closer to a form of piracy'.[25] On 29 October, two British-flagged ships, *Tsinan* and *Louise Moller*, were escorted by British warships up to the three-mile limit. This situation was somewhat unusual because the convoy arrived just as the Nationalist guard ship on duty was removing the Kiutoan light vessel and 'had towed it outside territorial waters'. Since the Nationalist ship was outside the three-mile limit, 'Ships were thus escorted past by His Majesty's Ship.'[26]

On 12 November 1949, however, the *China Weekly Review* reported a stalemate in which a Nationalist warship had intercepted *Tsinan* and *Woosung*, but had been prevented by the near presence of the frigate HMS *Alacrity* from escorting the two ships to a Nationalist blockade base. This proved to be a deadlock, since the Nationalist ship was safe only so long as it remained in China's territorial waters. The stalemate led to the conclusion: 'Shipping circles in Hong Kong believe that neither merchantman is in danger and that the action of the Nationalist vessel is the last serious effort to "blockade" the China coast'.[27]

The British naval policy toward the Nationalists firmed up between late October and early November 1949. By November it seemed more and more certain that the Nationalist government was on the verge of collapsing. As a result, the British government's recognition of the PRC was 'seemingly imminent'. Given this change, it was to be feared that British warships entering the PRC's territorial waters, even in defence of British ships, could be interpreted as a violation of China's sovereignty and so would be resented by the Chinese Communists. One telegram from the British consul in Nanjing even warned: 'I think it possible that such resentment in the minds of official individuals might prejudice the establishment of friendly relations with the new Government which we wish for', particularly since when 'recognition has been accorded, His Majesty's Navy will almost certainly not be permitted to enter Chinese territorial waters . . . except with the specific approval of the Chinese [PRC] Government, which is not likely to be accorded except in special circumstances'. Given this shift in perception in terms of which Chinese government – Nationalist or Communist – really controlled China's territorial waters, the message concluded: 'I venture to suggest that until recognition has been accorded and diplomatic representatives have been appointed . . . His Majesty's Ships should abstain from operating in Chinese territorial waters'.[28]

This position was confirmed on 19 December 1949, when representatives from the Admiralty, the Ministry of Transport, and the Legal Adviser's Department met in the Foreign Office to consider 'whether at the time of recognition these [naval] instructions could be modified so as to give more help to merchant shipping'. As a result of the meeting, however, it was decided that this modification could 'not be justified', in particular because: 'What shipping interests still needed was naval escort through territorial waters and no such support could be given without prior permission from the sovereign government, who would then be the Communists'. But it was further determined that once Britain recognised the PRC, then stationing British ships – in violation of the Nationalist blockade – off the mouth of the Yangzi River need 'no longer be called into question', and should Nationalist warships harass British shipping outside of China's territorial waters this 'could then be regarded as acts of piracy and lay them open to arrest'.29

Obviously, during this second period of transition the British government shifted from a position of full recognition of the inviolability of Nationalist China's territorial waters to one of qualified recognition, and then back again to full recognition, this time of Communist China's claim to sovereignty. This corresponded to changes in perception, as opposed to actual military and especially naval power, as to which government was the 'real' government of China. As British recognition of the PRC neared, British warships were warned once again to be more cautious about violating Chinese sovereignty.

Differing Anglo-American Views of the Nationalist Blockade

As well as conducting negotiations with the PRC government to end the blockade, a policy that clearly failed, the other country the UK had to be concerned about was the United States. The US government was always less ambivalent than the British about the blockade, perhaps because US trade with China was not as important at the time. Like Whitehall, Washington condemned the blockade initially, warning that 'the US government would take a serious view of any Nationalist attack upon American merchant ships running the blockade which, the US government wished to reiterate, it did not recognize as legal'.[30] On 24 December 1949, however, Washington actively supported the Nationalists when it warned American shipowners that their licences could be revoked if they attempted to run the Shanghai blockade.[31]

From 1949 through to 1971 – when Richard Nixon ended the US blockade policy as the first step in opening diplomatic relations with the PRC – Washington conducted an embargo against Communist China. Official US policy was not to cooperate with the Communist government in any way. During July 1949, the US Minister to China, J. Leighton Stuart, even explained to the British Consul-General in Guangzhou what he thought the American policy toward Communist China should be. His view was that the British and American investments in China were 'patently lost anyhow', and that the best policy would be to allow the Soviet Union to 'demonstrate, if they could, that they were able to give a communist China the assistance she will need'. Once Beijing realised that Russia could not possibly finance China's development alone, then 'China would have to turn to us again and we might then be able to come back on terms which would suit us'.[32]

This policy was confirmed by an Australian cable sent on 4 August 1949. After discussing the serious plight of the foreign community, the reporter stated, 'There is ground for hope that the Communist authorities, in the desire to see the port opened, will offer facilities for the Royal Navy to escort vessels the necessary distance inside territorial waters'. But, while the British government was prepared to provide protection, 'the United States of America will not be willing to do so for their policy is to evacuate their community and abandon their interests'.[33]

On 13 September 1949, US Secretary of State, Dean Acheson, met with Ernest Bevin, the British foreign minister, and argued that the US policy toward China would be to 'let the Chinese Communists learn by bitter experience that the position of a satellite of Russia had little to recommend it'. As part of this policy, 'Every effort should be made to show up any Soviet actions which were contrary to China's interests'. The US goal would be to await the 'development of a Chinese version of Titoism'. Bevin agreed with this, but because of the UK's extensive commercial interests in China, it was more important for the British to consider recognising the Communist government. Otherwise, Bevin 'thought there was a risk, if the Western Powers remained too obdurate, that the Chinese would be driven further into the arms of Moscow'. Acheson acknowledged that 'because of the much more extensive British interests in China the behaviour of His Majesty's Government was bound to be somewhat different from that of the United States Government'. Acheson was careful to state, however, that this was 'a difference in situation rather than a difference in policy, and as long as the objective of the two Governments remained the same he did not think it mattered greatly if there was some divergence on tactics'.[34]

In sharp contrast with the official British view that the blockade was illegal, *The American Journal of International Law* concluded in 1950 that 'within the limits of territorial waters unrecognized belligerents enjoy the right to prevent access of supplies to their domestic enemies'.[35] Beginning in November 1949, the US imposed an embargo on strategic materials to the PRC, eventually directed by the Coordinating Committee for Multilateral Export Controls (COCOM). In 1951, Congress adopted the 'Battle Act' to cut economic and military aid to allies and neutral countries that refused to support the embargo against the PRC. China

Committee (CHINCOM) controls were soon instituted that were even tighter, embargoing industrial machinery, steel mill products, and metal of all types.

With tacit support from the US, therefore, the Nationalist forces tightened the blockade through the summer of 1949, especially at Shanghai. On 27 June 1950, President Truman even ordered the Seventh Fleet to move into the Taiwan Strait, thereby delaying a planned Chinese invasion of Taiwan until 1951, then until 1952, and eventually it was cancelled altogether.[36] The US also became the major provider of aeroplanes and munitions to Taiwan, the very equipment which the Nationalists were using to bomb Shanghai. With American training, equipment and financial backing, it was hoped that 'the blockade of war goods, partially effective now, will grow more effective as the naval and air forces of Nationalist China are built up with US aid'.[37]

The British 'foot in the door' policy differed in many substantive ways from the American approach of having the PRC 'stew in their own juices'. On 20 April 1950, the Secretary of State for Foreign Affairs outlined four possible directions which the PRC's political development might take. First, China might become totally dominated by the USSR. Second, a Chinese form of Titoism might emerge, as a result of which the PRC would look to the West for potential allies. Third, the Communist regime might collapse and a new civil war begin. Finally, the PRC might single-handedly turn China into a great power. Of these, number four seemed most unlikely, and 'the odds against this at present appear to be considerable'. The British government was hoping to prevent China from becoming completely dominated by Russia, and to avoid option one, 'The present policy of His Majesty's Government is therefore to try to maintain contacts with China, for the reason that in the maintenance of contacts with the West lies the only hope of influencing China not to commit herself finally and irrevocably to the Kremlin'.[38]

In January 1950, the British government reiterated that the UK's and America's 'long term interests in the Far East go hand in hand', but on 'immediate tactics there may be some divergence'. In particular, the British wanted to put pressure on the Nationalist government to end the blockade, a naval operation that was merely strengthening the hand of extremists in Beijing so that 'any attempt to retain a foothold of Western influence in China will become increasingly difficult'. Only by retaining such a 'foot in the door' could the West help bring China back to a situation of 'normal relations and stability'; while without the West the extremists in China would be strengthened and China could fall completely under Soviet influence.[39] All British attempts to convince their American allies to adopt a more conciliatory policy toward Communist China ended in failure, however.

Britain Recognises the PRC

In preparation for the UK's recognition of the People's Republic of China, there was much discussion of what the British attitude toward the Nationalist blockade should be. Many British citizens in Shanghai assumed that the blockade would be broken immediately after recognition. There were good reasons, however, why this was not to be the case, including the Communist government's sensitivity about having a foreign 'imperialist' power infringe upon its territorial waters. Hence the UK's 6 January 1950 recognition of the PRC proved just one more step in a more lengthy diplomatic process to try to lift the blockade.

Prior to Britain's recognition, in a report of 29 November 1949 entitled 'Recognition of the Chinese Communist Government', the UK Chiefs of Staff had warned that while the 'withdrawal of recognition from the Nationalists would remove the basis in international law for their blockade of Chinese ports in Communist hands', initially 'there might be an increased risk of incidents between British and Chinese Nationalist forces'. They concluded, however, 'this risk will decrease as Nationalist resistance collapses'.[40]

Perhaps in anticipation of Britain's decision to recognise the PRC, on 12 December 1949, the Nationalists announced that they would lay mines to block the entrance to Shanghai. Full port closure would become effective on 19 December. As a result, the two main British shipping lines, Jardine Matheson and Butterfield and Swire, 'announced that they would give up their attempts' to run the blockade. This left only the American Isbrandtsen line, which continued its blockade-running operations in violation of a US government warning to 'American shipping that Shanghai was a danger zone to be avoided.'[41]

On 15 December 1949, the British consul in Nanjing (Nanking) reiterated that while the collapse of the Nationalist government would end the blockade, 'I do not think that this collapse will necessarily coincide with recognition'. Therefore, it was imperative to consider 'what attitude we propose to take regarding rump Nationalist Administration in Formosa after recognition has been accorded'.[42] In line with this approach, the policy of British warships towards Nationalist ships would remain the same, and recognition of the PRC 'will not … automatically give our forces full freedom to carry out retaliatory or preventive attacks on Nationalist naval and air bases.' In a clear break with former procedure, however, reconnaissance missions, mainly by air, of 'Nationalist naval and air bases' could now be conducted, but would have to be 'as unobtrusive as possible'.[43]

The British government recognised the PRC on 6 January 1950. The UK was the first major Western country to do so. Almost immediately afterward, on 10 January, the British Consul General reported that the Assistant Naval Attaché had conducted talks with the British shipping businesses, advising that British warships would probably be allowed to escort 'merchant ships inside territorial waters not under Communist control, provided the right approach were used and publicity such that the Chinese did not lose face'. In particular it was thought that 'the Communists would in any case have difficulty in raising objection to the proposal of escort of our merchant ships <u>as far as their batteries</u> particularly if it were included with a request for merchant ships to be allowed safe conduct past the batteries'.44

Yet, during the two to three months after the Nationalist announcement of the mining of the Yangzi River entrance there was no break in the blockade. During this time only one British vessel entered and then left Shanghai, while 114 air raids were conducted by Nationalist aircraft. A raid on 6 February 1950, in particular, included 17 Nationalist aircraft and led to extensive damage throughout Shanghai and as many as 1000 casualties, although reportedly none were British.[45]

Meanwhile, by mid-February 1950, the British Assistant Naval Attaché was reporting the results of a 13 February meeting between two 'British and American ex-officers with sweeping experience', and of their attempts to help the Chinese People's Navy sweep for mines blocking the passageway to Shanghai. He reported, however, that 'impression gained that Peoples Navy unwilling and incapable of action though they said they had been ordered to sweep mines by Peking'. Even if the mines could be removed, 'the real obstacle to breaking the blockade is the Nationalist guard ship', whose removal would take direct action by the British Navy.[46]

On 9 February 1950, a strongly worded letter was sent from Admiral of the Fleet Lord Fraser to Sir William Strang, Permanent Under-Secretary of State, condemning the Nationalist blockade in general, and recent aerial bombing of British property in Shanghai in particular. According to Fraser, 'the present position in which British ships of war watch one wretched little Chinese gunboat holding up our merchant shipping is intolerable … The action we could take would be to convoy our ships into Shanghai and either to blockade Formosa or, if air attacks do not cease, to attack their aircraft on the ground.'[47]

Almost at the same time that this message came in, the CinC Far East Station reminded the Admiralty on 14 February 1950 that any action by the British Navy, including driving 'Nationalist Patrols from the Yangtse', sweeping the area for mines, and in particular sending 'escort ships in the Estuary to afford some protection from air attack' would 'not likely to be effective unless Communists agree'. He also warned that any direct action by the British against the Nationalists might 'widen the rift with America caused by our recognition of the Communists.'[48]

The main holdup was that the PRC had not yet recognised the Consul-General in Shanghai, Hutchison, as the official representative of the British government. Hutchison had been ordered to proceed to Beijing to open discussions on how to terminate the blockade, but it was important to reiterate that, 'we cannot amend our Naval policy as far as intervention by HM Ships inside Chinese territorial waters is concerned until clarification of the Communist attitude has been achieved', and this 'cannot be done until effective diplomatic relations are established'.[49] Another memo to this effect re-emphasised that 'the major obstacle [to breaking the blockade] remains the attitude of the Chinese Communists', and that because of Communist antipathy to the West, 'Shanghai's importance as the economic and commercial capital of the Chine will very gradually be allowed to dwindle … I do not think that talks with the Chinese authorities about the blockade will prove particularly hopeful or helpful.'[50]

On 17 March 1950, the Foreign Office presented a paper to the Cabinet's Far East (Official) Committee examining the Chinese Nationalist Blockade and Bombing. After discussing all of the major factors surrounding this episode, and the

HMS *Concord* leading the Royal Navy's 8th Destroyer Squadron out of Hong Kong for exercises in the early 1950s. (A.L. PERRYMAN)

reasons why Britain wanted to retain its foothold in China, it concluded that, 'successful action to break the blockade should go far to produce a resumption of normal trade with Shanghai'. Not only would this 'avoid the immediate and grave possibility of the withdrawal of the British commercial community from China', but it would halt the 'failure of a major part of our China policy'. To make such an action a long-term success depended on 'the attitude and policies of the Chinese People's Government', and should Britain's allies, in particular the US, oppose such an action, then 'it would be necessary to reconsider the action proposed in the light of these comments and the situation then prevailing'.[51] What this meant in practical terms was that the Royal Navy was ordered not to use its naval assets to break the Nationalist blockade.

By the following month, however, the situation had notably worsened. The Secretary of State for Foreign Affairs was reporting that a number of British firms in Shanghai had decided 'that they must cut their losses and abandon their enterprises, with the probability that these will sooner or later be expropriated by the Communist authorities'. As a result of the Nationalist blockade, and of the PRC's unwillingness to help British merchants by lowering taxes and allowing them to fire redundant labourers, the 'total loss of a great proportion of British investment at Shanghai must therefore be faced in the very near future'. This policy of abandoning the British holdings in Shanghai impacted upon all foreigners still living there. And of the 1500 foreigners awaiting evacuation, about 400 were British citizens, with another 100 Indians and a handful of Australians and Canadians. It was the British government's intention to evacuate the last of its nationals as soon as possible, through the northern port of Tianjin, if necessary.[52]

In the final analysis, British recognition of the PRC government did not bring the hoped-for results. Although a British representative had been sent to Beijing following recognition on 6 January 1950, he was virtually ignored. By May 1950, official diplomatic channels had not been fully established, and there was still no British ambassador accredited to China. Talks on breaking the Nationalist blockade had not even begun, much less created a united Sino-British policy.

As a result of these delays, the British attempt to 'save' Shanghai by recognising the PRC proved a failure. This may have been an outcome of Beijing's strategy of making the Communist government appear surrounded by foreign imperialists. In August 1949, for example, Mao Zedong even challenged the US and the UK to blockade China, 'Let them blockade us! Let them blockade us for eight or ten years! By that time all of China's problems will have been solved.'[53]

Although the British government decided on 20 April 1950 that it should try to 'do what we can with the Chinese authorities as opportunity offers to try to bring about an improvement in conditions', if 'British firms nevertheless decide to close down, we should not deter them, but should advise them when closing down not to take any action which will prejudice their chances of obtaining at any rate some measure of compensation from the Chinese authorities'. Finally, the British government should 'facilitate to the extent open to us, the evacuation of British subjects who wish to leave' Shanghai.[54] The admission that all British nationals in Shanghai should be evacuated was the death knell for Britain's commercial interests in that city.

Conclusions

This study of the Nationalist blockade of China in June 1949, and in particular the blockade of the port city of Shanghai, shows the enormous impact that naval affairs can exert on political decisions. The vast majority of the major naval events during this period took place in the littorals, usually within the three-mile limit of China's territorial waters as recognised by Britain. Meanwhile, throughout the 1950s, Nationalist aeroplanes attacked targets on land, in the littorals, and on the high seas within the Taiwan Straits at will.

During July 1949, one British official based in Hong Kong was highly complimentary about the Nationalist blockade of China, concluding that 'the blockade is probably the only really effective piece of warmaking the Nationalists have carried out in the last couple of years'.[55] From their naval bases on Taiwan and on dozens of offshore islands, the Nationalists blockaded the Chinese coast from Shanghai down through southeastern China. Between 1950 and 1952, the Nationalists halted and searched some 90 ships heading for Communist ports, two-thirds of them British-flagged vessels registered in Hong Kong.[56] Furthermore, during 1954 and 1955, Nationalist bombers damaged or sank an estimated 54 blockade-runners, many of them British-flagged.[57]

For many British shippers, however, pro-Nationalist guerrillas based on the offshore islands controlled by Taiwan seemed little better than modern-day pirates.

As reported by one British official in 1952, 'indications are that most of the piracies are done by Nationalist guerillas not under the effective control of Taipei. The border line is obscure. In 1948 it was respectable for bandits to masquerade as Communist Liberation forces; in 1952 it is respectable for pirates to masquerade as the Nationalist Navy.'[58] Nationalist interference with ships involved in trade with the PRC continued through 1958, at which point the naval blockade ended for good.

In addition to the blockade's dire commercial and humanitarian effects, it had major political effects as well. As we have seen, one of the major motivating factors behind the UK's decision to recognise the PRC was the desire to negotiate a suitable method of breaking the blockade and reopening Anglo-Chinese trade, in particular between Hong Kong and Shanghai, so that British interests in Shanghai could be saved. Without the full cooperation of both the Communist government, on the one hand, and the US government, on the other, however, British recognition of the PRC proved insufficient to achieve this important goal. As a result, Britain's commercial interests in Shanghai were eventually lost.

One possible reason for this poor outcome was the major difference between London and Washington's policies. While Whitehall hoped that China could quickly be turned away from its pro-Soviet stance, the US government perhaps realised that this could not happen immediately and so was more interested in the long-term strategic impact of the Nationalist blockade. In this regard, Washington continued to support the blockade as merely one method of forcing the PRC to depend more and more on the USSR. According to British embassy personnel in Washington, beginning in summer 1949 and intensifying through early 1950, the US government's policy of allowing the 'Communists to stew in their own juices' included Washington's active 'encouragement of the present blockade measures'.[59]

With the Nationalist blockade in place, and with that blockade bolstered by the United Nations embargo of strategic materials, which the UK fully supported, the PRC had little choice but to make up its trade losses by turning to the Soviet Union. Before World War II, only 1 percent of China's foreign trade was with the USSR, but by 1957 this figure had increased to 50 percent.[60] Yet Beijing's overwhelming economic reliance on Moscow restricted its diplomatic leverage, so that China could not negotiate equitable relations. This may, in turn, have exacerbated border demarcation problems between the two Communist states, complicated the return of Manchurian ports and railways to Chinese sovereignty, and prevented the sharing of control of the international Communist movement.

Perhaps it was not a total coincidence, therefore, that the Nationalist naval blockade of the PRC ended during 1958, just as the first signs of Chinese 'Titoism' appeared and the country entered a new phase of political development that would eventually result in the 1960 Sino-Soviet split. At this point, the PRC indeed began to look to the UK and the US as possible allies, and Shanghai began anew its climb to economic prominence as one of Asia's premier cities.

In conclusion, while the Nationalist blockade may not have been completely effective operationally, it perhaps contributed significantly to the larger Anglo-American strategic goal of breaking the PRC away from the USSR. The UK-US 'carrot-and-stick' cooperative strategy may have been equally important to the success of this policy. Taken together, this shows how the careful application of sea power, especially in the littorals, can exert a significant influence upon political events on land.

12. Korea: The First Challenge for Australian Naval Aviation

Jack McCaffrie

28TH AUGUST 1948 marks the formal establishment of the RAN Fleet Air Arm with the commissioning of the 20th Carrier Air Group (CAG) at RNAS Eglinton in Northern Ireland. A little over three years later, the Fleet Air Arm was at war and the light fleet carrier HMAS *Sydney* and her CAG were deployed off the coast of the Korean peninsula. While this chapter deals with the deployment of *Sydney* in Korea it does so in the context of what came before and an aftermath that unfolded in unexpected ways.

By any measure, the establishment of the Fleet Air Arm and its rapid progress to being a highly efficient and effective operational entity, is a remarkable story. There were political, bureaucratic, financial, personnel and timing challenges, any one of which might have brought the process to a halt, especially in the early stages. That *Sydney* and her Air Group deployed to Korea and performed so well was the result of determination on the part of the RAN leadership, enlightened political decision-making – in the face of sometimes less than sparkling naval planning – and significant assistance from the Royal Navy.

Reflecting on the performance of the Fleet Air Arm in Korea – and indeed of the destroyers and frigates, which also distinguished themselves – the RAN would have had every reason to look to the future with confidence.

In the Beginning ...

Although other dates could be selected to represent the origin of the Fleet Air Arm, the one chosen for this chapter is 21 March 1944. On this date, the then Chief of Naval Staff (CNS), Vice Admiral Sir Guy Royle, RN, put an informal and unannounced proposal to Prime Minister Curtin's Advisory War Council. He proposed that the RAN take over and man eight or nine Royal Navy ships, which would include an aircraft carrier, one or two cruisers and six destroyers.[1] Admiral Royle emphasised the speed with which the transfers could take place, the essentially naval nature of the war in the Pacific and the expectation that the transfer of ships would be at no cost to the Australian government.[2]

Curtin agreed to discuss the proposal with Admiral Royle, but Defence Department Secretary, Sir Frederick Shedden, was most disturbed by the proposal and the approach taken. On the one hand, he suspected that Admiral Royle was attempting to steal a march on the other Service Chiefs in respect of post-war planning. On the other hand, Shedden saw the proposal and the way in which it had been brought forward, as a threat to him and his position.[3]

These issues apart, however, Shedden also opposed the proposition for other reasons. The government had significant difficulties in meeting competing workforce demands and Shedden did not want to add to them. He also believed that there were already sufficient naval forces available in the Pacific and Indian Oceans and that Australia's best post-war planning option would be to build up the Royal Australian Air Force to the maximum extent.[4]

Curtin agreed to raise the matter with British authorities during his May 1944 visit to the UK. Despite the pressure from the British side, on top of that coming from Royle and other members of the Advisory War Council, the Prime Minister would not commit. But Royle continued to battle on, despite setbacks and despite some of his tactics being questionable.[5] By early 1945, Curtin was prepared to ask British Prime Minister Winston Churchill to agree to the transfer of an aircraft carrier and two cruisers at no cost.

Circumstances had changed in the UK, however, and the British government indicated that the transfer could not be made at no cost and that in any event no cruisers would be available before late 1946.[6] Changes had also occurred in Australia, and Chifley, acting for the terminally ill Curtin, advised the British that personnel considerations (or was it really money?) prevented Australia from going ahead at that time.[7]

Still the RAN pleaded its case, as Chifley became prime minister and Royle was relieved by Vice Admiral Sir Louis Hamilton in mid-1945. Support was beginning to emerge within government for the post-war acquisition of a carrier as part of a balanced force. The Naval Board certainly considered naval aviation to have become the primary offensive arm of the Navy and that without it the RAN would cease to be a front rank navy.[8]

Soon after his arrival, CNS Hamilton had a study underway to determine the composition of a naval air arm. Coincidentally, a naval staff plan was prepared for the post-war Navy. It recommended a force of three light carriers, six cruisers, 32 destroyers and 50 escorts.[9] This kind of exuberance was also demonstrated by the other Services and was totally at odds with the financial restrictions which government would impose.

Hamilton had some success with both Chifley and Shedden, and in April 1946 was given approval to begin planning with the British Admiralty for the development of Australian naval aviation, subject to there being no financial commitment. Subsequently, in September 1946, in response to RAN inquiries, the Admiralty in September 1946 offered Australia two *Majestic* Class light fleet aircraft carriers for the building cost of one, which was £2,750,000.[10] Buy one, get one free! Then, after a significant amount of political and bureaucratic deliberation and a number of internal Defence studies, CNS Hamilton received provisional agreement to a two-carrier force that would have its aviation component provided by the RAN.[11] This latter issue had itself been debated at length and strongly opposed by Chief of the Air Staff, Air Chief Marshal Sir George Jones.

Defence funding was a major concern for the post-war Chifley government, and the Prime Minister had been reluctant to commit to the Navy proposal without comprehensive costing data. His fears were confirmed very shortly after the Cabinet's endorsement of the naval aviation plan on 3 June 1947. No financial contingency had been allowed in the acquisition process and within six weeks the Admiralty was tabling cost increases for the carriers and advising on the need for modernisation of the selected *Majestic* class ships to take emerging aircraft types.[12] So, from the very beginning naval aviation was hard pressed financially.

Formation and Workup

Early planning anticipated that the first carrier would be commissioned in June 1948 and that the second would follow in 1949 or 1950.[13] The RAN also expected that the Royal Navy would initially provide the majority of the aviation and air group personnel. The RAN also began its own training programs. The first Naval Air Pilots' Course began at the shore establishment HMAS *Cerberus* on 7 December 1947, and nine of the 14 starters graduated at RAAF Point Cook on 29 July 1949. They then undertook operational conversions in the UK, completed in April 1950, on both Firefly and Sea Fury aircraft.[14]

The second pilots' course began in August 1948 and the RN also loaned a number of Observers and Telegraphist Air Gunners to the RAN. Additionally, the first group of ex-naval (RN, RANVR and RNZN) and ex-RAAF aircrew joined *Cerberus* on 6 February 1948 for a six month orientation course before also travelling to the UK for further flying experience and to join either the 20th or 21st CAG.[15]

The Fleet Air Arm was established on 28 August with the commissioning of the 20th CAG at Royal Naval Air Station *Eglinton* in Northern Ireland. Four months later *Sydney* was commissioned in Devonport with Captain Roy Dowling, RAN, as Commanding Officer.

Air Group work-up in the UK entailed the full run of proficiency and operational flying, including deck landing practise in Royal Navy Fireflies and Sea Furies. Lieutenant Danny Buchanan, RAN, caused some excitement on 17 March 1949 while landing on HMS *Illustrious*. His Firefly missed the wires, jumped the barrier, destroyed four other Fireflies in the forward deck park and was itself also destroyed.[16] Nevertheless, no one else was hurt and as he himself walked away without a scratch, the landing must be recorded as a good one.

The 20th Carrier Air Group and *Sydney* sailed for Australia on 12 April 1949 and arrived in Jervis Bay on 25 May. Flying for the Air Group began from the newly commissioned Nowra Naval Air Station in June and there were two embarked periods between then and April 1950.[17] *Sydney* then returned to the UK to bring home the 21st CAG, arriving back in Australia in December 1950. This transport task meant that the ship and Air Group were unavailable to reinforce the Commonwealth effort in the Korean War which had broken out mid-year.[18] There had been some consideration of this possibility because of problems with the British carrier HMS *Theseus* and there was an expectation that the issue might be raised again.[19]

Indeed, by March 1951 the CNS, Vice Admiral Sir John Collins, had received a request from the First Sea Lord for *Sydney* to replace HMS *Glory* in Korea in late 1951. Consequently, in May 1951 the Navy decided to form an Air Group from 805 and 808 Sea Fury squadrons and 817 Firefly squadron and work-up plans began immediately. Most Air Group members were Royal Navy personnel but most of *Sydney*'s ship's company were Australian. The work-up was costly in aircraft with weather contributing to several deck crashes. Ultimately *Sydney* sailed for Korea, with the destroyer HMAS *Tobruk* in company, on 31 August 1951.[20]

En route to Korea, preparations continued for the forthcoming operations. They included rehearsal of bombardment spotting procedures, lectures on escape and evasion, planning conferences and small arms practice. Additionally, a number of officers went ahead of the ship to observe Royal Navy flying operations in Korea.[21]

The War and Australia's Commitment

How did Australia, and the RAN in particular, come to be involved in the war in Korea? Korea had been divided at the 38th parallel in 1945 by the US and the Soviet Union, essentially to facilitate disarmament of the Japanese.[22] Subsequently, neither the Soviet Union nor the US, even with the aid of a UN Temporary Commission on Korea, could agree on a government for the entire country. Elections were held in the South in May 1948 and were followed by the end of the US occupation of that part of the country.[23]

UN sponsored efforts to unify the two Koreas continued through 1949, with the active diplomatic involvement of Australia. Neither Kim Il Sung in the North nor Syngman Rhee in the South were inclined to cooperate or in any way compromise. By late 1949 border skirmishes were occurring and North Korean troops were infiltrating the South.[24] Then in May 1950, Australia agreed to provide two military observers to report on developments in Korea: Major F. Peach and Squadron Leader R. Rankin. They were on the border from 9 to 23 June and reported on 24 June that South Korean troops were not well placed and would probably retire if attacked. Nevertheless, they also reported no unusual troop movements in the North. On 25 June North Korea launched its attacks.[25] Australian official historian Robert O'Neill has claimed that Kim Il Sung may have determined that he would win quickly against the South and even if he did not, that Soviet help would reach him before the US could respond.[26]

Australia had a major problem in responding credibly to the subsequent calls for military assistance. There were severe defence personnel shortages, money was scarce and the Army had fewer than 1000 infantrymen in the regular force – and most of them were in the British Commonwealth Occupation Force in Japan, from which Australia wanted to withdraw anyway.[27] Additionally, RAAF No. 77 Squadron was in Japan and the frigate HMAS *Shoalhaven* was in Japanese waters.

Neither the Australian Services nor the Prime Minister wanted to become involved in Korea. Provision of forces to Malaya had been a consideration in May, and Britain was urging Australia to commit ground and air forces to the Middle East, which was seen as a higher priority in the event of any Soviet attack.[28] Yet the Supreme Commander Allied Powers in Korea, General Douglas MacArthur, had indicated that he wanted both No. 77 Squadron and the 3rd Battalion Royal Australian Regiment (3 RAR), the Army unit in Japan. Minister for External Affairs Percy Spender was keen for Australia to become involved as he had in mind a future security pact with the US. Eventually *Shoalhaven* was committed on 29 June, followed by No. 77 Squadron the next day and 3 RAR on 26 July.

On the naval front, the USN was not well disposed to respond at the beginning of the War. It had a few ships in Japanese waters but most of the 7th Fleet, by no means a powerful force, was in or around the Philippines. On the other hand the Royal Navy was much better placed. A major portion of the Far East Fleet, some 22 ships, including a light fleet carrier, was in Japanese waters – escaping the heat of Hong Kong and Singapore.[29] This force was placed under MacArthur's control.

For the RAN, *Shoalhaven* was to be relieved by the destroyer HMAS *Bataan*, but both ships initially remained in theatre. They were joined by Canadian and New Zealand units, all of them operating within the Royal Navy force. This British Commonwealth Naval Force formed the West Coast Support Group (TG 96.8), while the USN made up the East Coast Support Group (TG 96.5) and the 7th Fleet carrier force (TF 77) operated wherever it was needed.[30]

Shoalhaven was at first assigned to escort duties between Sasebo and Pusan and was relieved by *Bataan*'s sister ship, HMAS *Warramunga*, on 31 August 1950. The RAN had consistent problems in replacing ships in Korea, because of shortages of suitable vessels and of people.[31] Other tasking for the surface forces included support for amphibious operations and shore bombardment, often in the face of a significant mine threat and sometimes with the fear (ultimately unrealised) of a Soviet submarine threat.

From the outset of the War, carriers were in heavy demand, not least because USAF aircraft were in short supply and because NATO and domestic US demands limited the availability of reinforcements. Other reasons for the attractiveness of the carriers were the lack of suitable operating airfields in Korea and the limited time over target for tactical aircraft operating from Japan.[32]

Normally the USN carriers in TF 77 operated in the open Sea of Japan off the east coast of Korea. The British light fleet carrier force (TE 95.11) usually operated in the more restricted waters off the west coast. Their contribution in terms of aircraft numbers was not great, but did permit the maintenance of a continuous air offensive off both coasts. After December 1950 a USN light fleet carrier also joined the west coast operation, thereby allowing each of the Royal Navy and USN carriers to rotate on station about every 10 days.

In the first year of the war, United Nations (UN) air and sea power prevented almost all enemy movement at sea and fought off the first challenge from Soviet-built (and sometimes manned) MiG 15 jet fighters. With general control of the air, UN forces limited the North Korean movements to the night hours.[33] Generally, the carriers were not attacked but there was constant anti-submarine screening by escorts.

HMAS *Sydney* in Korea

Sydney, like the British carriers, operated a mix of Fairey Firefly and Hawker Sea Fury aircraft. The Firefly was a single engine, two-seat anti-submarine reconnaissance and strike aircraft. It had a maximum speed of over 350 mph and an endurance of over six hours. Weapons included four 20 mm cannon and two bombs of up to 1000 lb each.[34] The RAN had both Mk.5 and Mk. 6 aircraft, and because only the Mk. 5 had the cannon armament, took 12 of those aircraft to Korea with 817 Squadron, where they also carried two 500 lb bombs for strike missions.[35] For anti-submarine missions, depth charges were substituted for the bombs.

The Sea Fury was a single-engine, single-seat fighter bomber, the last piston-engined fighter to serve with the RN Fleet Air Arm and the only one to serve with the RAN. Its top speed was over 450 mph and it had a range of up to 1000 nm with two drop tanks. Its armament consisted of four fixed wing-mounted 20 mm cannon and it could carry a mix of either eight 60 lb rockets or two 1000 lb bombs. In Korea, the Furies usually flew with eight rockets, two drop tanks and had about 125 rounds per gun.[36]

Unlike the USN's Corsair aircraft, RAN Fireflies and Sea Furies were not armoured, although the Furies did get some armoured panels to protect the underwing coolers.[37]

During operations in Korea *Sydney* also embarked a Sikorsky S51 Dragonfly helicopter for the first time. With a crew of two, the Dragonfly had a maximum speed of 95 mph and a range of 300 nm. This helicopter was used for search and rescue, for plane-guard duties and inevitably for mail runs. At first the helicopter came from *Glory* but for most of the deployment it was provided by the USN. Colloquially it came to be known as 'Uncle Peter' because of its identification letters, 'UP'. The Dragonfly performed sterling service for *Sydney*'s aircrew during their Korean War deployment.

Under the command of Captain D.H. Harries, RAN, and with Commander V.A.T. Smith, RAN, as Executive Officer, Sydney's ship's company worked hard and kept the ship in as good a state as was possible under the conditions. Sydney did not operate as a flagship, although the Flag Officer Commanding the Australian Fleet, Rear Admiral J.W.M. Eaton, RN, did spend three days in the ship in January 1952. Rear Admiral Alan Scott-Moncrieff, RN – Commander West Coast Blockade Force – also spent time in the ship on occasion. Morale for the most part was very good, despite the cramped and often freezing quarters, awful winter weather and constant air, surface and sub-surface threats. Perhaps understandably, behaviour ashore was not always what was expected. Leave breaking was a problem and relations with the ships' companies of other UN force ships were also strained at times.[38]

Sydney had a single hydraulic catapult for aircraft launching and although free take-offs had often been used prior to Korea, the catapult was invariably used in wartime operations; not least because of the higher weights at which aircraft were operating. The catapult proved to be very reliable, with only one account of a failure, on 1 January 1952. Even then, 51 sorties were flown that day after the catapult was repaired.[39] Average time between launches was 42 seconds, although lower times were recorded. This was a level of flight deck performance which the Royal Navy carriers apparently found difficult to emulate.[40]

With a straight deck, *Sydney* was more prone to deck landing accidents than the later angled-deck carriers. Nevertheless, her record during the Korean deployment was very good. During the patrol which ended on 13 November, there had been only one deck landing accident in over 400 landings. The fact that no night flying was conducted undoubtedly contributed to the good figures. Pilots averaged one and a half sorties each day in the war zone, but acknowledging that not every pilot was always available to fly, most pilots flew two sorties on days on which they did fly.

With three squadrons totalling 38 aircraft embarked, *Sydney*'s flight deck and hangars were always crowded. Flight deck operations, never to be taken lightly, were invariably made more complex by the need to have 20 of the aircraft always parked on deck. These aircraft were manned and started for every flying event, so that any unserviceable aircraft could be replaced immediately.[41] After every recovery, unserviceable aircraft were parked forward or taken below for maintenance, while the serviceable ones were taken aft for fuelling and arming. At the end of every flying day, all deck-parked aircraft had to be heavily lashed in case of inclement weather. The handlers, armourers and maintainers worked hard under often very difficult conditions.

Flak damage to the wing of one of *Sydney*'s Fireflies. (RAN)

Aircraft maintenance was a never-ending task. On average, 38 percent of aircraft returned unserviceable – although the figure varied from 29 up to 66 percent.[42] All flak damage was repaired overnight.

Air Operations in Korea

Sydney operated mainly in the Yellow Sea and alternated on station with the US Ships *Rendova* and *Badoeng Strait*. At that time Operation STRANGLE was underway. This involved the UN Command air forces in interdicting the removal northwards of large North Korean supply dumps and the movement southwards of enemy combat supplies. The latter task proved more difficult because of the North Korean capacity to repair damaged supply routes.[43]

The carrier operating cycle was nominally nine days on patrol and nine days for return to Sasebo and replenishment there. Frequently, however, one of the on-station days was taken up with replenishment at sea, which provided a welcome break for aircrew. The missions flown by the Air Group can be grouped as strike, armed reconnaissance, close air support, naval gunfire support spotting, combat air patrol or anti-submarine warfare. There was also some time spent on convoy escort.[44]

Strike Operations

Strike operations were the major component of *Sydney*'s contribution to the War. Target location proved to be difficult, especially at first, because of excellent North Korean camouflage. As the operation continued, the most reliable targeting information was normally that provided by *Sydney*'s own air photography or visual reconnaissance.[45] Targets varied a great deal and included North Korean artillery positions, coastal shipping – including junks, troop concentrations, road and railway bridges, railway tunnels, buildings and supply vehicles including ox carts.

Much effort went into determining the best method for the Fireflies to bring down bridges, for which direct hits were necessary. Dive-bombing was tried at first, diving from 8000 ft and releasing at 3000 ft. Ultimately the aircrew fixed on 10 degree dive bombing at low altitude along the length of the bridges, in flights of four aircraft.[46] This was also the typical anti-submarine depth charging profile.[47] With experience, a flight of four Fireflies proved able to bring down three bridges in one sortie. However satisfying dropping bridges was to the aircrew, in most cases the North Koreans managed to repair them very quickly. This led to the following assessment by the Carrier Borne Army Liaison (CBAL) Officer:

> *It is of interest to note that the hard night-working Communists have already repaired the large hole we knocked in the rail bridge only three days ago. The Australian Country Roads Boards could well do with a few of these gentlemen!*[48]

North Korean gunners certainly enjoyed some success against RAN aircraft. Aircraft were hit by flak ninety-nine times and nine aircraft were shot down – seven of them Sea Furies.[49] The gunners did not use tracer and did react to changing aircraft operating patterns but, for the most part, aircraft were shot at only after they had passed over enemy positions. The flak could be astonishingly good and, at times, equally bad. There is one report of a Sea Fury inadvertently passing so close to a gun pit west of Chinnampo, that the pilot could see the gunner was missing a top left front tooth. Yet the gunner missed with everything![50] On the other hand the aircrew were extremely keen to press home attacks, and some consideration was given to limiting the minimum altitudes and number of attack runs. This became more of an issue as anti-aircraft fire increased.[51]

Other Operations

During the voyage to Korea much time was spent in practising naval gunfire support spotting techniques. The time was well spent as the subsequent results while in Korea were certainly impressive. The aircrews especially enjoyed working with the 'big gun' ships, such as the battleship USS *New Jersey* because of the much greater accuracy of their guns.[52]

Twenty percent of sorties, some ten each day, were flown as combat air patrol over the *Sydney*, because of the proximity of North Korean airfields. The extent of the self-defence effort led to a view that the carriers would be more effective if operating in pairs or in larger groups.[53] No *Sydney* combat air patrol ever intercepted an enemy aircraft.[54]

Normally, just one Firefly was assigned to anti-submarine warfare, because of the low level of threat. The same aircraft was also often used for surface contact identification and for search and rescue assistance.[55]

Sydney's Patrols

Sydney's first patrol began on 5 October 1951, as part of the operation to push the front back from the Han River and the approaches to Seoul.[56] She flew 47 sorties on each of the first two days, a rate considered quite high for a ship new to the theatre. During this first patrol, *Sydney* was ordered to join a special force under Rear Admiral Scott-Moncrieff on the east coast. She subsequently flew strikes against positions southeast of Wonsan but without clear details of the purpose of the overall mission. Later, the operation proved to have been a demonstration for the benefit of the CinC Far East Station, Admiral Sir Guy Russell, visiting from Hong Kong.[57]

A quite demanding, yet achievable, daily sortie rate was ordered by Captain Harries after he had determined what could be achieved on a continuing basis. He decreed that the planned daily rate of effort should be 54 sorties, spread over five events or waves. The shortness of the winter days and the distance to targets made any more than five events difficult to achieve. This followed some quite exceptional flying performances, such as that on 11 October, when 89 sorties were flown. On one occasion, late in the day, 31 aircraft were airborne at the same time.[58] This sortie rate was a record for the light fleet carriers, but was subsequently bettered by *Glory*.

The second patrol, which began on 18 October, was quite eventful. It involved attacks against railway bridges and tunnels as part of a major interdiction campaign, as well as strikes against enemy troops and coastal shipping. It was during this patrol that the first close support sorties were flown – for the British Commonwealth Division on 21 October. They sometimes included providing cover for South Korean troops and irregular forces and were most popular when ordered in support of Australian troops.

Although, the pilots really enjoyed this work[59] it was hazardous, and during the period 25-26 October, for example, three aircraft were shot down. Lieutenant Col Wheatley's Fury ditched off Chinnampo; Lieutenant Commander J.C. Appleby's Fury was badly damaged by flak but he managed to get it to Kimpo. Then Sub-Lieutenant Noel Knappstein's Fury crash landed on mud flats in the Han River estuary[60] There was an amusing end to this episode. The pilot was rescued by the frigate HMS *Amethyst*, but before being picked up he had removed some salvageable equipment from the aircraft. Then, showing commendable enterprise he sold the remains of the aircraft to some local Koreans. It was a very satisfied Sub-Lieutenant Knappstein who returned onboard *Sydney* with a thick wad of Korean notes. The Supply Officer duly changed them for him and presented him with the grand sum of one shilling and ninepence![61]

On 18 November *Sydney* left Sasebo in company with the cruiser HMS *Belfast*, the destroyer HMAS *Tobruk*, and several USN ships for Operation ATHENAEUM. This was to be a major bombardment of Hungnam on the east coast. In two days *Sydney* launched 113 sorties; of which 75 were against Hungnam coastal batteries and 38 were defensive combat air patrols for the ships. The operation was not notably successful, and *Sydney* returned to the west coast for operations in very poor weather. Returning to Kure at the end of this patrol on 30 November, *Sydney*'s Air Group had flown 966 offensive and 360 defensive sorties. Forty-three aircraft had been hit and four had been lost.[62]

In mid-December *Sydney* became involved in convoy escort, a task which the aircrew found intensely frustrating, mainly because it almost invariably proved to be boring. This followed a decision by the UN Command to exchange the operational tasking of the US Fifth Air Force and that of the west coast carrier force. During this particular patrol, before the change to convoy protection, much more flak damage was taken, with 25 aircraft damaged and five destroyed between 6 and 18 December. This was half of the aircraft lost to enemy action in the entire deployment.

Having spent Christmas in Kure, *Sydney* began her sixth patrol on 28 December, this time conducting more airstrikes in the Cho Do-Sok To area. Targets included North Korean artillery and junk concentrations. Sorties were also flown in support of South Korean troops and irregular forces.[63] On 4 January 1952 Rear Admiral Scott-Moncrieff personally briefed the aircrews prior to attacks on gun positions along the Yesong River, close to Panmunjom. Lieutenant Peter Goldrick, RAN, was wounded during this operation but managed to bring his aircraft back and land safely on *Sydney*. Navy Office was informed of the incident and in its reply to the ship displayed the kind of humour which we have all come to associate so readily with 'head office'. Navy Office directed that Lieutenant Goldrick's flying pay was to cease from the moment he had been wounded as he was then temporarily unfit to fly and might never fly again.

The seventh and final patrol began on 16 January. Bad weather and poor coordination between the blockading force and convoys affected the flying rate. Nevertheless, North Korean gunners still managed to damage ten aircraft. During this patrol, strike targets were changed from rail infrastructure to water towers on the basis that water towers would take longer to repair.[64] No flying was possible on the final day of operations, 25 January. Cold and the overnight snow had rendered all deck parked aircraft unserviceable. The five forward most Furies were almost totally covered in frozen salt water.

All in all, during the deployment *Sydney*'s Air Group flew 743 Firefly sorties and 1623 Sea Fury sorties.

The Carrier Borne Army Liaison Section

One of the most important behind the scenes organisations in the *Sydney* was the combined CBAL Section which was manned by officers and men from 71 and 72 Sections. It comprised two majors, one of whom, Major M.B. Simkin, had flown to Korea ahead of the ship, two WO2s and two corporals. En route to Korea, as well as conducting bombardment spotting practice, the CBAL Section also provided briefings on escape and evasion and Owen gun practice for Firefly crews.[65]

Once on operations, each of the majors briefed and debriefed alternate flying events. They also undertook target assessment, and each evening proposed a target list for the following day's operations to Commander (Air), the Air Group Commander and the Operations Officer. The Section also kept up to date situation and flak maps and maintained target dossiers and files of briefing photographs. A store of over 200,000 maps and charts was also maintained by the Section.[66]

Other specific tasks for which the CBAL Section was responsible included the provision of Army advice to the Captain through Commander (Air), the provision of all intelligence information to aircrew, and collection and assessment of all flak information.

Busy as they undoubtedly were, the Army personnel were not entirely oblivious to their surroundings. They developed a comprehensive handover process for replacement personnel, which included a set of 'How to survive the Navy' instructions. The advice contained in this document included the following points:

Never sign for anything. The Naval Supply Branch never forgets.
Enjoy your food. Eat ashore.
Treat all parts of the ship as though they have been freshly painted.
They probably have been.
If you should require a drink after the bar shuts, follow the trail of Air Group officers to cabin G87.

Other Challenges

Quite apart from the dangers associated with carrier aviation and low-level flying in the war zone, there were other matters that made life difficult from time to time. Of these, the weather was probably the most significant and frequent.

Sydney's first real taste of what was in store came on 14 October when all Commonwealth ships in Sasebo were ordered to sail, to ride out the approaching Typhoon Ruth. Some in *Sydney* wanted to remain in harbour, noting that Sasebo was surrounded by 2000 ft high mountains. Nevertheless, she did sail and met the full force of the storm with 18 aircraft lashed down on the flight deck. Chocks and fastening lugs were torn away from aircraft and the heavy movement collapsed some undercarriage legs. Minor electrical fires broke out in the ship as sea water shorted out ventilator fan motors.[67] Four aircraft broke free, one was washed overboard and many others were damaged.[68] The ship itself survived pretty well.

As the winter progressed, intense cold, low cloud and snow also caused problems. Snow tended to settle only on the forward and after sections of the flight deck as the relative warmth from the hangars melted it elsewhere. Aircraft often had to be thawed below decks, and engine oil in deck-parked aircraft often had to be diluted at night so that engines could be started in the morning. Propellers and wing leading edges were coated with an anti-icing paste, the residue from which made the deck very slippery. Cold also damaged propeller seals and coolant hoses.[69]

These conditions of course made working on aircraft on the flight deck extremely difficult. Fine work requiring the removal of gloves could frequently not be done for more than five minutes at a time. For the aircrew, of course, immersion suits were a constant necessity and map reading was a constant challenge over snow covered country. There were also days when the weather was simply too bad to fly; low cloud, wind and snow being the usual culprits.

Some of *Sydney*'s men rugged up for the Korean winter. (RAN)

Although it did not necessarily impact on individual aircrew, command and control was a periodic headache for the command. Much of the problem was cultural, with the RAN and RN ships generally unhappy with the degree of prescription and detail contained in USN operation orders.

There was also an underlying friction between Rear Admiral Scott-Moncrieff and his USN superiors; whether it was Vice Admiral G.C. Dyer or Vice Admiral C.T. Joy and which preceded *Sydney*'s arrival in Korea. More often than not there was an unfortunate level of mutual misunderstanding. Admiral Scott-Moncrieff was sceptical of the value of some US proposed operations against the risk to his ships. He was also mindful that the smaller navies, like Australia's and Canada's, could ill afford to lose ships in such operations.[70] The USN hierarchy, in contrast, felt that the Royal Navy had a half-hearted approach to the War and was prepared to allow the USN to take the risks so that British forces could be conserved.

During *Sydney*'s time in theatre the most significant of such issues arose with the UN Command's temporary exchange of duties between the US Fifth Air Force and the west coast carrier force. Vice Admiral Joy had argued for the switch on the grounds that it would be sensible for the naval aircraft to take over the convoy escort role between Japan and Korea if the Fifth Air Force took over the strike and ground support roles of the carrier force.[71] Admiral Scott-Moncrieff disputed the move because of the familiarity his aircrews now had with the terrain of north-west Korea, and because of the monotony associated with the convoy protection task. Through all of this Captain Harries was in full agreement with Admiral Scott-Moncrieff.[72]

The rescue of downed aircrew was another major challenge. Robert O'Neill has noted that the air war virtually ceased until downed aircrew were located and where possible, rescued.[73] *Sydney* made her own unique contribution to this process. The first element of it was the introduction of fluorescent panels, two feet square, coloured red and yellow. The panels could be laid out on the ground in ways that denoted the condition of survivors, the location of enemy troops and the intentions of the aircrew.[74] The second element was the development of a message bag which could be dropped to downed aircrew. Both innovations were taken up by other air forces in Korea.

Of the rescues conducted during the War, that involving Sub-Lieutenant N.D. MacMillan, RAN, and Observer First Class P.H. Hancox from *Sydney* stands out. The Firefly crew crash- landed in a paddy field and held off North Korean troops for some time, aided by other Air Group and No. 77 Squadron aircraft. *Sydney*'s rescue helicopter, flown by USN Chief Petty Officer A.K. Babbitt and Airman C. Gooding, took off at 1622 and arrived at the scene just as North Korean troops were moving in.

During an exchange of fire the helicopter picked up the two aviators and began the return flight at 1735.[75] This presented its own difficulties. In darkness, and with no instruments by which to fly, CPO Babbitt made for the nearest airfield and landed with the aid of trucklights. He was subsequently awarded the Distinguished Service Medal and the USN Navy Cross for what was the longest helicopter rescue transit over enemy territory during the War.

Effectiveness and Achievements

During her Korean War deployment, *Sydney* spent 122 days in Korean and Japanese waters. Of these 42.8 days were spent on flying operations; 11.7 days were lost to poor weather; 29.5 days were spent on passage or in replenishment; and 38 days were spent in harbour.[76]

Considered in purely financial terms it was a costly deployment but carrier air power was a necessary counter to North Korean land force predominance and to a lack of suitable air bases in Korea. Fortunately *Sydney* lost only three aircrew killed. The three, all RAN Sea Fury pilots, were Lieutenant Kenneth Clarkson and Sub-Lieutenants Dick Sinclair and Ron Coleman. Looked at in terms of damage to North Korean infrastructure and logistics support, and of the number of troops killed, the result is mixed. The official figures (and sources do vary a little) are:

3129 enemy troop casualties
66 bridges destroyed or damaged– many of them more than once
7 tunnels damaged
38 sections of railway destroyed
7 rail sidings destroyed
5 water towers damaged
1 locomotive destroyed and 1 damaged
139 rail trucks destroyed or damaged
2060 houses (troop shelters) damaged or destroyed
469 junks and sampans sunk or damaged
15 artillery pieces damaged or destroyed.[77]

Additionally, former RAN pilot Fred Lane claimed in a recent article that 234 oxcarts were destroyed, along with 16 ammunition dumps and seven fuel dumps.[78]

Battle damage assessment was quite difficult, and *Sydney*'s aircrew and CBAL Section staff took a notably conservative view of their achievements. For example, unless bodies could be counted, rockets had to be seen to hit known numbers of troops before kills were accepted.[79] That the approach was conservative was confirmed by 'Leopard', the US leader of an irregular ground force group. His 'on the ground' assessments indicated that RAN aircrew were underestimating their impact. That they were doing so had important ramifications in that it contributed to the US dissatisfaction with the performance of the Commonwealth forces. On the other hand, RAN aircrew were said to have been bemused by the American penchant for figures of often doubtful reliability.

Issues such as these may well have contributed to a degree of frustration that had emerged among the aircrew by the end of the deployment. A much more significant cause, however, may have been the inability to shut down the North Korean supply lines completely. This inability simply reflected the limits of air interdiction capability as observed in other conflicts, such as Italy in 1944 and subsequently Vietnam in the 1960s.[80] Some of the aircrew involved bemoaned the apparent trading of North Korean oxcarts and peasants for aircraft and aircrew. The significance of this criticism is apparent in the decision by USN and Royal Navy authorities to switch from tactical targets to strategic ones such as petrochemical installations and powerhouses. Only then, after *Sydney* had departed, did the North Koreans begin to show flexibility in the armistice talks.[81]

Yet while there may have been some frustration at their inability to close down North Korean supply lines, the RAN aviators should have been satisfied with their efforts. They created significant problems for North Korean logistic support and ensured that most, if not all, such movements took place only at night. They also provided ground troops with consistently accurate close air support. Probably unknown to them at the time, at the political level *Sydney* and her Air Group enabled Australia to avoid having to commit additional ground forces to the conflict.[82] For their efforts they were awarded four Distinguished Service Crosses and one Bar, one Distinguished Service Medal, ten Mentions in Dispatches and two US Legion of Merit awards. As the CBAL Officer, Major Simkin, noted in his official diary: 'Australian Naval Aviation can be justly proud of its first major war effort'.[83]

The Aftermath

The Korean War heralded a new Australian naval construction program, announced in February 1952. It was to include six River class frigates, one replenishment ship, three boom defence ships and four coastal minesweepers. It was also to include modernisation of the cruiser HMAS *Hobart*, the three Tribal class destroyers, and conversion of the 'Q' class destroyers into fast anti-submarine warfare frigates.[84]

The Fleet Air Arm was not forgotten in the program. HMS *Vengeance* was offered on loan to the RAN while the second carrier, HMAS *Melbourne* was being modernised. Additionally, Sea Venom and Gannet aircraft were ordered in 1951 and 1952 respectively. Schofields airfield also became HMAS *Nirimba* on 1 April 1953, with two reserve air squadrons, a repair base and the Captain (Air).[85]

The personnel element of this program was designed to have the Navy grow from 10,252 PNF (Permanent Naval Force) and 4943 citizen force members in 1950 to 17,000 PNF and 10,000 national servicemen. The closest it got to that target was a strength of 14,144 PNF and 7398 citizen force members in 1953.

Circumstances, however, conspired to curtail the program. Hostilities ceased in Korea in 1953 and the focus of Australian defence efforts changed from the Middle East – a Commonwealth commitment – to Southeast Asia. As well, high inflation in 1954 saw the imposition of cuts in the Defence program. These resulted in cancellations and delays to building and modernisation activities.

Melbourne nevertheless entered RAN service in 1955, together with her Sea Venom and Gannet aircraft. Even so, *Sydney* was not modernised and soon ceased to be a front-line aircraft carrier. Additionally, a mere four years after *Melbourne* had entered service, seven years after Sydney had returned from her successful wartime deployment (and a mere 11 years after the Fleet Air Arm had been established) the government decided that carrier aviation would have no further place in Australia's defence and that the carrier and its aircraft would be paid off in 1963. A reprieve was subsequently granted, but for the carrier and fixed-wing aviation it lasted only until 1982-83. Clearly operational excellence and strategic value were not enough on their own to guarantee organisational prosperity.

13. Virtual War in the 'Ice Jungle': The Changing Nature of Shore, Littoral, and Terrain

Gary E. Weir

MUCH OF THIS BOOK seeks to examine the history of joint and combined power projection across the littoral.[1] In this world of shoreline, beach, and barely submerged terrain, forces can easily define the nautical and terrestrial geography. They see the strategic goals, the space to occupy, and the assets to seize if capability and planning are up to the task. During the Cold War, submariners faced a similarly daunting mission, in a world where the terrain, turned 180 degrees, hung inverted above their heads and an immediate and very restrictive three-dimensional geography placed severe limits on their normally freewheeling open ocean tactics.

Unlike the geography at Gallipoli, Normandy, or Okinawa this potential battlespace changed constantly, altering almost daily any assumed defensive or offensive solution. In our own post-Cold War world, this battlespace continues to change, suggesting new deep water and littoral configurations that will present significant problems for the future of diplomacy and naval planning. The significance of the Arctic during the Cold War emerged from two very simple truths. Firstly, the shortest distance between major strategic targets for the two hostile, nuclear-armed superpowers was the great circle route across the top of the world. Secondly, in that region, the environment set tactical limits on naval operations as strict as, and remarkably similar to, those imposed by any traditional amphibious or expeditionary effort.

Reflecting on his work in the 1970s, the father of naval applied Arctic science in the United States, Dr. Waldo Lyon, recalled the value of placing acoustic arrays in the ice cover, providing scientific and operational intelligence regularly via satellite. Building on that observation, he noted that:

> *... we finally demonstrated with the arrays [that] the Russians were using the Arctic ... one of these arrays tracked [a contact] right across, from Franz Josef Land, [it] came across to the top of Greenland, went back at 14 knots, just the right speed, [and] disappeared on the other side. That's what this array caught. That caught attention. It was from that point on, then, which was in about the middle seventies ... everybody believed then that we'd better be watching the Arctic Ocean, so that was the beginning of ... what you'd call the Arctic Cold War era.*

> *During that time, particularly in '73, I had [USS]* Hawkbill *and [USS]* Seadragon *in the Bering Sea, and we played a war game, with … practice torpedoes. The game was on, two of us in the shallow water under the ice, and that's when we learned … no way … we don't know how to do this.*[2]

During the Cold War, the naval conflict at the top of the world belonged primarily to the submariners. With the advent of nuclear power in both the American and Soviet fleets, the ice cover in the Arctic became what one Soviet Victor II driver called the *armour* above his head that offered both protection and a challenge to his survival.[3] For the American submarine force, exercising under-ice capability by voyages to the Pole or deep penetrations into the Norwegian and Barents Seas on special operations changed roughly thirty years ago into the far more aggressive posture of the US Maritime Strategy. Frequent exercises demonstrated that American submarines could successfully and regularly venture under the ice and into protected Russian waters, deliberately refining their ability to bring any conflict home to the Soviet Navy. The Maritime Strategy transformed the Arctic environment conceptually from a natural scientific laboratory and a region of occasional and exceptional operational activity into a possible battlespace on par with the northern Pacific or the GIUK (Greenland-Iceland-UK) gap. In this battlespace, the incomparable environment reigned supreme, strictly limiting the options available to opposing submarines, limiting strategic choices, and dictating the only tactics possible.

In recent years the post-Cold War Arctic environment has once again redefined the possible. During the confrontation between the North Atlantic Treaty Organisation (NATO) and the Warsaw Pact the ice 'armour' limited both access to the region and its potential. In recent years the dramatic recession of the Arctic ice cap during the summer months has opened new commercial possibilities and dramatic new strategic questions. Data gathered in large part by American submarines in cooperation with the National Oceanographic and Atmospheric Administration (NOAA) has enabled scientists to suggest with a high degree of certainty that the Arctic ice will begin to disappear completely each summer by the end of this century. The process has already begun. The summer recession is already dramatic and obvious. For countries in Asia, the summer Arctic would provide a very valuable short route to the eastern coasts of North and Central America, as well as to the commercial possibilities of Scandinavia and Western Europe. Economic potential exists which makes it almost certain that naval forces will follow. Can study of the Cold War experience provide a productive basis for an Arctic naval strategy over the next century? Will receding ice define the shoreline, and littoral areas that emerge from regions previously denied to surface vessels? How do we project power into these new open areas? As long as ice remains, the submarine will play a role in any expeditionary foray, in the far north.

In response to the success of regular American deep submerged penetrations into their protected northern waters, during the Cold War, the Soviet Northern

Fleet looked for ways to secure the very limited routes available to their nuclear-powered ballistic missile submarines (SSBNs) transiting from the Kola Peninsula to their launch stations. Employing locations in the Norwegian or Barents Seas protected by seasonal ice and Soviet nuclear-powered attack submarines (SSNs), as well as sending their SSBNs into the Arctic Ocean for natural protection within range of their American targets made strategic sense. As the Maritime Strategy pushed American forces forward, the Soviets looked to the Arctic to maintain their internal lines of communication, to achieve effective deployment, and to ensure the safety and potency of their submerged deterrent.[4]

A Soviet Navy Delta III, SSBN, 23 May 1984.
(US DEPARTMENT OF DEFENSE)

Both adversaries had already taken their first steps in the Arctic. The US Navy reached the North Pole with USS *Nautilus* (SSN 571) in 1958 and continued to engage in scientific and operational projects in the far north with both the 571 boat and USS *Skate* (SSN 528) to understand both the capabilities of the vessels and one of the most demanding submarine environments in the World Ocean. By 1962, the Soviet November class nuclear submarine *K-21* made the first Russian Arctic voyage, surfacing in a man-made polinya[5] opened by a four-torpedo volley. On 17 July 1962 the first Soviet nuclear submarine, *K-3*, took the Russians on their initial trip to the North Pole under the command of Captain First Rank L. Zhiltsov, who became a Hero of the Soviet Union. In September 1963 the Soviet ballistic missile submarine *K-178* cruised from the Pacific Ocean across the Arctic to the Northern Fleet, travelling 617 nautical miles submerged,[6] emulating a feat already accomplished by a number of American nuclear submarines, including *Nautilus* and *Skate*.[7]

These activities simply constituted a beginning. From 1949 onwards, addressing the Soviet challenge successfully, in the Arctic and elsewhere, occupied a very high place on the US Navy's list of strategic priorities.[8] For nearly a half century after each superpower began making regular trips to the North Pole, American and Soviet submariners engaged in an intense and regular *virtual war* on many fronts far below the surface of the World Ocean. A period referred to by Admiral James Watkins as an era of violent peace.[9]

In the early 21st century, employing the phrase 'virtual war' to describe this conflict seems false and immediately conjures up unwelcome images of video games, computer graphics, and child's play. The more adult view, evident in the current historical literature, suggests that these two Cold War superpowers engaged for a third time in a battle for the Atlantic, naturally following the path paved by the major naval powers of the two World Wars. This view holds that after 1945, the hunt simply occurred at greater depth, at greater speed, with more powerful and innovative weapons and techniques, and eventually on board nuclear submarines. All of this took place with the constant threat of nuclear Armageddon lurking in the background via missile or nuclear torpedo.[10]

Actually, the Cold War's constant and dangerous submarine joust barely resembled past undersea confrontations, if at all. With the constant threat of mutually assured destruction (MAD) placing tenuous but strict limits on both the use of weaponry and the traditional expectations of battle, Cold War submariners formulated strategy and tactics, and calculated victory and defeat in a very different way. Since using even conventional offensive weapons could easily precipitate horrible and almost uncontrollable geopolitical consequences, undersea warriors measured victory in terms of surveillance, detection, and constant monitoring. If you knew the enemy, his vehicle or ship, his location and capability, and could follow or 'shadow' him without betraying yourself, you claimed victory by Cold War standards. The object of this dangerous and difficult exercise had little to do with destroying the target in an encounter that might escalate to jeopardise the human race, but rather with redefining the conflict in terms of surveillance, detection, submerged capability, and destructive potential. With this victory achieved, a national leader might anticipate any threat with sufficient time to destroy an adversary's ability to act. *Control* over the opponent became the operational objective and precise knowledge of him became the means to that end. Underwater acoustics, early and reliable detection, and capable submarines would provide the keys to a more easily recognised victory if the nuclear nightmare began to unfold. Without that terminal event, surveillance, detection, and control provided the desired advantage, and for nearly 50 years became synonymous with success.

The undersea conflict also provided the only stage on which the two major players regularly met in face-to-face armed confrontation. While their soldiers glared at one another across the checkpoints between East and West in Europe, fought in the occasional combat episode during the air war in Korea and in various capacities during the Vietnam War, only under the surface of the World Ocean did the two leviathans risk daily encounters and possibly deadly confrontation in their mutual effort to achieve the desired control. Thus the submarine Cold War evolved into a high-tech, high-risk venture, that involved the best and brightest practitioners on both sides, each one seeking to control the other via seamanship, technology, enhanced listening, and stealth.

When asked about the pursuit of these goals, Vice Admiral Anatoli Shevchenko recalled with his larger-than-life enthusiasm a particularly wild, high speed chase he conducted with his Victor II boat, *K-513*, in December 1978. He encountered an 'American colleague' commanding a *Sturgeon* class attack boat in the North Atlantic. The contest lasted for over six hours at submerged speeds approaching 26 knots.

In this case, Shevchenko decided to fall in behind the American without trying to disguise his presence, feeling sure that his opposite number had already detected him. He decided to deprive the American of any stealthy illusions while driving him away from the contact point to ensure the relative privacy of ongoing Soviet fleet exercises. Suddenly, Shevchenko's opponent decided to lose his pursuer and resume his patrol, so he accelerated away at high speed, doubtless hoping to double back more quietly at some future point without benefit of a Victor class tail. Unfortunately for him, Shevchenko found the prospect of a chase interesting. Listening for the sound of the *Sturgeon*'s screws loudly departing at over 24 knots, Shevchenko waited for the burst of speed to end without accelerating himself. While he waited, his sonar plotted the American's course and speed. When the acoustic signal seemingly went quiet, *513* took its turn at high speed, along the same bearing and course, for the same amount of time. The boat would then slow down to permit the sonar to reacquire the American, who would then attempt to elude at high speed once again. While accelerating, neither submarine could detect the other due to the turbulence generated by their high-speed screws, internal machinery sounds, and increased flow noises along the hull. Some of the high-speed bursts would last as long as 20 to 30 minutes. After a burst, Shevchenko would then slow down to five knots to give the sonar a chance to work. Once the American realised the effort to elude his pursuer had failed, the chase began again.

Since the sonar could not 'hear' anything with the boat moving at high speed, the officers in the control room took time to relax. This chase stood out in Shevchenko's memory not only for its challenge, its speed and near recklessness, but also for a tradition initiated by the boat's political officer. After two hours of chasing had passed, he had the cook reward each hard working rating in the sonar room for his intense devotion to the chase. While their boat sped after the American and attention to the momentarily deaf passive sonar could relax for a few moments, the galley provided each rating with a bowl of cream-cheese, topped with sour-cream, and crowned by a magnificent strawberry! So while the submarine rushed after its adversary, the acoustics team sat back and enjoyed this decadent treat laced with colour and vitamin C. As soon as their boat slowed, they went back to work.

During this bizarre episode, a well-known quieting and acoustics expert, Captain First Rank Anatoli Kolmakov, came into the sonar room. He joined Shevchenko's crew for this deployment on board *513* to complete an on-board submarine sonar qualification course. Upon entering, he saw everyone enjoying their strawberry-topped delicacies. He smiled and commented that everything seemed well in hand. Shevchenko's navigator Sergey Efremenko responded very seriously by saying, 'We are chasing an American submarine!' Kolmakov laughed, never took him seriously, and left the compartment. Many years later, Shevchenko recalled with characteristically energetic gestures and a broad smile that:

> *Two hours later he came into the sonar room again and saw us eating the cream cheese and he was told the same thing. When he left the post the men laughed! Kolmakov went to the navigators' room, made sure that it was true that we were chasing the US submarine and said 'An amazing tradition! I would never believe that the symbol of surveillance aboard a submarine is a strawberry with sour-cream!*[11]

The new nature of undersea warfare, its object and tactics, placed control of the Soviet Union's growing submarine fleet at the top of the US Navy's list of strategic priorities. The control sought by each side depended completely on knowledge of the environment alongside a mastery of submarine technology. Combining the latter with intimate knowledge of the ocean made detection and identification possible, and with that, control over the Soviet submarine force.[12] Nowhere on Earth was this more important or more difficult than under the Arctic ice.

As a result, regular, early and accurate detection of Soviet submarines became every bit as important as developing the capability to guide a ballistic missile to its target. Oceanographic research during World War II began to reveal the physical nature of the ocean in ways never before imagined, enabling underwater acoustics to emerge as the vital eye of an ultimately victorious American submarine force and as the key to identifying an approaching or lurking threat. During the two decades following the War, ocean research sponsored largely by the Bureau of Ships, the submarine force, and the Office of Naval Research (ONR) improved passive sonar and signals analysis, conceived the deep ocean surveillance system called SOSUS and installed its first generation arrays, initiated the operationally-oriented Long Range Acoustic Propagation Program (LRAPP), and laid the foundation for the surface towed-array sonar system (SURTASS).[13]

These projects and systems eventually provided a striking advantage for the US Navy. For example, by the time SURTASS went to sea in final form in 1984, the Soviet Northern Fleet already understood its importance to the battle for control. After an encounter with two American *Los Angeles* class boats in 1989, Captain Evgeni Vanin's evasion with *Komsomolets* (*K-278*) provided him with a brief respite, only to have another threat emerge. His sonar supervisor reported intense active sonar signals from surface vessels close aboard. In the control room, Vanin consulted first with his executive officer Oleg Avanesov and then his navigator Michael Smirnoff. Studying the charts, they chose a new course for the boat just in time. The active sonar 'pings' grew even stronger, requiring *Komsomolets* to evade. When the signals again began to increase in strength three hours later, Vanin estimated the distance to the surface vessels and, satisfied with the margin of safety, decided to take a look. He came shallow and broke the surface with his periscope to determine more precisely his location and to record the radio signals of the pursuing ship. His crew went to battle stations in response to the general alarm as Vanin placed his pursuer in the periscope crosshairs. Increased magnification revealed the persistent American as one of the US Navy's *Stalwart* class. Vanin realised that this specialised vessel carried the SURTASS system. As Captain Boris Kolyada recalled,

> *Vanin commented openly that 'If the SURTASS system vessels are here, it's difficult for us to stay undetected'.*
>
> *He turned to his diving officer and issued a quick order. 'Emelyanov, estimate the best depth for evasion. If it goes on like this an [American P-3] Orion [aircraft] will be here soon. We definitely need to maneuver away from the region.'*
>
> *'The course and depth have been confirmed' snapped navigator Smirnoff.*
>
> *'Our radio transmission to base was received,' reported communication officer Alexander Volodin.*
>
> *Vanin turned back to Emelyanov and simply said, 'Dive'.*[14]

Even further north, in the Arctic, operations under the ice would depend absolutely on environmental knowledge, the detection possibilities it provided, and the ability to use it to best advantage.

As early as 1973, when he played his makeshift war game with *Hawkbill* and *Seadragon* under the winter ice covering a portion of Bering Sea, Waldo Lyon had already come to a disturbing conclusion regarding the unique confluence between tactics and the environment under the Arctic ice:

> *We learned then that whatever submarine was stopped in the ice cover, he had such an enormous advantage over the one that was [searching], there was no chance for the [latter]. He was going to be hit. [The attacker in the ice cover is] listening . . . all around. All he has to do is drop his weapon in the water, and the one that's coming, is in the open water underneath. [The searcher is] trying to find [possible threats] up in the ice cover. If [the Attacker] shoots, [the searcher] has a long time before he even gets close enough to shoot his weapon… and not much water to maneuver in to avoid [an incoming] weapon. The advantage is enormous for the one that's stopped in the ice. It's unfair.*[15]

In February 1992, within the context of the late Cold War Maritime Strategy, Lyon commented in the US Naval Institute's *Proceedings* magazine that:

> *Warfare in the ice is like jungle warfare. The sea ice canopy becomes the jungle in which the submarine must live, work, and fight – not just transit. The sea ice canopy is highly variable. Viewed from below it is upside down, rolling hills of ice: a 'badlands of ice buttes and blocks, scattered or piled on one another;*

canyons between massive ridges; flat ice planes and open cracks and lakes. It is dynamic, changing under thermal stress, and is always in motion under stress of wind and current.[16]

Discussing submarine encounters under the Arctic ice as if they resembled open ocean patrols struck Lyon as absurd. Naval officers and scientists with Arctic experience reacted strongly to the notion that this region would ever become the site of regular submarine operations on a large scale. In his career-long struggle to keep his Arctic Submarine Laboratory in San Diego open and working, on this issue Lyon had encountered every reaction from professional interest and concern to complete incredulity. Submariners treated the notion of Cold War operations under the ice either as fantasy or as a terribly deadly business akin to walking into a confined space with your vitals already in a sniper's crosshairs. Very few had given serious thought to the Arctic as a battlespace until submerged navigation improvements opened up debate on the tactics of under ice naval operations in the face of a capable enemy.[17] As Lyon observed, looking back over twenty years of debate:

*…in the late seventies, we started bringing this point up. If you've got warfare coming up in the Arctic Ocean, we don't know anything about how to fight. [In] these 637s [*Sturgeon *class] we can't evade. To go into the ice cover to evade, you've got to come to a complete stop, and I mean stop, be sure that you're zero speed, and then you must go up vertically against the ice at twenty to thirty feet a minute? Let's just start thinking... You're going to avoid a weapon? No way. It's going to take you twenty or thirty minutes to make that maneuver. Otherwise you're going to wreck the submarine… So the point was, 'All right, we'd better start doing something about learning …how to do warfare in the ice.' So Rickover then proposed, 'Well, let's build an experimental submarine, one that has a little sail and hardened, so that we can go ramming into the ice cover to evade the weapon, or let's modify a* Skate *and try it.'*

Lyon and Admiral Hyman G. Rickover understood both the importance of the problem and the peculiar nature of the Arctic environment. The submariners found it necessary to consider the 'terrain', and factor its characteristics into their operations. The environment would impose its own discipline on any tactical and technical solution under consideration, in much the same way as the shoreline, the beach, and the shallows influenced success or failure in the projection of power by surface and air forces.

The Soviet submarine force found itself simultaneously experiencing the very same growing pains. While accustomed to demanding and dangerous transits by both diesel and nuclear boats across the northern rim of the Russian landmass, operating under the Arctic ice fell into a completely different category.

One of the Russian Navy's current experts on Arctic operations, Vice Admiral Anatoli Shevchenko, first experienced the far north at the same time that the

Americans began to realise the strategic options presented by the Arctic in a world of longer-range ballistic missiles and aggressive naval strategies. Shevchenko qualified for command in 1974, less than a year after the trial war game by *Hawkbill* and *Seadragon* in the Bering Sea, and after a staff assignment he took command of *K-513*, a Victor II class fast attack submarine. Shortly after a first deployment took his boat to the Mediterranean, the deputy commander of the Northern Fleet submarine flotilla, Rear Admiral Eugene Chernov presented him with an unwelcome surprise immediately upon his return. Impressed with the *513*'s crew, the admiral intended to send them north to explore the Arctic, their boat's capabilities, and themselves. It had become plain to Chernov that the Arctic would soon assume an importance far beyond the symbolic achievement of reaching the Pole or the inspiration of a fascinating natural laboratory.[18]

When asked how he prepared for the assignment, Shevchenko immediately recalled reading George Steele's book *Seadragon: Northwest under the Ice* in 1966, four years after its publication in the United States. At that point he began to appreciate the demands of the Arctic through the eyes of this American submarine commander. Now he went back to that book and, with good reason and renewed purpose, looked for the spoken and unspoken detail that might emerge from Steele's account.[19] Weeks later, when his *K-513* arrived at the Pole after a successful transit marred only by a slightly bent forward dive plane, Shevchenko announced to his crew that as of 1832 hours on 31 August 1979 they had arrived in 'the land of *Nautilus* and *Seadragon*'.

In the later stages of his long career, Waldo Lyon argued that neither the United States nor the Soviet Union ever completely developed under-ice capability. American 688i fast attack submarines with their low and forward-mounted control surfaces, as well as the latest Sierra and Akula class Soviet/Russian boats, can operate with considerable agility in the various basins of the Arctic Ocean and have little trouble finding polinyas for surfacing. But even with this capability, the two navies have never thoroughly addressed the deadly endgame predicted by Lyon in his analysis of the Arctic operational environment, the realities of the ice cover, and the need to seize and master the battlespace via naval power projection. Both sides quickly discovered Lyon's meaning when he stressed the importance of a regular presence and being there first. A 1989 RAND report noted the growing habit of the Soviets to adapt their submarine design,

> *… for just this type of environment. All of the large Soviet* Typhoon *class SSBNs are based in the Kola region and are thought to be the first submarines designed particularly for under-ice operation. These submarines exploit their design by using a technique called [by the US Navy] 'ice picking' in which they quietly drift for months while resting immediately below the surface of the ice.*[20]

The term 'ice picking,' used by the RAND analyst, comes from a DARPA, through-the-ice surveillance device, under development in the mid-1980s, called 'Ice Pick,' which first appeared in an airdrop version late in the Cold War.[21]

Crew members stand by the sail structure of USS *Seahorse* (SSN-669), which has broken through the Arctic ice off Alaska, 25 March 1990.
(US DEPARTMENT OF DEFENSE)

The ice cover provided the stealth the Soviet/Russian Navy always sought to make their SSBNs, perhaps some of the quietest submarines in the world, virtually undetectable on station. They also still carried torpedoes designed for breaking through the ice in case they could not immediately emerge through the cover in the event of a quick launch order.[22] While its submarine force did not miss the advantages of this practice, the US Navy did not have the same geographic disadvantages that obliged the Soviets to look to the Arctic for strategic solutions. Thanks to advanced surveillance capability and a much friendlier geography, the Americans had a wider variety of answers to their tactical needs. For the Russians, the ice cover offered greater safety, proximity to target, and tighter internal lines of operation and communication.

However, the Arctic provided neither side with a complete advantage or disadvantage. Like the dense Vietnamese jungle that proved so formidable in the 1960s and 1970s, the natural Arctic environment severely limited the tactical options available in the region of closest physical proximity between the Cold War adversaries. Neither side could exercise the desired control with certainty in the Arctic and both had to accept the limits imposed by nature. The Arctic became a battlespace both too dangerous to ignore and too terrible for combat operations save under the most dire circumstances: a perfect metaphor for the MAD-ensured Cold War standoff in which both navies found themselves trapped for the better part of a half century. For the submariner, without the greatest care, this environment could be every bit as demanding and tyrannical as the nuclear technology that drove their amazing boats.

In recent years the environmental tyranny of the far north has confirmed that the only constant in earthly experience is change. Given what scientists now predict about the Arctic, the Soviet and American confrontation in that area during the Cold War may become far more important than either power expected. Bringing to the attention of the general public widely-held scientific conclusions already familiar to the US Navy, journalist Daniel Glick recently reported in *National Geographic Magazine* that:

The annual breakup of sea ice off the coast of Alaska happens weeks earlier than it once did. Since 1978 the area of perennial Arctic sea ice has decreased by 9 percent per decade, but the degree to which the ice has thinned has been harder to assess. Scientists using submarine sonar data documented a 40 percent thinning in the past 30 years. Others put the estimate at about 15 percent. Some predict that the sea ice could be absent in summer by 2100.[23]

Thanks to the unclassified SCICEX submarine ice exercises in the Arctic in the 1990s, as well as research conducted by means of satellites, icebreakers, and via the SCAMP acoustic ocean floor mapping project at Columbia University, scientists have concluded that the configuration of the Arctic ice and the region's general climate will soon pose amazing possibilities as well as fresh potential threats. According to one scientific publication,

In ten years' time, if melting patterns change as predicted, the North-West Passage could be open to ordinary shipping for a month each summer. And the Northern Sea Route across the top of Russia could allow shipping for at least two months a year in as little as five years. The new routes will slash the distances for voyages between Europe and east Asia by a third, and open up new fisheries. The resulting boom in shipping could lead to conflicts, as nations try to enforce fisheries rules, prevent smuggling and piracy, and protect the Arctic environment from oil spills. To complicate matters, Russia and Canada consider their northern sea routes as national territory, while the US regards them as international waters.[24]

As these changes occur, geological time suddenly shrinks to the span of a single human life. The environmental lessons as well as the limited Arctic capabilities developed during the Cold War will soon become the foundation upon which naval strategists build their operational reaction to these developments. Powers like Japan, Russia and China will naturally look closely at the possibilities presented by these dramatic new circumstances and their evolving plans will naturally involve their military forces. Thus, within the next few years the increasing appearance of Russian, Chinese and European forces at the top of the world seems very likely. In the near term, they can do so only by following in the footsteps of *Nautilus*, *Skate*, and *K-21*. However, a 2001 Office of Naval Research-sponsored symposium examining the operational challenges posed by these developments began by assuming that the Navy would have to expand its operations in the Arctic by 2050 and that over the next 20 years the volume of sea ice in that area would decrease by 40 percent. There was also an understanding that within five to ten years there was the strong possibility that the Sea of Okhotsk and the Sea of Japan would remain ice-free year around.[25]

For a time, their shared Cold War experience and under ice capability will give the United States and Russia an advantage in the region. That will not last for long. As the ice recedes further the acoustic detection equation will change dramatically. Wind will drive up the ambient noise levels in an open sea while mixing the surface layer with warmer air, affecting the acoustic ducting near the ocean surface. For a time new littoral regions will emerge in proximity to ice cover that will present problems of expeditionary warfare very difficult to imagine, and made more threatening by an under ice aspect. In addition, as the surveillance problem becomes more complex, the population of possibly hostile forces will likely increase, given the attractiveness of the region as a sea route and as a centre of virtually untapped natural resources. As the United States Arctic Research Commission put it, 'In summary, melting of sea ice in the Arctic will turn it into a conventional open-ocean ASW environment, with none of the advantages it now affords to an adversary strategic submarine'.[26]

There is every reason to suspect that the pursuit of *control* may very well resume once again, only now in the far north, driven by the natural strategic advantages presented by the Arctic's location as well as the economic promise of easier access to valuable world markets. In the process we shall have to constantly revise our assumptions about littoral, shoreline, and terrain and revisit the possibility of an undersea component to expeditionary warfare.

14. A Maritime Approach to Joint and Coalition Warfare: Three Case Histories

David Hobbs

THROUGHOUT THE PERIOD OF the Cold War the Western navies placed great emphasis on anti-submarine capability in an open ocean environment. Fortunately, they retained the ability to dominate the land and the air above the littoral as the three examples detailed in this chapter illustrate. These cases were seen at the time as aberrations but today, more than forty years later, the operations seem very familiar and stand as examples of a capability that has not only survived but now has over-arching importance.

Lebanon-Jordan 1958

The unsatisfactory conclusion of the Suez Crisis in 1956 left the Middle East unstable, and the consequent weakening of British influence in the region enabled the expansion of communist political influence. The United States attempted to fill the vacuum with the 'Eisenhower Doctrine', which offered to 'protect the territorial integrity and political independence' of nations 'requesting aid against overt armed aggression from any nation controlled by international communism'.[1] King Hussein of Jordan dismissed his government in April 1957 when it threatened to seek Russian aid and, supported by loyal Bedouin troops, he appealed to the USA for help. In order to avoid 'that chaos Moscow hopes to exploit'[2] the Sixth Fleet cruised off Beirut in a regional display of strength and the US government offered King Hussein $10 million in aid which he accepted at once.

In February 1958 the situation became more complex when Presidents Nasser of Egypt and Kuwatly of Syria announced the union of their countries as the United Arab Republic (UAR). Days later Jordan and Iraq, both denounced as traitors to the Arab cause by Nasser, formed the Arab Federation with King Feisal as its head although each ruler retained sovereignty within his own domain. Between the two Arab combinations, the Lebanon with its partially Christian population remained unattached. There was a substantial opposition party, supported by the UAR, that criticised President Chamoun of the Lebanon for taking US aid which was compared unfavourably with the Soviet aid which was, by then, pouring into the UAR. On 13 May 1958, the Lebanese Foreign Minister informed the US, British and

French ambassadors that his government would ask for military assistance if his country were attacked.

Consequently both the British Mediterranean Fleet and the US Sixth Fleet were brought to a high state of readiness and deployed units to the eastern Mediterranean. In order to preserve secrecy, the movement of the Amphibious Warfare Squadron from Malta to Cyprus was declared to be part of the NATO Exercise 'Medflex Fort'. The same cover was used to explain the move of the aircraft carrier HMS *Ark Royal* with its battle group to the eastern Mediterranean from Malta 'with all convenient despatch as soon as possible under cover plan with utmost secrecy'[3]. The Royal Navy maintained a carrier force, varying in strength and composition from time to time, in the Eastern Mediterranean for the next five and a half months save for a short period at the beginning of July. HMS *Eagle*, flying the flag of Vice Admiral A.N.C. Bingley, Flag Officer Aircraft carriers (FOAC), replaced *Ark Royal* in the Eastern Mediterranean and FOAC assumed command of British maritime forces.

British and US contingency plans were drawn up in expectation of a formal request to defend the Lebanon against aggression. Given the code name Operation BLUE BAT the plan anticipated US Marines landing over the beach to seize Beirut airport after which British troops would be flown into it from Cyprus. Air support for both would be provided by British and American aircraft carriers and their air groups. In early June 1958, the United Nations Assembly discussed the Middle East situation and arrived at a consensus that the use of force would be unnecessary. Unfortunately, FOAC shared this opinion and after three weeks at sea at 12 hours' notice for action, he sent *Eagle* back to Malta at the end of June for a 14-day self-maintenance period (SMP).

The British fleet carrier was, thus, off station when the crisis broke on 14 July 1958. With ruthless speed and efficiency, the Iraqi Brigadier Kassem and his 19 Brigade murdered the young King Feisal and his family, the former Regent Abdul Illah and Nuri-es-Said the pro-British prime minister. A day later President Chamoun made an urgent appeal to the USA for military aid and President Eisenhower agreed to land US forces 'to protect American lives and, by their presence, to encourage the Lebanese government in defence of Lebanese sovereignty and integrity'.[4] On 16 July King Hussein asked the British Chargé d'Affaires in Amman for military aid and the British government agreed to provide it. Operation BLUE BAT was set in motion by the Sixth Fleet as a purely American venture to defend the Lebanon. London and Washington agreed that any international criticism of the intervention would best be shared and so the British airlift of troops was transferred to Amman in Jordan, a task given the name Operation FORTITUDE. It was to involve 16 Parachute Brigade Group to be flown from Cyprus in Royal Air Force (RAF) transport aircraft. Unlike the Lebanon, however, Jordan had no Mediterranean coast and political permission would have to be obtained for the aircraft to overfly neutral territory.

Eagle sailed from Malta with all despatch but suffered from boiler defects, which limited her speed and delayed her arrival in the operational area until after the first attempt to airlift troops. This took place on 17 July and followed a route from Nicosia airfield to Cape Pyla, thence through the American Amphibious Objective Area, to Lod in Israel, over which they turned east, for Amman. As the leading aircraft approached Amman, however, they were ordered to turn back by the British government because no permission had been obtained to overfly Israel. The first five aircraft went ahead and landed Brigadier Pearson and his small advance party, but the remainder turned back and returned to Cyprus. An argument between the British and Israeli governments eventually led to permission being given for the airlift to resume on 18 July, the day on which *Eagle* returned to the operational area south east of Cyprus. Her air group comprised 803 and 806 Naval Air Squadrons (NAS) equipped with Sea Hawk day fighter-bombers, 894 NAS with Sea Venom night and all-weather fighters, 813 NAS with Wyvern strike aircraft, 849A NAS with Skyraider airborne early warning (AEW) aircraft, 814 NAS with Gannet anti-submarine aircraft and two search and rescue (SAR) helicopters.[5] Just before sunset on 18 July, *Eagle* recovered 802 NAS Sea Hawks that had flown out from the UK to reinforce her air group via Dijon, Pratica di Mare, El Adem and Akrotiri in 36 hours. The British carrier group took up an operating position some 50 miles to seaward of Haifa, well to the south of the US operating area.

The first airlift had fighter cover from the RAF near Cyprus and from USN combat air patrols (CAP) in the amphibious area but flew on alone near Israel. A conference was held at the Joint Headquarters at Episcopi in Cyprus to which FOAC's Chief of Staff was flown from *Eagle*. It was agreed that from 18 July, RAF fighters would provide day escort near Cyprus with the limited number of aircraft available. Fighters from *Eagle* would defend the transport aircraft from the point where they entered the American air defence sector in the amphibious area to the point where they went 'feet dry' over Israel.

The airlift to move 16 Brigade to Jordan lasted five days and throughout this period Eagle's air group protected it. By day, two sections of Sea Hawks patrolled near the transport route and a third maintained a CAP over the radar picket HMS *Salisbury* which was stationed 50 miles south of *Eagle* between the Task Force and the Egyptian coast. By night, one section of Sea Venoms was maintained on patrol to give cover to both the airlift and the fleet if necessary. Skyraiders maintained a radar barrier over *Salisbury* with instructions to detect fast warships, referred to as 'E-boats', and low flying aircraft heading toward the force. Gannets flew on an armed anti-submarine patrol 40 miles square around the main force against the threat posed by Soviet submarines based in Albania. During the critical five days, *Eagle*'s aircraft flew 388 operational fighter sorties and Gannets and Skyraiders a further 112. Unfortunately, the RAF transport aircraft were not fitted with IFF Mark 10, making them difficult to identify positively in what was often a crowded air plot. Two hundred and eight aircraft were intercepted and identified, 121 of which proved to be allied transport aircraft. Of the remainder, seven were UAR military aircraft, 14 were UAR civil aircraft and seven were Israeli military aircraft. Fifty-nine others were miscellaneous civil aircraft on their lawful business[6]. For part of this period, there was an American airlift flying US citizens out of Beirut, following a route parallel to the British one but inshore of it. This, too, was protected by *Eagle*'s aircraft until 22 July when the movement of allied transport aircraft by day ended and the task force withdrew to Cyprus to rest, replenish and prepare for further action.

The US operation in the Lebanon was on a larger scale but also relied primarily on maritime presence for its successful conclusion. An Army Rapid Reaction Force (RRF) comprising two battle groups based in Germany had been nominated for the defence of Lebanon. But there was only sufficient air transport in Europe to move one at a time, and then by a long circuitous route avoiding communist territory to Adana in Turkey, the nearest US base to Beirut. The tactical fighters and strike aircraft allocated to support BLUE BAT were based in the USA and faced a trans-Atlantic, trans-European flight to deploy to Adana. They also needed large numbers of transport aircraft to carry their maintenance personnel, support equipment and ammunition. The US Sixth Fleet comprised two strike aircraft carriers and one anti-submarine carrier and three amphibious squadrons, each with a battalion landing team of about 1800 Marines. One of the later was south of Cyprus, only 12 hours steaming from Beirut and was ordered to land its men at Beirut at 1500 on 15 July, when President Eisenhower would announce US intervention publicly. The Marines were supported by 11 armed Fury fighters and Skyraider strike aircraft from USS *Essex* which had left their carrier off Greece and refuelled at a British airfield on Cyprus so as to be over the landing beach on time. The Marines secured Beirut airfield an hour after landing, and the remaining battalions landed as they arrived on 16 and 18 July. A fourth USMC battalion flew from the USA in Marine Corps aircraft and landed at Beirut on 18 July. The US Army and Air Force units began to arrive from 19 July but were not effective for some weeks until their vehicles, ammunition and support equipment arrived by sea. *Essex* continued to fly support missions until 20 August, supported from 18 July by USS *Saratoga*. Other carriers were sailed from the USA to back up the two ships committed to the operation in the eastern Mediterranean.[7]

The allied moves stabilised the situation in the Lebanon and Jordan without fighting, although the forces ashore would have been hard pressed if attacked. Carrier task forces were, therefore, maintained in the eastern Mediterranean until the crisis passed and troops were withdrawn in October. Sixteen Brigade in Jordan was lightly equipped and, had the Jordanian Army's armoured corps rebelled against its government, would have been totally reliant on carrier-borne aircraft for close air support. The Commander in Chief (CinC) Mediterranean, Admiral Sir Charles Lambe, signalled the Admiralty concerning the vital presence of the carrier: 'nobody else can give the Army in Jordan much help'.[8] The presence of aircraft

HMS *Albion* leaving for Lebanon, 1957. (DAVID HOBBS)

carriers, capable of providing a dominating presence over the eastern Mediterranean and the landward theatre of operations can, therefore, be said to have underpinned the whole allied operation.

The protection of the airlift and the allied troops once they were deployed was not the only example of how the Royal Navy influenced events on land in 1958. The threat of revolution was present throughout the Middle East and warships were involved in moving Royal Marines and troops to protect British and Commonwealth interests in Libya, the Persian Gulf and the Arabian Peninsula. The activities of the strike carriers have been highlighted but other aircraft carriers showed the versatility of type by ferrying troops and their vehicles over large distances. HMS *Albion* was working up in the UK in early July when it was decided to use her to transport 42 Royal Marines' Commando to the Mediterranean. She disembarked her air group and entered Portsmouth Dockyard to have 1000 extra ringbolts welded to her flight deck to allow the Commando's vehicles to be lashed on deck. *Albion*'s air group flew out to rejoin her in Malta and she was able to relieve *Eagle* for a period in the operational area before resuming her planned deployment to the Far East Fleet. Another light fleet carrier, HMS *Bulwark*, was in Mombasa in mid-July and she, too, made a significant contribution. She flew strike missions against rebels in Oman, ferried an Army battalion from Kenya to Aden, from where it was landed in Jordan at Aqaba and, on her way back to the UK, she too relieved *Eagle* for a spell. Even so, during the first 95 days after *Eagle* left

Devonport, operations in the eastern Mediterranean meant that she was only in harbour and able to give leave on 17 occasions.

Throughout the crisis, three British and three US aircraft carriers played a key role in enabling the military measures that stopped the spread of revolution and brought a period of stability to the Middle East. They achieved their aim, but more to the point, neither the UAR nor any other potentially hostile power succeeded in deterring them from achieving it. Two quotations sum up the Royal Navy's attitude after its performance in Operation FORTITUDE. The FOAC, Vice Admiral Bingley, emphasised in his Report of Proceedings (ROP) that 'it is of the greatest importance that our slender allowance of aircraft carriers should be kept up to a state of maintenance and training which will ensure that they are available for the many demands which they alone can meet'. Admiral Mountbatten, the First Sea Lord, went further when he stated that 'the Navy is completely mobile, flexible, versatile. And so, you see, it has got an enduring and vital function.'[9]

Kuwait 1961

On 19 June 1961, the British government formally restored to Kuwait the right to manage its own foreign affairs. It gave an assurance, however, that assistance against aggression would be given if Kuwait were to request it in an emergency. A week later, on 26 June, the Iraqi dictator Brigadier Kassem laid claim to Kuwait and by the evening of 28 June the UK Chiefs of Staff ordered that forces allocated to Operation VANTAGE, the contingency plan for the defence of Kuwait, should be brought to early states of readiness. They did not, however, give the executive order to implement it, an omission that was to cause confusion in the early stages of the operation. Fortuitously, *Bulwark*, which had recently been converted from a fixed-wing strike carrier into a 'commando' carrier (LPH) was at Karachi with 42 Royal Marines' Commando and 848 NAS equipped with Whirlwind helicopters embarked. The Admiralty ordered her to proceed 'with despatch' directly to Kuwait.[10] Other naval deployments followed quickly. The strike carrier HMS *Victorious* and her destroyer screen, which were on passage from Singapore to Hong Kong, were ordered to make a fast passage to the Gulf.

The Emir of Kuwait formally requested British protection on 30 June 1961 and the CinC Middle East, Lieutenant General Sir Charles Harrington, assessed that his minimum requirement was for two infantry battalions, two armoured squadrons and two RAF Fighter/Ground Attack (FGA) squadrons in place by dawn on 2 July. Operation VANTAGE orders anticipated a four-day build up in Kuwait of British forces which were airlifted from Cyprus and the UK. The Chiefs of Staff's earlier failure to order the implementation of the contingency plan had already disrupted its execution, however. As in 1958, the rapid movement of troops by air was entirely subject to obtaining clearance to overfly 'neutral' countries along any of the three possible routes. The shortest, at 3500 nautical miles, needed permission to cross Turkey and Iran. An intermediate route, at 5000 nautical miles, needed

Whirlwinds from 848 NAS operating at Aden. (David Hobbs)

permission to cross Libya and the Sudan. The longest route, at 7000 nautical miles, needed permission to cross Nigeria and the Congo. None of these permissions had been requested and all needed a minimum of three days for diplomatic clearance under normal circumstances. This the Turks refused to give when asked on 30 June and the first aircraft had to fly from the UK to Aden via Libya and the Sudan. British troops in Cyprus earmarked for VANTAGE were, therefore, effectively isolated. The situation was partially resolved by the reinforcement of the Middle East air transport force by aircraft deployed from the RAF's Far East Air Force and the Royal Rhodesian Air Force.

42 Commando was deployed ashore to Kuwait airport during the forenoon of 1 July, 24 hours after the formal request for aid. Later that day, the first of two squadrons of Hunter FGA 9 aircraft followed them from Bahrain together with transport aircraft, which flew in their maintenance personnel, support equipment and stocks of ammunition. On completion of this task, the transport aircraft joined the airlift of military units. The CinC Middle East juggled the land forces under his command to provide a second infantry unit alongside 42 Commando. To achieve this, two companies of the Coldstream Guards were ordered to move from Bahrain and the 45 Royal Marines Commando and the personnel of an armoured car regiment were ordered from Aden to Kuwait. By the evening of 1 July, only the heavy armoured strength met the goals of the VANTAGE plan. This was due to the stockpile of tanks held in the Bahrain-based Amphibious Warfare Squadron's Landing Ships Tank (LST), the stockpile of tanks in Kuwait itself and the fortuitous presence in theatre of two half-squadrons of Dragoon Guards who were carrying out a routine change-over of standby personnel.

The Guards arrived on the evening of 1 July and took over the defence of the airfield and other rear areas from 42 Commando. Reinforced by the Royal Marine Detachment from the frigate HMS *Loch Alvie*, the Marines moved up to the border to join the Kuwaiti Army in its defensive positions on the Mutla Ridge. The frigate itself stood by to provide naval gunfire support for the troops ashore, the only artillery available at the opening phase of the crisis. There were no military radars in Kuwait and so all air warning and fighter direction was carried out by *Bulwark*, which managed to maintain a non-stop air plot in its operations room, albeit with gaps in cover caused by the proximity to the land. The RAF Hunter units in Kuwait lacked sufficient numbers to allow a standing CAP and so were held at alert on the ground, being launched when necessary to investigate suspicious contacts. Interceptions had to be carried out by visual sighting of the target since the Hunters had no air-intercept radar. The air defence task was made more difficult by the lack of modern IFF in both the Hunters and the RAF transport aircraft flying into Kuwait. A pre-positioned air-transportable Type 787 radar station was flown into Kuwait from Bahrain early in the operation but did not become operational until 16 July, when it was finally set to work by RAF technicians flown out from the UK.

Bulwark remained at anchor in Kuwait harbour until dusk on 2 July when she sailed for Mina al Ahmadi where arrangements to refuel her had been made. Thereafter, until 12 July, the Commando Ship followed a daily routine of remaining at sea during the night to reduce the risk of any attack being mounted by Iraqi fast patrol boats (FPB) known to be based in the Shatt-al Arab, then returning to anchor off Kuwait during the day. Throughout the crisis, *Bulwark* provided direct logistic support to the Royal Marines as well as maintaining command and control of the air defence organisation. In addition to functioning as a large heliport, she also functioned as a 'fuel dump', 'magazine' and 'supply depot' for items as diverse as toilet paper, barbed wire and replacement combat clothing. Her helicopters were able to move water, fresh bread and other creature comforts up to the Mutla Ridge on a daily basis in stark contrast to the Army formations. These, having been flown in at short notice with only the equipment the men could carry, lacked both logistic support and, to a large extent, the vehicles with which to move it. Nothing highlights the difference between air and sea transport as starkly as a close study of the first week of Operation VANTAGE.

45 Royal Marines Commando was flown into Kuwait from Aden on 2 July, together with the Hussars to operate the armoured cars delivered from Bahrain by Army LSTs. The Turkish government had a change of heart and agreed to permit night overflights of its territory from 1 July. It was not until the evening of 2 July, however, that the first aircraft left Cyprus with elements of the 2nd Battalion the Parachute Regiment. Even then, air movement was complicated by the fact that movements personnel and technical reinforcements who would normally have been early arrivals in theatre to 'activate the route' were at first left behind in Cyprus to get fighting troops into Kuwait. Their absence was another casualty of the failure to

give an executive order to execute Operation VANTAGE at the appropriate time, and was to cause problems at the airhead. The movement of 2 Para took two nights with the result that the battalion was not at full strength until 5 July by which time another infantry battalion had arrived from Kenya. The latter relieved 42 Commando on Mutla Ridge on 7 July and a second battalion from Kenya relieved 45 Commando two days later.

Another arrival on 2 July was the headquarters ship of the Amphibious Warfare Squadron, HMS *Meon*. She berthed alongside the new Kuwait Jetty and established herself as a Joint Services communication centre and relay ship. Virtually no Army or RAF communications support existed ashore, but with her complement of Royal Navy communicators and 601 Signal Troop Royal Signals embarked *Meon* established links between Task Force 317, the forward airfield and 24 Brigade Headquarters. In effect, she became the focal point for all VANTAGE communications and Flag Officer Middle East (FOME) repeatedly stressed her importance in his ROP.[11]

Other elements of the Amphibious Warfare Squadron made significant contributions from 2 July including the delivery of freight, heavy stores, ammunition and the armoured cars from the Bahrain stockpile by HM Ships *Striker*, *Bastion* and *Redoubt*. The CinC Middle East and the General Officer Commanding Middle East Land Forces also set up their Joint Headquarters on 2 July having moved, with their staffs, from Aden to Bahrain in order to be nearer the scene of action.

Victorious and her task force arrived on 9 July, immediately taking over the tasks of air defence and close air support. On 10 July the RAF Hunter squadrons withdrew to Bahrain where they were able to return to their normal support arrangements. *Victorious*'s air group comprised 803 NAS with the relatively new Scimitar strike fighters, 892 NAS with Sea Vixen missile-equipped all-weather fighters, 849B NAS with Gannet AEW aircraft and 825 NAS with Whirlwind anti-submarine helicopters[12]. By this time, the Intelligence appreciation of the situation had come firmly to the conclusion[13] that Iraqi military movement was defensive in character although concerns remained that 15 July, Iraq's National Day, might see offensive action. *Victorious* was flying the flag of Rear Admiral J.B. Frewen, Flag Officer Second-in-Command Far East Station (FO2FES) and, from 10 July he assumed duty as Commander Task Force 317 from Captain R.D. Franks, RN, of *Bulwark*. *Victorious* was able to settle into a cycle of four days flying followed by two days replenishment and maintenance. In the event of an Iraqi attack during the latter, she could have returned to operational status at short notice. After studying reports from this phase of the operation, Rear Admiral R.M. Smeeton, FOAC, commented that 'all reports show that the Royal Navy improvised a workable air defence organisation which made the best use of the available ships, aircraft and equipment'.[14]

The high ambient temperatures and low surface wind in the Persian Gulf, together with *Victorious*'s short catapults, combined to limit the Scimitar's capacity for ground attack. The aircraft were unable to launch with fuel in their drop tanks and sorties were limited to 40 minutes, mostly spent at low level. In view of the short distances involved this was acceptable. The primary offensive weapons were 2-inch rockets and 30-mm Aden guns. The limited ability to carry bombs was reduced in effect by a UK government ban on their use for counteroffensive action. On 31 July HMS *Centaur* arrived from the Mediterranean to relieve *Victorious* and Admiral Frewen transferred his flag there. She was supported by destroyers that moved into the Gulf from the both the Mediterranean and Far East Stations. This force continued to provide strike, air defence and naval gunfire support for the Army in Kuwait until it was withdrawn on 15 August when the threat of invasion had passed.

As the forces withdrew from Kuwait, both the FOME and FO2FES drew attention to the need to improve communications. As the flagship, *Victorious* had handled 80 percent of the Task Force signal traffic and had barely been able to cope.[15] Admiral Frewen noted that the VANTAGE communications plan was designed for an intensive operation over a short timespan and was over-complicated for the protracted, uneventful operation that, in fact, eventuated.[16] The plan was simplified on 28 July, and thereafter *Centaur*'s statistics showed an average of 151 messages a day passing through her bridge wireless office compared with the 352 a day handled by *Victorious* in early July. Difficulties were experienced in dealing with freighting RFAs (Royal Fleet Auxiliaries) used for replenishment at sea and with War Department LSTs owing to their lack of cryptographic material. FOME wrote a comprehensive report on the communications aspects of the operation and concluded that the lack of an established joint organisation and procedure was seriously felt ashore.[17] FO2FES drew similar conclusions and, after commenting on the inter-service muddle in the early stages with regard to call-signs and address groups, he recommended that a Joint Chief Signals Officer should be appointed in future, responsible to the overall commander of the operation.

Overall, Operation VANTAGE was successful in that it deterred an Iraqi invasion of Kuwait and responded to the Ruler's call for help. The emergency had long been foreseen, plans for its implementation had been kept constantly under review and brought up to date and yet it did not unfold as envisaged. There are few better case studies of littoral, expeditionary warfare.

East African Mutiny 1964

Zanzibar and Kenya were granted independence from British colonial rule in November 1963. Tanganyika had been independent since 1961. All three had democratically elected governments committed to peaceful progress. Their small national armies had evolved from the former King's East African Rifles. On 12 January 1964, however, the African population of Zanzibar revolted against the Arabs and Asians who held power on the prosperous island and murdered a large number of them. The British High Commissioner became trapped in his residence and unable to communicate with London.

On hearing the news, the British CinC Middle East, Lieutenant General Sir Charles Harington, ordered preparations to protect British citizens. The survey ship HMS *Owen* and the frigate HMS *Rhyl* were ordered to stand by Zanzibar and Rear Admiral J. Scotland, FOME, ordered the aircraft carrier *Centaur*, off Aden, to embark 600 men of the 45 Royal Marines Commando. No alongside berth was available and the carrier had to embark the men, 24 Land Rovers and 70 tons of stores and ammunition while anchored a considerable distance offshore. She also embarked five ferret armoured cars and personnel of the 16/5th Lancers, hundreds of cans of vehicle fuel (MOGAS) and two Belvedere helicopters of 26 Squadron RAF plus their aircrew, maintainers and spares. She retained her full air group on board. This comprised 892 NAS with Sea Vixen strike fighters, 849B NAS with Gannet AEW aircraft and 815 NAS with Wessex helicopters.[18] From the time of the order to embark to that of sailing took just 13 hours, and *Centaur* left Aden at midnight on 20 January with the destroyer HMS *Cambrian* in company. Employed on military duties in the Radfan, 45 Commando had no recent amphibious warfare experience.

The commando stores were piled on the starboard side of *Centaur*'s flight deck, forward of the island in an area known as Fly 1. The vehicles were parked alongside the extreme starboard edge of Fly 1 leaving the port catapult clear to launch aircraft should it prove necessary. The Belvederes were too big to strike down into the hangar and were parked at the after end of the flight deck. Space was kept on deck for the marines to keep fit. A plan, known as 'Sardine Stations' was evolved for the operation of fixed-wing aircraft. Marines, vehicles and as much of the stores as possible were struck down into the hangar and the six Wessex and two Belvedere helicopters were launched to hover alongside while the fixed wing launch and recovery were in progress. Thankfully, the plan did not need to be put into practice.[19]

On 20 January 1964, men of the Tanganyikan Rifles mutinied in Dar es Salaam and pillaged the city, terrorising the population. Two days later, African troops mutinied in Uganda and British troops were flown in to restore order at the request of the Ugandan government. In Kenya, President Kenyatta asked the British government for support from the British Army units that remained in the country and these quelled a mutiny by the 11th Battalion of the Kenya Army on 24 January. At sea in *Centaur*, Captain O.H.M.StJ. Steiner, RN, her commanding officer and Lieutenant Colonel T.M.P Stevens, RM of 45 Commando set up three planning teams that were to cover Intelligence, Assault (including Logistics) and Communications. With no knowledge of how their forces would be used, the teams concentrated on essentials and prepared simple, flexible procedures that could land marines in the Wessex helicopters, from which the sonar sets had been removed to make space. On arrival off Dar es Salaam, *Centaur* received instructions from the CinC Middle East to 'Disarm Tanganyikan Army Colito Barracks soonest'. President Nyerere asked the British government for military assistance at 2300 hours on 24 January. At midnight, Brigadier Sholto-Douglas, the former commander of the Tanganyikan Army, was taken off shore by boat and reported a second night of

HMS *Centaur* with Belvederes on deck. (DAVID HOBBS)

rioting, looting and killing in Dar es Salaam. By then it was 0100 hours on 25 January and swift action was clearly necessary to save the lives of innocent civilians. Captain Steiner decided to take Colito Barracks at dawn, only five hours away and, to avoid courting delay, he acted on the authority he had been given without seeking advice from London. The essence of his plan was speed, surprise and an overwhelming display of force, in effect the 'shock and awe' tactic used again by the coalition allies in Iraq in 2003. Marines would be landed in a sports field right alongside the barracks; *Cambrian*'s guns would be used together with strike aircraft if necessary. It was essential to minimise casualties among the mutineers and their families and so *Cambrian* was to lay down a diversionary bombardment over an uninhabited area north of the barracks, no easy task in a tideway with no forward observer. The plan was deliberately kept simple and a briefing was held at 0330 hours. The helicopter flying program was the only element that was written down.

At 0610 hours on 25 January, only hours after British aid was requested, Wessex helicopters of 815 NAS landed Z Company of 45 Commando at Colito Barracks.

There was an initial show of resistance but after an anti-tank missile was fired into the guardroom, the mutineers surrendered. Many had run into the bush and were shepherded back by a hovering helicopter with the naval pilot brandishing his pistol out of the window! At 1100 hours *Centaur* moved into Dar es Salaam and requisitioned tugs and lighters to get the vehicles and stores ashore and to land her own Royal Marines detachment. Armed parties were sent to Taboora, 340 miles inland, where the mutinous 2nd Battalion was disarmed, while Sea Vixen fighters of 892 NAS flew overhead in case they were needed. Other mutineers laid down their arms at Natchingwea after British troops were flown in from Kenya by the RAF, protected by Sea Vixens from *Centaur*.

By a happy coincidence, 26 January is Australia Day and the Australian High Commissioner in Dar es Salaam held an official reception to mark the occasion and the return of peace to the streets. *Centaur* showed that there are many ways in which a warship can actively influence events on land by parading her Royal Marines' Band for the occasion. Later an air display was organised over the city. Within hours of President Nyerere's call for help, the mutiny was over, order had been restored and the authority of his government re-established. Only two persons, both in the guardroom when the missile hit it, were killed. *Centaur* and *Cambrian* sailed shortly afterwards, to continue their general service commissions in the Far East Fleet.

Lessons

In his ROP on the Jordan-Lebanon Crisis of 1958, the CinC East Indies highlighted the fact that the numerous standing Joint Operational Instructions (JOI) had been written in isolation. Thus, when more than one had been ordered into effect, the air transport lift written into them had to be split resulting in delays and confusion. He recommended future consideration of the worst case scenario, especially in the Middle East, where several JOIs might have to be implemented at once. On the purely naval aspects of the operations, however, he noted that 'it is most gratifying to see how few of the many mistakes which occurred at the time of Suez and again at the onset of the Muscat/Oman rebellion have been repeated. I have no doubt that before long we will, by the continual practice that we have, have a crisis where the Navy makes no mistakes.'[20]

The Kuwait Crisis, three years later, proved to be an excellent test of Britain's ability to deploy its limited forces to deal with a 'brush fire' outbreak. The emergency had been foreseen many years before and operational plans for dealing with it had been maintained. No other contemporary emergency complicated the issue and yet the system was found wanting in some of its basic assumptions. Superficially, air transport had seemed a prompt way of delivering military power to a trouble spot. In practice, this proved not to be so for a variety of reasons. The flights themselves were too easily thrown into disarray, in this instance by the failure to order a timely D-Day and the lack of early overflight rights. More importantly, when airlift troops did arrive, they were not combat fit and had no logistic support. Even when the airlift got underway, its achievements cannot have been very satisfactory since the Commanding Officer of 42 Commando described his unit as 'the only fully equipped unit ashore for at least seven days'[21]. The vital contribution of *Bulwark*, her commando group and the need for carriers to provide overland air defence, were the key elements of the operation. Without them, Operation VANTAGE could have been a disaster if serious opposition had been encountered.

The quelling of the East African mutinies was an excellent example of the Navy's ability to achieve an effect by using capabilities latent in sea control. The determined and professional approach shown by *Centaur* is an example of the broad spectrum of aircraft carrier capability in littoral warfare. The surprise achieved in the assault on Colito Barracks illustrates the need for aggressors to consider the sea dimension and the potential consequences they might suffer if they seek to provoke it.

The three case histories have similarities. They all used both air and sea mobility, and allow direct comparisons to be made between the two forms of transport. The first was a UK/US coalition operation; the second was a planned intervention by UK joint forces; and the third a rapid reaction to a crisis by an improvised UK joint force brought together in only hours for the purpose and based on an aircraft carrier. Despite difficulties of organisation, all were successful. In all three cases, the projection of power over the land and in the air from ships at sea proved to be the enabler that ensured success, while airlift proved disappointingly inadequate and did not deliver the military force ready to fight with the logistic support it needed. Most significantly, none of the potential adversaries proved able to deter navies from projecting and sustaining power into the regions they had sought to attack.

Twenty-First Century Reflection

These operations continue to have relevance today. Critics who do not understand maritime power have assumed that with the end of the Cold War and the absence, at least temporarily, of peer competition, navies have become irrelevant. Nothing could be further from the truth. Today's likely mission scenarios are shaped to react to three very real international influences and the fleets that respond to them would be familiar to those who planned and took part in the three case histories.

The three influences include the need to combat unconditional terrorism, the responsibility to protect populations subject to genocide or ethnic cleansing and protection of the oceans for peaceful use. All have extensive maritime implications and the first two translate into a mission of 'strategic raiding'; the temporary interposition of force to achieve a permanent effect before it withdraws.[22] Afghanistan and Iraq are recent examples of the first of these and Western fleets have thrown a 'big, blue blanket' over the land from task forces at sea. The third is straightforward and requires sustained presence and long reach, best provided by aircraft embarked in ships to guard the oceans and their resources and to provide emergency relief. The second needs more explanation.

In 2004 a United Nations Study Group, charged with looking far into the future of the organisation, endorsed the emerging tenet that national sovereignty may become conditional once a certain high threshold has been crossed. When it has been, there is a collective responsibility to protect populations exercisable by the Security Council. A 'just cause threshold' is deemed to be genocide, ethnic cleansing or large-scale loss of life which sovereign governments have proved powerless or unwilling to prevent.[23] The study also concluded that under certain, clearly defined circumstances the protection could be anticipated before, rather than be consequent upon, the actual loss of life. The R2P mission, as it has become known, has a recent and very successful example in Operation PALLISER, the British rescue of Sierra Leone and the failed UN peacekeeping force deployed in it during 2000. Study reveals a host of similarities between that operation and this chapter's three case histories.

The fundamental underpinning of the expeditionary capability in the histories and PALLISER was the task force containing large aircraft carrying warships capable of operating naval aircraft. These could gain control of an area on land as well as at sea and exploit it with both fixed and rotary-wing aircraft types. Both the carriers and other ships carried military forces with their vehicles, ammunition and logistic support. They arrived ready to fight if necessary or poised 'over the horizon'. They achieved surprise by using dimensions the likely antagonists proved unable to oppose. Naval air supported operations many miles inland and not for the first time. There was no need for a prolonged build-up to create a base from which to operate, even if it had been possible to do so. Air transport from a distant base could do none of these things, but provided valuable support once the first lodgement ashore had been achieved.

Forty years ago, the operations described were thought of as aberrations against the background of a global Cold War. They are now seen as demonstrations of a naval capability, which the Royal Navy and many others have included in their arsenal for over a century and are now being re-structured to perform better. There is a clear and long-term future for new ships and aircraft that will continue to project naval power over the land and in the air as well as on the sea. Britain's future carrier and its cousins in the USA, Australia, France, Spain, Italy, Brazil and India may be products of the 21st century, but they will be ships entirely familiar to the men who achieved success in these littoral case histories.

15. The RAN and Air Mobile Operations: The RAN Helicopter Flight Vietnam

David Farthing

AFTER ITS CRUSHING ELECTION victory in December 1966, the Holt Government began openly to canvass further involvement in the Vietnam War. Among other problems, the United States was acutely short of helicopter pilots and one of its pressing requests was for support in this area. The RAN was the only Australian service which conceded that it could spare some pilots – although not without consequences in other areas – and it was initially proposed that these pilots would be deployed with RAAF 9 Squadron which commenced operations in Vietnam in June 1966. This was not, however, what the US wanted. It wanted the Australian Navy pilots to operate with the US Army and, after much discussion, it was agreed that the naval pilots would fly with the 135th Assault Helicopter Company of the US Army (to be known as the EMU's: not as a flightless bird, but, as an Experimental Military Unit) under US operational control. The original preference of the Australian Chiefs of Staff, that the RAN pilots would only operate within the general area of the Australian Task Force in Phouc Tuy province, also went by the board.[1]

The RAN then took a quite remarkably sensible step in the training process and sent the Officer in Command (OIC) designate of the RAN Helicopter Flight Vietnam (RANHFV) for an in-country reconnaissance. This far-sighted action enabled the implementation of an effective training program to equip the Flight for the unique experience of operating within the massive and often baffling structure of the US Army. The success of this ongoing program was clearly demonstrated to me, as OIC third contingent RANHFV when, after we had been in country for a few weeks, one of my junior sailors said to me 'it's just like they said it would be'.

Thus, in October 1967, the first contingent of the RAN Helicopter Flight arrived at Vung Tau. The make up of the Flight was eight pilots, four observers, four aircrewmen, 24 technical sailors and 6 other personnel. With minor adjustments, this structure was preserved for the four contingents which served in Vietnam until June 1971. The Australian OIC became the Executive Officer of the 135th Assault Helicopter Company, and other members of the Flight assumed positions commensurate with their rank and experience. Inevitably, because of the enormous difference in training and experience of the Australian regular service personnel when compared with those of the typical short service US serviceman, most of whom were conscripts, the majority of the key leadership positions in the Company, both flying and maintenance, were filled by Australians. This had good and bad consequences: the 135th was undoubtedly the elite aviation company operating in the Delta and whenever a crisis arose the cry went out 'call for the EMUs', but the political constraints imposed, increasingly rigidly, by the Australian government resulted in operational limitations for the Company in some circumstances.

The sign proclaiming the presence of the 135th Assault Helicopter Company at its base at Bearcat. (RAN)

This problem was highlighted starkly by the operations into Cambodia in May 1970, which the EMUs were nominated to lead. This operation, an excellent example of the strategic use of air mobile operations, was unfortunately in a geographical area into which Australians were forbidden to go. I was tempted to ignore the constraint, but the order was reinforced the night before the operation by a message from the Australian Embassy. My Flight Commanders and I then had the forlorn experience of watching our Company head off for this most important operation without us. The problem was overcome after Day 1 of the operation, which incidentally was most effective in reducing the enemy war-making capacity in the Delta, by returning the EMUs to tasks wholly within Vietnam and allocating other Companies to continue the task in Cambodia. It reflected nevertheless an appalling piece of political cowardice.

Air Mobile Operations

The major task of the 135th Assault Helicopter Company was the provision of air mobile capability to the Army of the Republic of Vietnam (ARVN) in the Delta of South Vietnam, designated as IV Corps. The aviation company was not organic with the units it was supporting and thus operations carried out by the EMUs could, hypothetically, just as easily have been launched from a naval platform as from our distant airfield. A typical daily commitment of one Command and Control aircraft (Charley Charley), eight troop carrying helicopters (Slicks) and three Gunships could have been provided from a light carrier.

To provide this force, the Company had thirty Slicks and eight Gunships. Typically, on any given day, 50 percent of these aircraft would be unavailable due to battle damage and major maintenance.

Unfortunately, the aims of air mobile operations to use technical superiority to gain a force multiplier effect and to achieve tactical (and occasionally strategic) surprise were sadly corrupted in Vietnam. Too often the helicopter was used as a tool of convenience, as an airborne truck, to make the soldiers' lives more comfortable and to avoid the hard slog of occupying territory.

Our standard operation in the Delta was for the Flight to depart its base at Bearcat an hour before dawn, fly 100 miles to the Delta, pick up an army unit and conduct a combat assault, usually reasonably close to their base. This operation, depending on the opposition and the numbers involved, would take from one to four hours. Then the Flight would revert to stand-by before extracting the same army unit before dusk and returning them to their base. Thus the soldiers returned to a 'warm lee' for the night, but any territory gained during the day was immediately ceded back to the enemy. As a result, the long suffering peasants in the Delta hated the government forces, who pillaged them during the day and the Viet Cong (VC or colloquially 'Charley'), who pillaged them during the night, with equal intensity. This was a major factor in the eventual disaster. When in mid-1970 all the military indicators showed that the war in the Delta was being won, the political ground had long since been lost: the 'hearts and minds' of the people were gone.

It was not all bad news. In 1970 overall command in Vietnam was assumed by General Creighton Abrams who, at his first meeting with his operations staff, announced 'Gentlemen, this fucking rocketing is going to stop'. At this point, the practice of ground forces retiring at night into enclaves surrounded by earthen 'berms', where they were at the mercy of every raggy-arsed guerilla with a two-inch mortar or a primitive rocket, ended, and a policy of aggressive patrolling did, indeed, stop the rocketing.

At about the same time, the commanders of the 7th and 9th ARVN Divisions were replaced by much more aggressive successors. As a consequence, although the routine of dawn insertions and dusk extractions continued, other tactics were adopted which were much more conducive to the concept of air mobile operations.

To overcome the problem of the VC owning the night, we started night 'hunter-killer' operations. This involved a tactical unit of one Slick rigged as a flare-ship, Charley Charley rigged with a powerful searchlight and a twin 50-calibre machine gun and two Gunships. This was intensive flying, all at low level and often terrifying, but it was remarkably effective. The enemy, who had been used to moving in the dark with impunity, suddenly found himself being illuminated and engaged by rockets and guns at close range. We also began flying night combat assault missions and these, too, could be rather frightening, especially since there were often no pathfinders on the ground. Flying in an eight Slick formation into a pitch-black paddy field was a good test of skill and nerve but, again, it was effective and the result was that 'Charley no longer owned the night'.

We also commenced operations with a detachment of US Navy SEALs – special forces who were a bunch of ruffians you would not want to bump into on a dark night, or allow to consort with your teenage daughter. But, they were very good at their job and remarkably effective. Their modus operandi was counterterrorism based on operational intelligence. They would identify a house, for example, in a village where a VC leader was thought to be residing and then use a small EMU element of Charley Charley and four Slicks to launch a surprise raid. After a very low level approach, designed to minimise the warning given by the noise of the helicopters, the Charley Charley would pop up and identify the target with smoke, and the Slicks would then surround the target. We did not always succeed – target identification was the most difficult part – but we had sufficient successes to more than justify the effort, and intelligence feedback indicated that both these and the night operations were having a significant effect on enemy morale. Above all, we were at last taking the initiative away from the enemy.

We also worked with the SEALs when an Intelligence Officer on General Abrams' staff, defying explicit orders, went for a joyride in an Army Mohawk fixed-wing aircraft and had to eject near the mouth of the Mekong after the aircraft was hit by enemy fire. He was promptly captured by the VC and, as you would expect, all hell broke loose and maximum effort was made to retrieve him before he could be interrogated. The chase was absolutely implacable and, after three days, the VC decided that he was simply too hot to hold and, sadly, they killed him. It was a bit like a Hollywood Western with a tragic outcome: a constant series of leap-frogging insertions and an equal number of frantic evasions.

The Aircraft

Our main aircraft was, of course, the ubiquitous Bell Iroquois, UH 1H 'Huey' (or UH 1C in the case of the Gunships). A wonderful aeroplane; robust, easy to fly and easy to maintain. With its big doors on both sides it was ideal for the insertion of an infantry platoon, and eight Hueys could lift a company. On the easy to fly aspect, it should be noted that there is a much greater differential in capability between individual pilots in helicopters than you would expect to find with fixed-wing aircraft. My best pilot could lift about 1000 pounds more than my worst pilot; this in an aircraft with a nominal lift of 4000 pounds. This can be an important consideration when planning special operations.

In Vietnam we flew two-hour sorties interspersed with refuelling with the engine running. Spending fourteen hours airborne in a day was not unusual. For the first assault with full fuel, the Slicks would be too heavy to hover and would have to slide along the runway for a hundred metres or so until they gained translational lift. If there was only a short flight to the first landing, the aircraft would still be too heavy to hover, requiring a zero-zero landing which was a test of pilots' skills; we lost a few aircraft that way.

The Huey also had some disadvantages, the main ones being vulnerability and noise. All helicopters are vulnerable to heavy weapons, although modern helos have much better protection than the Huey. One or two hits from 50-calibre machine gun rounds would cause a Huey to break up in the air: a sure way to ruin the whole day. More importantly in the Vietnam context, it was also vulnerable to small arms fire. The doctrine in Vietnam was that you were immune to small arms fire above 1500 feet, although Lieutenant Commander Pat Vickers, RAN, was killed by an AK47 round above that so-called safety height.

Towards the end of my tour, when I was conducting an assault in a supposedly secure area of eastern Kien Hoa province, a VC soldier emerged from a small haystack, armed only with an AK47 rifle, and shot down five of the flight of eight aircraft. To add insult to injury, he got away. I hope that his boss gave him a medal! In this case he was acting alone and we suffered no casualties, but had the enemy been present in force it could have been a major disaster. As it was, all five aircraft had to be hooked home by Chinook heavy-lift helicopters.

A further vulnerability of the Iroquois was only apparent in the wet season when the rice paddies were flooded. The protective 'stinger' on the rear, which prevented the tail rotor from striking the ground, was not effective in water and we lost a few aircraft through this cause.

An insertion of ARVN troops by the RAN Helicopter Flight Vietnam. During four years of intensive operations, the flight suffered the highest casualty rate of any RAN unit and earned more than half of the honours awarded to naval personnel for Vietnam service. (RAN)

The Huey was very noisy. We have all heard the Bell two bladed rotor noise on the Iroquois and Jet Ranger – a very distinctive sound and audible for miles in the right atmospheric conditions. This noise was inimical to the air mobile aim of surprise and, although we adopted various tactics to minimise it, we never overcame the problem.

Finally, on vulnerability; it is my view that enemy capability in the form of heavy weapons, and particularly simple guided weapons, would have made helicopter operations as practised in Vietnam impossible or, at least, prohibitively expensive. It is for more current aviators to comment on modern helicopters, but I observe that if you can disrupt the rotating parts of even the most sophisticated helicopter, it will quickly beat itself to death.

Command and Control

The 135th Helicopter Company was part of an enormous military machine which, it is ironic to note, was not ultimately successful. We were part of the 1st Aviation Brigade which boasted 3000 aircraft and 30,000 men. I understand that this was the largest aviation brigade in history.

Our chain of command came through Group commanded by a colonel, and Battalion commanded by a lieutenant colonel. Each day's tasking came from Battalion, but as the task was crystal clear we had remarkably little interference in the actual operations. The only problems occurred when we became engaged in a major battle, which might quickly involve 1000 troops on the ground, immediate support from artillery and a flight of AH 1H Huey Cobra Gunships and, a little later, support from the US Navy OV10 Broncos, based at Can Tho, and then, if the action continued, from a flight of USAF Phantoms. On one occasion we even received support from the 16-inch guns of USS *New Jersey*; everyone gave them a very wide berth indeed!

All of this considerable force was fairly routine for Charley Charley and his embarked Ground Commander to manage until the 'goofers' (Navy parlance for sightseers) arrived. We would be going about our business without too many problems when suddenly we would be confronted with a near mid-air collision with another Huey or two. There were the Battalion and/or Group Commanders arriving on the scene for a share of the action. I had been warned by my predecessor of this hazard and was prepared, but it was still pretty alarming to miss another Huey by only a few yards, and the passengers in the back were not too impressed either. After a while I would impose strict height separation and things would settle down, but they always wanted to be close to the action!

As a national commander, I had a separate chain of command to the Commander Australian Forces Vietnam (COMAFV) through the Naval Liaison Officer in Saigon, an RAN commander. I only used this line once during my tour, which would seem to indicate that the command arrangements in this unique situation were remarkably successful.

Personnel

We were inclined to be critical of our American comrades, but had to remind ourselves constantly that we were a tiny, albeit highly trained element in an enormous conscript army. When I arrived in the country in September 1969, there were 550,000 American troops in country and over 1,000,000 in the ARVN. The entire Australian force was about 8000.

Nevertheless, many of the age-old lessons were reinforced. Most important, of course, was that there is no substitute for thorough training and that experience is nice too! US Army Aviation had expanded so quickly that training was often sketchy, and even many of the senior officers had very little aviation experience. Most of the US pilots were young Warrant Officers who would arrive with, typically, 100 hours total experience. Training for two vital emergencies – tail rotor failure and engine governor failure – had ceased in the rush to save time in the training program and the result was that either of these not uncommon emergencies resulted in the loss of the aircraft and sometimes loss of life. On one occasion a young Regular Army Captain was posted to the Company as our Flying Instructor, a vital position in any aviation unit, but doubly so when so many of your aviators are straight out of flying school. Inquiries revealed that our new Instructor had only 125 hours in total. This situation caused the only real argument in my time with the EMUs: I said that he did not have sufficient experience to instruct (observing that none of the Australian pilots had less than 1000 hours), but my American CO did not agree. Sadly, the new Instructor managed to kill his first student the next day and the CO was sacked for something which was really the fault of the system.

The eight Australian pilots (actually nine because two were rotated due to serious wounds) flew an extraordinary amount: over 12,000 hours in the course of the year. The Deputy Australian Commander, Air Commodore C.H. Spurgeon, sent for me and gently told me that what I was doing was dangerous, that we were flying too much. But we did not have one accident or incident involving an Australian pilot other than those directly caused by enemy action. We operated under a nominal limit of 140 hours per pilot per month, which could and did go up to 160 hours in case of necessity. Individual pilots reacted in different ways to this demanding regime, and I found that management required individual assessment rather than a blanket rule. The record speaks for itself.

Our Technical Sailors also performed heroically in arduous and primitive conditions. It was always hot and either muddy or dusty and there was only rudimentary aircraft hangarage, but we always had enough aircraft to meet our daily mission (sometimes only just). Junior Sailors found themselves supervising major maintenance and repair tasks that would have required a Senior NCO in peacetime, and our Senior NCOs provided quite outstanding leadership throughout the Maintenance Platoon. We often had aircraft forced down by enemy action, but very rarely through any maintenance inadequacies.

Conclusion

What did we learn from this rotten war which we probably should not have fought but, having fought at such enormous cost, should have made damn sure that we won? The answer is 'not much'. We affirmed the professionalism of the Fleet Air Arm, as well as the ability of our people to adapt to new and unexpected challenges, and we contributed a small chapter of successful cooperation with our most important ally. But as for incorporating any of the lessons into future planning, the page is blank. I discussed this aspect with Lieutenant Commander Neil Ralph, OIC of the First Contingent RANHFV (later Rear Admiral Ralph, Deputy Chief of Naval Staff) and his response was that every time he tried to incorporate some of the lessons into planning the response from all three Australian services was that 'they did not want to know'.

And so we are left with the propensity for ignoring the lessons of history which is not, of course, uniquely Australian. In Vietnam the Americans wilfully ignored the lessons of the French experience. When I raised this point with a senior American he replied: 'the lessons have no relevance to us; we are better than the French'! When I read Bernard Fall's great book, *Street Without Joy*, I wept at the repetition of the French folly. I have read and re-read David Halberstam's superb analysis of the politics of the Vietnam War, *The Best and the Brightest*, and I have no confidence that our political masters will ever stop making the same mistakes, discarding unpopular advice and ignoring the lessons of history.

About four weeks after arriving in Vietnam, at the end of a long day's flying, my two young Flight Commanders, Peter Clark and Dick Marum, asked if they could have a word. Over a cold VB (Victoria Bitter) they expressed their concern 'that the war could not be won the way it was being fought'. This was from two young officers, brilliant pilots, but with no other military background, and a conclusion I was already fretting about myself. It coincides with Halberstam's observation that the majors and lieutenant colonels in the field knew only too well that the war was not being won, but that knowledge never made it back to the halls of power.

I would like to think that this volume will do something to preserve and examine the lessons of history, but I note that none of our political masters attended the 2005 conference. In the 1980s, Kim Beazley, then Minister for Defence, said to me 'if I can get one or two members of Cabinet to show any interest in Defence matters I feel that I have had a victory'. I fear that the situation has not changed.

16. The Type 42 in the Falklands War: The Influence of the Air Warfare Destroyer on Joint Operations

Mark Schweikert

FROM 2013 THE RAN plans to acquire three Air Warfare Destroyers (AWD) at a cost of some A$8 billion. Based around the US Navy's Aegis combat system, the new AWDs' integrated and highly capable weapon and sensor systems will be able to detect and destroy hostile aircraft and missiles at extended ranges and in all weather. Offering direct protection not only to themselves, but also to friendly forces at sea, on land and in the air, the AWDs will have a far wider impact on joint Australian Defence Force (ADF) operations than is generally realised. Much of this benefit will be delivered through advances in force networking, which will both enhance the combat effectiveness of individual units and provide deployed commanders with the tools to make better and faster decisions. By using the cueing data provided by remote assets, for example, the AWD will be able to engage targets well beyond its own sensor horizon. Yet outside the ADF, blinkered and outdated attitudes prevail, many of these connectivity advantages are not recognised, and not a few defence analysts have dismissed the AWDs as 'inflexible, [and] of dubious operational relevance'.[1]

The ship type known as the destroyer has undergone many changes over the years and, although it would be wrong to consider them all to be cut from the same cloth, issues of flexibility and relevance have rarely been concerns. Originally conceived as a small, fast vessel to defend fleets of dreadnoughts from torpedo boats, by the end of World War II the destroyer had become a capable battle unit in its own right. In the post-war era the term was often attached to vessels more specialised in the anti-air role, but the destroyer always remained a flexible asset, sufficiently versatile to take on virtually all the tasks that a navy might require of it. Certainly, the modern Australian term 'AWD' falls far short of an accurate description of the planned vessel's capabilities across the operational spectrum. It does, however, highlight a warfare role that smaller vessels are generally unsuited to master due to both the size and technical requirements of the task.

Recognition of the operational advantages of deploying a highly mobile, integrated, air warfare system are not new and, during the Cold War, Western navies introduced a succession of technological advances to deal with the multiple threats and increasing pace characterising modern air defence. But rather than actually coping with massed Soviet air attacks, the potential of such a system was first seen in the Falklands War of 1982. Although more than 25 years have passed since the end of this conflict it still provides a valuable case study, simply because in no other recent conflict have naval forces faced an equivalent level of air threat. Moreover, the British recovery of the Falklands after the Argentine invasion was almost entirely a maritime affair. It entailed a long-range, joint, expeditionary mission requiring the Royal Navy to establish sea control in a littoral environment.

In today's uncertain security situation such expeditionary missions are becoming increasingly common, and experience suggests that it will only be a matter of time before the ADF is faced with a determined air threat. The part played in the Falklands by the British 'AWD', the Type 42 destroyer, offers pointers to the problems that the ADF may need to overcome, and to the advantages it may gain, through the future capabilities provided by its own AWD. Despite significant losses to air attack in 1982, the contribution of the Type 42 to the Falklands campaign was far from inconsequential and, as this chapter will show, it had a far more significant impact on the outcome of the war than is generally understood.

Background to the Conflict

The Falkland Islands lie some 400 miles off the Argentine coast and approximately 8000 miles from the UK, deep in the South Atlantic. The islands have changed hands several times since the 18th century, with Britain's colonial administration dating back to 1833. Despite the somewhat longer standing Argentinean claim to sovereignty, there is no doubt that the local population see themselves as steadfastly British.[2] Yet in the lead up to the 1982 conflict, the British Foreign and Commonwealth Office treated the islands more as a forgotten outpost of the old Empire, one needing to be cut loose. Indeed, in 1980 the UK's Cabinet Defence Committee proposed finally to solve the issue through a leaseback arrangement with Argentina. Although such a proposal was acceptable to the Argentine government, it was not greeted warmly by the Falkland Islanders, who also faced a proposed Nationality Bill that threatened to exclude them from British citizenship.[3]

Further adding to this display of indifference was the 1981 UK Defence White Paper, *The Defence Programme: The Way Forward*. Sometimes referred to as the 'Nott Review', after Minister for Defence Sir John Nott, the White Paper articulated a refocusing on Britain's NATO (North Atlantic Treaty Organisation) responsibilities by placing priority on nuclear forces and land-based airpower, at the expense of distant British territories and drastic cuts to the Royal Navy. Apart from reductions in the numbers of destroyers, frigates, support ships and personnel, the new aircraft carrier, HMS *Invincible*, was to be sold to Australia and the older carrier, HMS *Hermes*, to be decommissioned early. The Royal Navy's two large amphibious assault ships, HM Ships *Fearless* and *Intrepid*, were also to be decommissioned, while (of immediate interest to the Falkland Islanders) the Ice Patrol and Falklands guard ship HMS *Endurance* was to be decommissioned after the 1981-82 season.[4] The Governor of the Falkland Islands, Sir Rex Hunt, later recalled the dismay with which this news was received, because the 'islanders perceived immediately how this would be seen from Buenos Aires (a factor that appeared to have been totally ignored in the decision-making process in Whitehall)'.[5]

Hunt's fears were evidently well grounded, for the leader of Argentina's military junta, General Leopoldo Galtieri, reputedly remarked that a British military reaction to an Argentine invasion 'was scarcely possible and totally improbable'.[6] Without doubt, there was little expectation in Buenos Aires that the British would carry through with a difficult and costly military campaign. In reaching this conclusion the Argentine leadership had made three strategic assumptions, which on the surface all seemed quite plausible. Firstly, the evident British lack of interest in the islands meant that they would not fight to regain them. Secondly, the Americans would not take sides, for fear of pushing the Argentines towards the Soviets. Finally, the Soviets would veto any UN Security Council Resolution (UNSCR), given the large quantity of grain they bought from Argentina.[7] As it turned out, the Argentine invasion on 2 April 1982 resulted in UNSCR 502 calling for Argentina's immediate withdrawal, with the United States declaring its unqualified support for British actions, and a British Task Force steaming more than halfway to the theatre before Argentine preparations to defend the Falklands had really begun.[8]

Argentina had originally planned to invade the Falklands under Operation AZUL (BLUE) on 15 September 1982,[9] a date with a number of advantages. Not only would the fierce South Atlantic winter have passed, but the world would have seen Argentine diplomatic efforts to find a solution exhausted, thus gaining Argentina the moral high ground. Furthermore, Argentina's armed forces would have been far better prepared. By September, all 14 Super Etendard aircraft and 14 AM-39 Exocet anti-ship missiles on order from France would have been delivered to the Argentine Naval Air Arm, and pilots should have been both carrier qualified and proficient in the offensive use of this anti-ship capability. The Army's conscripts, meanwhile, many of whom actually deployed to the Falklands with just 45 days experience, would have had another six months of training.[10]

Added to these advantages, an invasion later in 1982 might have left the Royal Navy to respond without either aircraft carriers or assault ships, and with no ship on the spot to provide local knowledge and intelligence. As the foreword to the 1983-84 edition of *Jane's Fighting Ships* declared, 'the entire operation [to recover the Falklands] could have been nullified had the requirements of the cuts proposed in June 1981 taken place'.[11] But an incident involving Argentine scrap iron merchants on the island of South Georgia in late March 1982 led the junta to gamble on a move sooner rather than later. Operation ROSARIO replaced AZUL and the date for the invasion was advanced.

The few British troops stationed in the Falklands, having demonstrated their willingness to fight, quickly surrendered in the face of overwhelming Argentine strength. But this reflected military prudence rather than British resolve. Three days after the invasion the largest task force in recent history left the UK, and the achievement in mounting what the British termed Operation CORPORATE soon became obvious to all:

> *In the space of seven weeks a task force of 28,000 men and over 100 ships had been assembled, sailed 8,000 miles, effectively neutralised the Argentine navy and fought off persistent and courageous attacks from combat aircraft which outnumbered its own by six to one. This in itself was no mean feat, but the task force then put ashore 10,000 men on a hostile coast while under threat of heavy air attack; fought several pitched battles against an entrenched and well supplied enemy who at all times outnumbered our forces; and brought them to surrender within three and a half weeks.*[12]

Type 42 Development

Nearly every warship design is the result of a compromise of varying, and at times, conflicting demands and limited resources, and the Falklands War proved yet again that the perfect warship has yet to be built. Despite the ongoing Cold War threat, it is clear that inter-service rivalry, financial considerations and political perceptions still had more to do with warship design than contemporary naval thinking or experience. The production of the Type 42 destroyer came about as a stopgap measure following the 1966 cancellation of two major UK warship programs: the CVA-01 or *Queen Elizabeth* class aircraft carriers and their purpose designed air defence escort, the Type 82 destroyer. Designed around the new Sea Dart area air defence missile system, and incorporating a sophisticated ADAWS (Action Data Automation Weapons System) computerised combat information system, the Type 82s were to represent the second generation of British guided-missile destroyers (DDGs). But with the class reduced to only one unit, HMS *Bristol* commissioned in 1973, the Royal Navy faced the pressing problem of how to replace the older and aptly named Sea Slug missile fitted in the first generation DDG, the County class. Hence, by default, the Sea Dart system became the Type 42's *raison d'être*.

Briefly, the Sea Dart missile is guided by homing in on radar energy reflected from the target and generated by a powerful fire control radar on the ship, in this case one of two Type 909 illumination radars. This means that should the target manoeuvre below the ship's radar horizon, it may be lost to the system. Sea Dart has a range of approximately 40 nautical miles and a maximum speed of Mach 3.5. Australia played a big part in Sea Dart development. Trials of the new British missile were conducted at Woomera in South Australia between 1965 and 1968 with Australian scientific help under a bilateral testing agreement.[13] Another trial was conducted at Woomera in 1975 where Sea Dart performed well against sea-skimming missiles. Employing vacuum-tube technology, the system nevertheless had a reputation for poor reliability. A Sea Dart 2 program, announced in 1977 to improve missile performance, was cut back before production began, and then killed as part of the Nott Review.

The Type 42 began life as the private venture Type 3009, designed by the British company Vickers.[14] Preliminary design studies during 1966 indicated that each Type 3009 would cost £19-£20 million. This was nearly twice what the Royal Navy could afford for the number of Sea Dart systems it needed at sea so the new destroyer immediately faced a tight and inadequate budget. In this environment, accountants and politicians became naval architects, and to them this often meant small was better. The Type 3009 design was shortened, its beam reduced, weight and space savings were made everywhere and cost cutting measures subjectively applied. Even the size of the Sea Dart magazine was reduced to save money, with space for just 22 missiles, thereby allowing only 11 targets to be engaged in salvo firing mode. The cumbersome magazine arrangements also required the missiles to be loaded onto the launcher in two stages, improving safety but potentially reducing the rate of fire. Damage control and survivability measures were other areas targeted for savings, and the ships were so small that there was no space or weight margin for later fitting of major new weapons or systems.

Notwithstanding these limitations, the Royal Navy accepted the altered design, and in October 1968 signed a contract for six warships.[15] The name ship of the class, HMS *Sheffield*, was laid down in 1970. Despite the cost cutting, the Type 42s ended up being 97 percent more expensive than planned, as well as five years late.[16] The destroyer's dubious claim to fame was that it was the smallest hull that could embark a Sea Dart launcher. Further contracts followed for four improved Batch II Type 42s in 1976, but only with the order for four stretched Batch III ships in 1978 were many of the design problems addressed. These latter were constructed to the original and slightly enlarged concept of their chief designer.[17] Ironically, Argentina became the Type 42's only foreign customer when it ordered two Batch I ships in 1970. One was built in the United Kingdom, the other under licence in Argentina.

The air search radar is one of the most important elements of an air defence system. But here again the choice reflected a cost saving philosophy. The Batch I ships mounted a Marconi Type 965P long range air-search radar, a slightly improved version of a 1950s 'bedstead' design then widely deployed on most Royal Navy ships

The Batch I Type 42 destroyer, HMS *Glasgow*. The dual Sea Dart launcher is clearly visible behind the 4.5-inch gun. The large antenna forward belongs to the Type 965P long range air search radar, while two Type 909 fire control radars are within the fore and aft radomes. (GRAZEBROOK COLLECTION)

from frigates upwards.[18] The Type 965P had a range of some 150-200 nautical miles and a horizontal beam width of 12 degrees. Its greatest deficiency was the lack of a moving target indicator (MTI) function in the processing suite, which would have allowed the operator to detect and track air targets at low level or against a land background. Only with the building of the Batch II and III Type 42 variants was a much-improved radar, the Type 1022, introduced. This radar incorporated a digital MTI function, frequency agility, pulse compression and a beam width of only 2.3 degrees, providing good performance in the presence of surface clutter and better detection of sea-skimming missiles.[19] The improved combat system on the Batch II, ADAWS 7, also gave them the ability to engage aircraft targets 15 seconds earlier than the Batch I ships, which had the older ADAWS 4 system.[20]

The somewhat arbitrary reduction in the dimensions of the original Type 3009 design left the Batch I and II Type 42s with very poor sea-keeping capabilities. The commander of the Falkland's task force, Rear Admiral 'Sandy' Woodward, had once commanded *Sheffield* and well understood their limitations:

> *I knew from my own time in* Sheffield *that the 42s were not as fast as they should be, that they were unreasonably slow in a short swell, with their bows slamming into the waves rather than splitting them each side cleanly. Even in those days, it had seemed to me the waves were landing on the foredeck too steeply and too short, breaking downwards, on to the Sea Dart System. The consequences had not been obvious then ... (I resolved to look into this further … It transpired much as I feared: the bow had 'happened' rather than been designed …).*[21]

The consequences revealed in 1982 included the tendency for salt encrustation to interfere with Sea Dart operations. On 12 May HMS *Glasgow* found herself unable to engage attacking Argentine A-4 Skyhawks off the Falkland's capital, Port Stanley, due to a salt-encrusted micro-switch on the launcher, the result of a big sea the night before. This gave the launcher computer an error message and prevented it from accepting the missiles. Later HMS *Coventry* was unable to fire on an Argentine C-130 Hercules which closed to within eight miles of the ship, due to dried salt jamming a missile flash door.

Wartime Performance

As both sides deployed the Type 42 destroyer in the conflict, it is of interest first to recall the role played by the two lesser known Argentine vessels. In the opening stages of the assault on the Falklands the Argentine Marine Commander used ARA *Santisima Trinidad* as his amphibious command and control centre.[22] *Santisima Trinidad*, which had only recently returned from missile firing trials in the United Kingdom, also launched over 100 commandos in inflatable boats towards the shore as a precursor to the main invasion force. Following the initial British surrender, the second Argentine Type 42, ARA *Hercules*, stayed behind to provide air traffic control at Stanley airport.[23]

Once the British task force was clearly on its way, the two Argentine destroyers reverted to their primary role of providing close air defence protection for their sole aircraft carrier, the elderly ARA *Veinticinco de Mayo*, in particular while she prepared to launch an air strike against the British task force on 2 May. During this approach, *Hercules* nearly shot down a British Sea Harrier fighter, which was on a reconnaissance mission to find the Argentine fleet. The Harrier had remained radar silent until the last moment, and when he switched on his radar the pilot found himself less than ten miles from the Argentine task group. Promptly illuminated by *Hercules*' Type 909 tracking radar, the pilot immediately broke away and dived low to deny an engagement opportunity for Sea Dart. *Veinticinco de Mayo*'s air strike, meanwhile, could not be launched because the carrier could not find sufficient wind.

A Royal Navy Sea Harrier of 809 Squadron carrying Sidewinder and drop tanks returns to HMS *Hermes* following an operational sortie during the Falklands War. (RAN)

Even after the virtual withdrawal from the conflict of the Argentine surface fleet – in the aftermath of the British sinking of the cruiser ARA *General Belgrano* – the Type 42s continued to be active in home waters. *Santisima Trinidad*, for example, was used as a signals intelligence asset. As it was fitted with the same system, it attempted to hack into the Royal Navy's networked datalink, and also attempted to jam these signals from the safety of the Argentine coast.[24] But it was the British who were to gain even greater benefits from the Type 42's versatility. The destroyer's multi-faceted ability to launch helicopters for anti-shipping strikes, conduct accurate shore bombardment, direct fighters, support special forces, conduct force networking and deny airspace to the Argentine Air Force in turn multiplied the flexibility, adaptability and lethality of the limited number of ships and aircraft available to the task force.

The Royal Navy initially deployed three Batch I Type 42s to the Falklands: *Sheffield*, *Glasgow* and *Coventry*. The Argentine Air Force and Naval Air Arm assessed that these vessels would not only be their major threat, but also their greatest obstacle in attacking the more important aircraft carriers and amphibious assault ships. It followed that if the Argentines could take out the air defence pickets then the war could be won, and consequently the Type 42s become their main target.[25] Having two Type 42s of their own, Argentine pilots were able to use them to develop tactics to play to the class's weaknesses, and were well aware of the advantages of flying ultra low or with land behind them.

Given the Type 42's inherent handicaps it may come as no surprise that of the 117 Argentine aircraft destroyed during the Falklands War only eight definitely fell victim to Sea Dart.[26] Even adding those kills by other systems in which the Type 42 directly contributed does not greatly increase the proportion. On the figures alone, one might be forgiven for questioning the value of the Type 42, particularly given that two of these vessels were sunk by air action and another badly damaged. Measuring air denial capability purely in terms of the engagement of individual aircraft, however, takes insufficient account of other and often more important effects. As the authors of *Air War Over the South Atlantic* correctly observe:

> *to judge the effectiveness of an air defence system solely on the number of aircraft it shoots down is to miss the essential point. The primary purpose of air defences is to protect targets; if in the process of securing that aim enemy aircraft are shot down, that is a bonus.*[27]

The task force relied for air defence upon a mix of systems, but without doubt the Type 42s provided effective protection, not least because as a priority target they diverted Argentine attention away from more vulnerable and higher-value units. In boldly carrying out this duty the Type 42s undoubtedly suffered. *Sheffield* was the first major casualty. On 4 May she was part of a three destroyer picket line 30 miles 'up threat' from the two task force carriers when two Argentine Super Etendards attacked with Exocet. *Glasgow* gave the first warning of the attack, which she detected about 25 miles out. Attempts to acquire the sea-level targets with the Sea Dart system failed and at 12 miles both aircraft launched their missiles. One Exocet failed to find a target and splashed harmlessly at the end of its run, but *Sheffield* sighted the other missile about two metres above the sea, less than a mile from the ship. Seconds later it hit the destroyer amidships. The missile's warhead did not explode but its remaining fuel ignited, spreading dense smoke and starting major fires. Despite valiant attempts to save the ship *Sheffield* had to be abandoned and sank five days later while being towed to South Georgia for possible salvaging.

Intelligence provided to Britain by France indicated that the latter had delivered only five Exocets to Argentina, meaning that just three remained, but it was also known that Argentina was vigorously searching the black market for more.[28] Without an organic Airborne Early Warning capability the British task force was already severely handicapped. And with *Sheffield*'s loss the task force had graphic evidence of the weakness of its air defences in the face of the Super Etendard-Exocet combination. Rear Admiral Woodward well understood that a successful Exocet attack on one of his carriers would spell disaster: 'Lose *Invincible* and the operation is severely jeopardised. Lose *Hermes* and the operation is over[29]'. This realisation forced the fleet to operate to the east of the Falklands, at what was thought to be the limit of the Super Etendard's range.[30] This also meant, however, that the Sea Harrier's time on combat air patrol (CAP) over the future amphibious bridgehead would be limited. This was not an ideal situation when attempting to project land forces ashore.

In an attempt to mitigate the Type 42's weakness in defending against low-level attacks, where possible Woodward teamed each destroyer with a Type 22 Sea Wolf missile armed frigate to act as a 'goalkeeper'. Intended as a self-defence system against the sea-skimming missile threat, Sea Wolf was specifically designed for short range, high speed, low-level interceptions. The frigate's air search radar also had an MTI function which could be used close to land, and the detected enemy track data-linked to the destroyer for Sea Dart interceptions.[31] If Sea Dart was still unable to engage then Sea Wolf would form a second defensive layer.

The combination of Type 42 and Type 22, sometimes colloquially referred to as a 'Type 64', nevertheless had mixed results, as evidenced by the attacks on *Glasgow* on 12 May. That afternoon the destroyer and the Type 22 frigate HMS *Brilliant* were acting as an anti-aircraft 'missile trap' as well as bombarding the airport off Stanley, when *Brilliant* detected an Argentine raid approaching overland. *Glasgow*'s Type 909 directors picked up the four Argentine Air Force A4 Skyhawks but, as already noted, salt encrustation prevented the Sea Dart missiles from loading. With the aircraft less than a mile away *Brilliant* fired three Sea Wolf missiles, scoring two direct hits and causing the third aircraft to fly into the sea. The fourth aircraft released its bomb early and this ricocheted off the water and over *Glasgow*'s hangar.

Twenty minutes later the Sea Dart problem had yet to be solved when the ships again detected an approaching A-4 raid, the four aircraft flying so low that their windscreens became encrusted in salt making it difficult to see.[32] But rather than adopting a direct approach, these aircraft began weaving to upset the aim of the ships' upper-deck gunners. *Brilliant*'s Sea Wolf system, designed to deal with a non-manoeuvring missile, also became confused, and automatically trained the launchers fore and aft. With no time left to use the system's alternate television guidance, *Brilliant* could do nothing to stop the attack. One 1000 lb bomb skipped over the Type 22, but another hit *Glasgow* amidships, just above the waterline.[33] The bomb passed through the destroyer without exploding, but the internal damage reduced her ability to provide anti-air cover and she was withdrawn early when reinforcements arrived. In the meantime, with *Coventry* his one undamaged Type 42, Admiral Woodward could not risk her on further offensive missions.

The British amphibious assault at San Carlos on East Falkland, beginning on 20 May, signalled the end of the precursor phase of maritime operations and the shift to direct support of the land forces. The assault achieved complete tactical surprise and allowed vital time for defences to be set up before the inevitable Argentine response. Air attacks began at midday on 21 May and faced a multi-layered air-defence designed to achieve a high attrition rate on Argentine aircraft. Sea Harriers on CAP provided the outer layer. The second layer consisted of a Type 42/Type 22 'missile trap' positioned off the northern entrance to Falkland Sound. The third layer of defence was a group of three or four frigates inside the Sound using every gun and missile they possessed. Finally, the anchorage was protected by short-ranged missiles on the assault ships and on shore, together with small calibre guns of every description.[34] The hilly local terrain also offered some protection to the anchorage, but at the same time deprived the force of radar warning during the last 50 miles of an aircraft's approach.

Between 21 and 24 May more than 40 Argentine aircraft were shot down, but their attacks were courageous and persistent, sinking two frigates in the third layer, HMS *Ardent* and HMS *Antelope*, and damaging another six ships. Despite their losses, on 25 May the Argentines made another major effort. British defences continued much as before with *Coventry* and the Type 22 HMS *Broadsword* providing the 'missile trap' outside Falkland Sound and controlling the Sea Harrier CAP. The first raid of the day turned away after losing its leader to *Coventry*'s Sea Dart, while the second returned to base upon realising that the destroyer had vectored the CAP to intercept. A third raid also lost its leader to Sea Dart. To this point the trap had worked as intended, with *Broadsword* making the first detection and then handing tracks over to *Coventry* either for Sea Dart engagement or fighter interception. Five aircraft had been accounted for in 24 hours, but this success now brought about a deliberate Argentine response.

Shortly before 2PM four Skyhawks were detected by the British some 100 miles out, flying at medium altitude. The formation split into two pairs, and dived down to low level. On re-detection the CAP attempted to intercept the first pair, but could not close to a firing position before reaching the edge of *Coventry*'s Sea Dart engagement range. The destroyer's captain ordered the Sea Harriers to haul off but, with less than a minute's flying time to go, found that his Sea Dart System could not acquire the targets as they hugged the water. *Broadsword*, meanwhile, continued to track the targets in, but her Sea Wolf system became confused by the explosion of *Coventry*'s gunfire, and would not fire automatically. Although damaged by the ships' guns the two aircraft continued their attacks, each releasing two 1000 lb bombs. Three bombs missed, but the fourth bounced off the sea and struck *Broadsword* without exploding.

The second pair of Skyhawks appeared moments later, and once more the Sea Harriers had to be warned off as they approached the Sea Dart engagement envelope. The Sea Dart system, however, again failed to acquire and at the critical moment *Coventry* passed in front of Broadsword preventing her using the television aiming display to engage with Sea Wolf. Three of the four bombs dropped hit *Coventry* with devastating results and the destroyer sank within 30 minutes. *Coventry*'s loss was undoubtedly a severe blow to the British but, of more importance to the overall operation, Argentine air attacks had failed to stop or delay the amphibious build-up and the vital landing ships had not been seriously damaged. By the afternoon of 25 May the beachhead was in good order and the assembled troops ready to march out.

Fortunately for the task force's air defence, reinforcements had by this point arrived in the form of the two Type 42s, HM Ships *Cardiff* (a Batch I) and *Exeter* (an improved Batch II) and the sole Type 82, *Bristol*. Fitted with the same sensors and combat system

The Type 22 frigate, HMS *Broadsword*. Her armament includes four Exocet surface to surface missiles in single cells mounted forward and two six-barrelled Sea Wolf launchers. (GRAZEBROOK COLLECTION)

as the Batch I Type 42s, *Bristol* suffered from similar deficiencies. *Exeter*'s performance on the other hand is quite instructive. Equipped with improved electronic surveillance equipment, better radars and the enhanced ADAWS 7 combat system, *Exeter* was the first to raise the alarm during an Argentine Super Etendard attack on the task force on 25 May. Positioned only some 60 miles north east of Port Stanley, the task force was covering the passage of the important stores ship *Atlantic Conveyor* to San Carlos. In a notable display of networked operations *Exeter* used her datalink to pass her track information to the fleet, enabling all ships to deal better with the threat. These measures were effective to the extent that both Exocets launched at the task force were successfully decoyed by one or other of the radar reflective chaff clouds deployed by the defending ships. Unfortunately, however, as one of the Exocets emerged from under the chaff its seeker head re-acquired another target, Atlantic Conveyor. Not having its own self-defence systems and outside the range of those of its escorts, the merchant ship was hit on the port quarter. Once more the missile failed to explode, but again the fires which broke out brought about the subsequent loss of the ship.

Exeter was again in the thick of the action on 30 May, when her ability to engage fast low-flying targets was well tested. With only one Exocet remaining, the Argentine Air Force used a bold tactic to maximise its effect. In a joint mission, two Super Etendards were joined by four A-4C Skyhawks to attack the British task force, which was again east of Stanley. Following a circuitous, low and silent approach to avoid early detection the Super Etendards popped up to use their search radars to locate the British ships. As their targets could not be distinguished from a number of rain showers in the area the aircraft descended again to further close the range. *Exeter* nevertheless intercepted the transmission, and on receipt of this warning all ships turned to meet the threat. When, some three minutes later, the Super Etendards took another look they were immediately detected by *Exeter* and *Cardiff*, and the former locked her Sea Dart system on to the contacts. Making no attempt to identify his targets, the Argentine leader fired his Exocet at a range of 21 miles leaving the four A-4s to follow it in while the Super Etendards turned sharply for home. The Exocet, which headed for the frigate HMS *Avenger* acted as a pathfinder for the Skyhawks, which had no search radar of their own. *Exeter* fired a two Sea Dart salvo at the A-4s, and a third missile at the Exocet, while Avenger engaged the targets with guns. Two A-4s were destroyed, at least one to *Exeter*'s Sea Dart, and there is strong evidence to suggest that she also shot down the Exocet. In any event, no British ship was damaged.

Argentina's air forces were essentially spent after 30 May and unable to conduct more attacks on the ships at sea. At this point *Exeter* was repositioned by Woodward to conduct air blockade and bombardment operations off Stanley. After this she was moved to the amphibious bridgehead at San Carlos, assuming responsibility for air defence coordination of the transport area.

Effect on Enemy Operations

Despite their flaws, it is clear that the Type 42s had a remarkable influence on Argentine air operations. As Admiral Woodward later described:

> *we did not know … of one fundamental factor which was going to dominate the thinking of the Argentine aviators: their high regard for the effectiveness of the British medium-range surface to air missile system, Sea Dart. And it caused them to decide against using the middle and upper air, to get below Sea Dart at all costs. This left them with only very low flying.*[35]

In effect the Argentine tactic to counter Sea Dart, while initially successful, also had a number of unexpected and crucial tactical consequences. To begin with, attacking aircraft dropped bombs from such a low level that the safe arming device on the bomb's fuse usually didn't have time to arm. Many bombs, while accurately delivered, failed to detonate and either passed through the targeted ship, bounced over it or lodged inside. Approximately 15 ships were hit or narrowly missed by bombs that failed to explode due to low-level release. Had the Type 42 destroyers not been present, and their capabilities weighing on the minds of the Argentine pilots, then bombs would have been dropped from a more appropriate height, with potentially devastating results. An example of this was seen at Fitzroy on 8 June when A-4s attacked and bombed the two Royal Fleet Auxiliary landing ships, *Sir Galahad* and *Sir Tristram*. Without effective air defence the Skyhawks could bomb from the correct height, causing terrible loss of life on the two ships.[36]

Bringing Argentine air attacks down to low level also had the effect of forcing them into the engagement zones of other air defence systems such as Sea Cat, Sea Wolf, Rapier, Blowpipe and guns of all calibres. Together these weapons accounted for 44 aircraft, which might otherwise have used higher altitude avoidance.[37] Finally, the need to fly low meant that Argentine pilots could generally not fly at night. This acted as a force multiplier for the small force of British Sea Harriers as their maintenance could be conducted at night and their pilots rested. As the task force's senior Sea Harrier pilot noted in his book *Sea Harrier at War*: 'In spite of the air supremacy established by the task force fighters, it was still not possible for us to achieve airspace denial for twenty-four hours a day'.[38]

There was another operational benefit to denying the middle and upper air to the Argentine Air Force: its long-range reconnaissance and strike assets were essentially sidelined. When, on 7 June, the Argentines pressed their luck and launched four Learjet reconnaissance aircraft to photograph the British positions in San Carlos, a Sea Dart from *Exeter* shot down the lead aircraft at over 40,000 ft. The primary Argentine bomber, the Canberra, was in like manner unable to act in its anti-shipping role, and was found to be of limited use against land targets. Although towards the end of the war the Canberras made some night bombing raids on British land force positions around the hills of Stanley, these missions achieved very little, and on 14 June another Sea Dart from *Exeter* shot down a Canberra over Mount Kent.

Finally, there was the indirect influence exerted by the Type 42 on Argentine ground operations, in particular its potential to interdict the logistics supply line. Air defence destroyers placed close to Stanley airport for bombardment missions effectively shut it down, even when not directly active. Sea Dart's threat to air traffic was first seen on 9 May when *Coventry* fired a missile at a C-130 Hercules approaching Stanley airfield. The first Sea Dart missed, but a second is thought to have claimed two A-4s acting as fighter escorts by detonating between them.[39] By maintaining a Type 42 within Sea Dart range of the airfield whenever possible, Admiral Woodward severely constrained the Argentine defences. During May and June, instead of the hundreds of transport flights into Stanley needed by the defenders, the Argentine Air Force managed just 33.[40] These flights brought in only 434 tons of supplies, insufficient to sustain an army of 11,000 troops cut off from home in cold weather and already experiencing shortages. Although the British forces ashore were outnumbered, a better logistics system and secure lines of communication ensured that the balance remained tipped in their favour.

Conclusion

During the Falklands War the Royal Navy's Type 42 destroyer was able to produce a result greater than the sum of its parts. The Type 42 was not fully networked, but neither did it work in isolation. Forming part of a sophisticated layered air defence, the destroyer severely constrained Argentine air operations and in conjunction with other systems helped win the air war in the South Atlantic. Fielding systems far in advance to those of 1982 and better able to integrate the overall management of a task group's combat assets, the ADF's proposed AWD will also be a significant force multiplier. Unfortunately, critics of the AWD tend to look at the ship in isolation, and often seem unable to appreciate the complexities of modern networked and layered air warfare. This is not to suggest that the ships will be perfect, or that current and future threats will not pose a risk, but this is a reality that navies have long come to accept. Asking why everyone in the room was so gloomy on the morning after *Sheffield*'s loss, the Chief of the UK Defence Staff, Admiral of the Fleet Sir Terrance Lewin, reminded his audience: 'there's no point in having these ships if you are not prepared to lose them'.[41]

PART IV

Naval power projection in the contemporary world

17. Sea Power and Peace Operations: The Naval Contribution to UNITAF, Somalia, 1992–93

John Connor

THE END OF THE Cold War has had two significant effects on naval strategy and operations. The first is that the demise of the Soviet Navy and the absence of any other rival to the United States as a blue water naval power means that navies have moved away from planning for open ocean warfare and now concentrate on the application of sea power in the littoral – the area of land which can be influenced from the sea. The second is that the end of Cold War rivalry means that the great powers that hold permanent membership and veto powers in the United Nations Security Council (UNSC) are now more likely to agree to the deployment of peacekeeping forces, resulting in an increase in the number of these operations. These two developments are related. Few peace operations take place in open sea, so naval peace operations will almost always be examples of the application of sea power in the littoral. This chapter will examine the role which sea power has played in peace operations in three ways. Firstly it will attempt to define what constitutes a 'peace operation'. Secondly it will point out briefly that the history of naval peace operations includes participation in what is arguably the world's first peace operation in 1920 and what is definitely the United Nation's first peace operation in 1947. Finally, it will analyse the deployment of the multinational naval component of the Unified Task Force – Somalia (UNITAF) in 1992 and 1993 as an example of sea power ashore.

Peace Operations

In arriving at a definition of 'peace operations' – known simply in popular usage as 'peacekeeping'[1] – it is first necessary to come to a definition of 'peace'. In the context of peace operations, peace means more than simply the absence of violence. It implies a belief that peace is the normal state of human existence, and that the maintenance of world peace is a responsibility of national governments and regional and international organisations. This meaning of peace is a relatively new understanding. For example, Social Darwinists of the late nineteenth and early twentieth centuries believed that conflict between nations was the natural order and war was inevitable. The concept of an international community which should be committed to maintaining world peace was only formalised with the creation of the League of Nations following the carnage of World War I. Even then, it was not until the European empires decolonised following World War II that the majority of the world's population achieved, however imperfectly, some form of national sovereignty.

It is difficult to talk about peace operations before the 20th century, and before the creation of international organisations dedicated to maintaining world peace. The best examples prior to the 20th century of naval operations for humanitarian purposes are the Royal Navy's anti-slave trade patrols. Established following the British government's abolition of the slave trade in 1807, the Royal Navy was continually deployed to the west coast of Africa in the first half of the 19th century, and the east coast in the second half, to interdict slave shipments.[2] It has been claimed that French naval interventions in the late nineteenth century had a basis in 'respect for human rights with the intention of integrating "barbarous" countries into the concert of civilised nations on a footing of diplomatic equality and reciprocal exchange'.[3] In reality they were based on national ambition. French naval operations in Vietnam did not result in that country gaining international equality but rather in it becoming a French colony.

Current peace operations are mounted in one of three ways. The first is as a fully-fledged UN mission, such as the longstanding presence of the UN Force in Cyprus. The second is with a multinational force sanctioned by the UNSC but not organised by it. In some cases these operations are designed to get a force on the ground as quickly as possible, and are later replaced by a UN operation once the latter has been organised. For example, the Australian led International Force East Timor (INTERFET) arrived there on 20 September 1999, just a week after the Indonesian government announced that it would accept peacekeepers in East Timor. INTERFET then handed over to the UN Transitional Administration in East Timor on 23 February 2000. In other cases, such as that of the Maritime Interception Force which enforced UN sanctions against Iraq from 1990 to 2003, there was no replacement by a UN force. The third type of peace operation is organised outside the UN by regional organisations or groups of nations such as the African Union's Darfur Integrated Taskforce deployed to Western Sudan.

UN operations are often classified according to the sections of the UN Charter which outline the means by which international disputes can be resolved. Chapter VI of the UN Charter details peaceful dispute resolution, so a 'Chapter VI operation' is a peace operation which aims to assist parties to end a conflict through peaceful negotiation. Chapter VII of the UN Charter gives the Security Council the power to 'take such action by air, sea, or land forces as may be necessary to maintain or restore international peace and security', so a 'Chapter VII operation' refers to the use of coercive measures to enforce peace.[4] These categorisations can be overly proscriptive, however, and may not reflect the complex reality of actual missions. Dag Hammarskjöld, the UN Secretary-General from 1953 to 1961, described peace operations as coming under 'Chapter six and a half' because they 'fall short of the provisions of Chapter VII [but] at the same time they go beyond purely diplomatic means of those described in Chapter VI of the Charter'.[5]

Article Two of the UN Charter enshrines the concept of national sovereignty, stating that, 'Nothing contained in the present Charter shall authorize the United Nations to intervene in matters which are essentially within the domestic jurisdiction of any state'.[6] During the 1990s, this idea of national sovereignty was tempered by the development of the concept of 'humanitarian intervention' to justify interference in a state's internal disputes if the population is suffering grievous human rights abuses. This was the justification for deploying UNITAF to Somalia. The concept was enunciated most clearly in a speech made by British Prime Minister Tony Blair on 22 April 1999 during the Kosovo War. Blair defined what he termed 'a new doctrine of international community' and stated, 'We cannot turn our backs on conflicts and the violation of human rights within other countries if we want still to be secure'.[7] Critics of humanitarian intervention argue that it is applied inconsistently, and that it remains a cloak for military action aimed at advancing national interest and neo-colonialism. As the American political commentator Noam Chomsky points out, NATO's concern for the human rights of Kosovar Albanians, which led it to fight the war against Serbia in 1999, should logically have led the alliance to intervene also in Turkey to protect the Kurds. But as Turkey was a NATO member and – unlike Serbia – did not challenge US hegemony, its human rights abuses were ignored.[8]

How then should 'peace operations' be defined? Military doctrine documents generally agree in the broad definition of peace operations, even if they differ as to how the types of operation should be categorised. The French doctrine of 1995, *La doctrine militaire française pour les operations de maintien de la paix*, talks of three types of peace operations. These are 'peacekeeping', an intervention carried out under Chapter VI of the UN Charter with consent of all parties and where have hostilities have ended; 'peace restoration', a Chapter VII intervention to return a country in civil war to peace where no aggressor has been designated; and 'peace imposition', a Chapter VII intervention to re-establish peace against a well-defined aggressor.[9] The RAN's *Australian Maritime Doctrine* also defines three types of peace operations but in slightly different terms. These are 'peacekeeping', where forces are deployed under 'conditions which are agreed by all the belligerent parties'; 'peace enforcement', where 'one or more of the belligerents have not consented to intervention by international forces and coercive action may be required'; and 'peace building', where a country or region is reconstructed following the end of a war.[10] Despite these minor variations, all doctrines agree that peace operations can involve a wide range of means, up to and including high intensity warfare.

Naval Peace Operations

Peace operations almost always include the forces of more than one nation, and are approved by a regional or international organisation in an attempt to avoid the suggestion that the intervention is aggressive or colonialist in intent. For example, there is a world of difference between HMAS *Adelaide*'s deployment to the Solomon Islands in 1927, and HMAS *Manoora*'s operations in the same archipelago in 2003. In 1927, the Australian cruiser was sent to Malaita on a punitive expedition to force the island's population to accept a tax imposed on them by British colonial officials.[11] In 2003, the deployment of the Regional Assistance Mission to the Solomon Islands was agreed to by the Solomon Islands government, endorsed by a meeting of Pacific Islands Forum foreign ministers, and included personnel not only from Australia, but also from Fiji, New Zealand, Papua New Guinea and Tonga.[12]

The involvement of a group of nations or a regional or international organisation may give legitimacy to a peace operation, but for practical reasons the naval contribution to a mission may sometimes be delegated to a single nation on behalf of the larger grouping. From 1966 to 1975, for example, the Royal Navy enforced the UNSC's oil embargo against Rhodesia following the White minority government's Unilateral Declaration of Independence by patrolling the Indian Ocean to prevent oil entering Rhodesia through the port of Biera in Mozambique, then a Portuguese colony.[13] In the same way, the Italian *Marina Militare* has provided the Coastal Patrol Unit for the Multinational Force and Observers (MFO) since the creation of this mission in 1982 to monitor the Egyptian-Israeli border as part of the peace agreement between these two nations.[14]

The majority of peace operations have been mounted since the end of the Cold War at the beginning of the 1990s. Navies have been involved in peace operations, however, since the very origin of this type of deployment. Naval personnel formed the majority of the military contingent in the first mission that can be clearly defined as a 'peace operation'. This took place in the aftermath of World War I when an International Commission, established by the Treaty of Versailles, administered Schleswig in Germany. This region, annexed from Denmark following a war in 1864, was inhabited by both Danish and German speakers and the International Commission organised two plebiscites in northern Schleswig in 1920 to allow the population to choose whether they would remain part of Germany or join Denmark. Vice Admiral T.D.L. Sheppard, RN, commanded the force sent to maintain order. This consisted of the British cruiser HMS *Carysfort*,[15] destroyers HM Ships *Seawolf* and *Sesame* and the tugboat *St Anne*, the French cruiser *Marseillaise* and destroyer *Téméraire*, the 1st Battalion, the Sherwood Foresters and the *22éme Bataillon de Chasseurs Alpins*.[16]

The concept of using carrier-borne aircraft in peace enforcement operations has a surprisingly long history also. When Italian dictator Benito Mussolini invaded Ethiopia in October 1935, there was the initial possibility that the League of Nations – precursor to the UN and created at the end of World War I – might take military action against Italy under the terms of the League of Nations Covenant. No action was taken in the end, but in late 1935, while the British government was considering what to do, naval planners drew up options for war against Italy. If this peace enforcement operation under the auspices of the League of Nations had taken place, it would have included an attack by aircraft from HMS *Glorious* on the main Italian naval base at Taranto.[17] This plan became the basis for the actual attack on Taranto on 11 November 1940.

The very first UN peace operation included a naval officer. Commander Henry Chesterman, RAN, deployed on 14 September 1947 with three Australian Army and Air Force officers. They were to serve as military observers for the UN Consular Committee during the Indonesian War of Independence.[18] These early naval experiences of peace operations should be remembered, and studied for the lessons they can teach for current and future operations.

The Naval Component of UNITAF

The UNITAF deployment to Somalia in 1992-93 provides a useful case study of the application of sea power in a particular form of peace operation: humanitarian intervention in a state with almost no existing infrastructure.

The Somali people inhabit a large area of East Africa which came under the colonial control of four nations: Ethiopia, Italy, France and Britain, during the late 19th century.[19] In 1960, the Italian and British colonies of Somaliland became independent as Somalia. Major General Mohamed Siad Barre took power in a military coup in October 1969, and Somalia was drawn into the Cold War due to its strategic position at the entrance to the Red Sea. At first, Somalia received Soviet military and economic aid in exchange for consenting to the construction of military bases. But when a Marxist government was established in Ethiopia in 1977, the Soviet Union shifted its support to this larger and more influential neighbouring country. In the 1980s, the Americans took advantage of the Soviet withdrawal, providing economic and some military aid to Somalia and gaining access to the former Soviet port facilities and air bases.

By the beginning of the 1990s, Siad Barre was losing his grip on power. The US Government withdrew aid in 1989 following his brutal repression of clans in northern Somalia. The main opposition groups united to defeat Siad Barre in October 1990, and by January 1991 he had fled the country. The clan-based civil war which followed the ousting of Siad Barre coincided with a drought that caused poor harvests and food shortages. In July 1992, humanitarian organisations launched a worldwide appeal for aid for Somalia, and the UN authorised an emergency airlift of supplies. With no functioning system of law and order, factions took the bulk of the food the aid agencies had shipped into the country and prevented its distribution to those who needed it. The food shortage became a famine and about 300,000 Somalis died.

The UN sent military observers to Somalia to monitor a ceasefire between the two main belligerents: the militias led respectively by Ali Mahdi Mohamed and by Mohamed Farah Aidid, who were fighting for control of the Somali capital of

Mogadishu. This mission was named the UN Operation in Somalia (UNOSOM). In September 1992, Pakistani troops were added to the deployment to provide security at Mogadishu's port and airport. UNOSOM had neither the mandate nor the resources to protect food distribution, and the relief effort faltered as the security situation worsened. Between May and November 1992 only ten ships carrying humanitarian aid had entered the port of Mogadishu. When a relief ship was fired upon in late November, they stopped coming at all.[20] Television images of the famine and the actions of the unrestrained militias combined with lobbying by humanitarian relief organisations to put pressure on US President George H.W. Bush to intervene.[21] On 25 November 1992, the US Government announced that it would lead a force to enter Somalia and enable aid agencies to distribute humanitarian relief. The UNSC gave this force, the Unified Task Force–Somalia (UNITAF) the robust mandate to use 'all necessary means' to carry out this task.

UNITAF at its peak consisted of 37,000 personnel, of whom 21,000 were Americans, with the remainder coming from 20 different nations.[22] The land force was deployed and sustained ashore by a large multinational naval force. Between December 1992 and May 1993, a force of over 50 naval ships, mostly American but also including ships from Australia, Belgium, Canada, France, India, Pakistan and Turkey, operated in Somali waters.[23]

The decision by the US and other governments to deploy a force to Somalia was made at very short notice. The US Central Command informed I Marine Expeditionary Unit, which provided the bulk of the initial force, of the possibility of being deployed to Somalia on 20 November 1992, only five days before the US Government publicly announced the intervention.[24] President Bush could deploy UNITAF swiftly because the US policy of forward deployment meant that the USS *Ranger* Carrier Battle Group, the USS *Tripoli* Amphibious Ready Group and pre-positioned supply ships were already in the Indian Ocean. The UNITAF operation began only three months after the publication of *From the Sea*, the joint USN/USMC white paper which called for 'a fundamental shift away from open-ocean warfighting on the sea towards joint operations conducted from the sea'. An opinion piece published in the *Navy Times* on 21 December argued that the UNITAF deployment was a perfect embodiment of this new post-Cold War doctrine.[25] The US Navy had also had recent operational experience in Somali waters. In January 1991, the US Ships *Trenton* and *Guam* were despatched – from the vast fleet assembled for the 1990-91 Gulf War – to evacuate foreign nationals from war-torn Mogadishu during the dying days of Siad Barre's regime.[26]

The speed with which the UNITAF deployment arose made it difficult for Canadian military planners to keep up with the pace of events. The Canadian government had already committed a battalion to UNOSOM. This meant that when HMCS *Preserver* sailed from Halifax on 16 November 1992 it carried quantities and types of ammunition appropriate for troops taking part in this UN operation, where they were only to use deadly force in self-protection. Much of the ship's cargo space was taken up with building materials for a camp. But by the time the ship arrived off Mogadishu on 5 December, the government had decided to participate in UNITAF with its stronger rules of engagement. The Canadian battalion therefore now needed items that had not been originally loaded in *Preserver*, such as larger stocks of ammunition and combat engineering supplies.[27]

On the night of 9 December 1992, Marines of the Special Purpose Marine Air-Ground Task Force came ashore at Mogadishu. They secured the airport, which enabled troops to be flown in direct from the US, and the port. The Center for Naval Analyses report on UNITAF describes graphically the situation which the Marines encountered at the devastated port:

Trash and excrement covered the piers and warehouse floors. Two derelict tugs were tied up at the pier. Local relief agencies had built a perimeter wall of containers to keep looters out. In some areas, this container wall was three high, nearly 25 feet. Derelict cranes, forklifts, and vehicles littered the operating areas. Warehouses were half full of bags of grain and half full of litter.[28]

Peace operations will often be conducted in states where the port facilities have limited capacity due to lack of economic development, or where existing infrastructure has been destroyed in war. In the case of Somalia, Mogadishu was the only port in the country that could receive ocean-going shipping immediately.[29]

US Navy divers completed an initial survey of Mogadishu port on 10 December. That evening Rear Admiral James B. Perkins, USN, Commander of the Maritime Prepositioned Force (MPF), decided to begin unloading MPF ships the following day. On 11 December the US Navy's Amphibious Construction Battalion One and

HMCS *Preserver* replenishes HMAS *Tobruk* and Commander Naval Forces Somalia, USS *William H. Standley*, February 1993. (RAN)

the US Marines' 1st Force Service Support Group (Forward) disembarked to find that the port lacked any administration, harbour police or dock workers. As Perkins later commented, 'We knew we'd offload ships, but we didn't know we'd run the port. Our intelligence was not particularly good.' He was required to establish a de facto port authority immediately.[30]

With the assistance of two tugboats from Mombasa chartered by US Military Sealift Command, the MPF ships disgorged their cargo without incident. But as the Center for Naval Analyses report points out, this unloading took place, contrary to established doctrine, before the port area had been secured. This exposed the ships – 'strategic assets of tremendous value' – to potential risk.[31]

The port's limitations hampered operations. The first transports carrying cargo for the US Army arrived in late December only to find that their draft was too deep for them to offload. It was not until 19 January 1993 that a US Army Prepositioning ship successfully docked in Mogadishu.[32] Mooring restrictions and the amount of traffic in the port meant that *Preserver* never went alongside during its long deployment off Somalia.[33]

When the Australian contribution to UNITAF was being planned in December 1992, Maritime Headquarters believed that the Somali port infrastructure would be unable to deal with the volume of shipping required to support the operation. It was proposed instead to supply the Australian contingent across a beach near Mogadishu or Kismayo. This would have required the LSH (Landing Ship Heavy) HMAS *Tobruk* to embark two LCM8 landing craft or two LARC amphibious vehicles. Loading this extra equipment in a ship of *Tobruk*'s size would have significantly limited the amount of space for stores and equipment to support the Australian battalion being deployed. Fortunately, assurances were received that US forces would provide ship-to-shore capability if this was required and the amphibious equipment was not loaded.[34] In the end this US assistance was unnecessary as *Tobruk* was able to use the roll-on roll-off berth at Mogadishu.

The position of this roll-on roll-off berth was yet another constraint on the use of the port. The berth was near the port entrance, and when ships docked with their stern at the wharf, they almost completely blocked the harbour entrance. The difficulty in using the roll-on roll-off berth was compounded by the prevailing winds in Mogadishu during the period of the UNITAF deployment. From late December to March, Somalia experiences strong north easterly dry monsoon winds. These made it difficult to carry out the Mediterranean moor manoeuvre required to come into the berth. When *Tobruk* went alongside for the first time on 20 January 1993, she had to deal with a 35 knot wind.[35] These high winds also caused problems with the operation of ship-to-shore fuel and water lines.[36]

Despite these difficulties, Mogadishu became, under UNITAF control, one of the busiest ports in Africa. During the first 35 days of operations to 15 January 1993, the 300 Seabees and 700 Marines unloaded 123,950 tonnes of equipment, supplies and food, 6668 vehicles and 96 helicopters from 48 naval and merchant ships.

Almost 20 million litres of fuel were pumped ashore. Most importantly, given the rationale for the operation, humanitarian relief supplies made up 37 percent of total port operations during this period, including the unloading of 14 aid ships carrying 39,700 tonnes of grain. On 15 January, port operations in Mogadishu were transferred from the US Navy and Marine Corps to the US Army's 7th Transportation Group.[37] In the southern Somali port of Kismayo, Belgian naval personnel from the supply and command ship *Zinnia* re-opened the port for military shipping and humanitarian relief.[38]

President Bush's special envoy, Robert Oakley, a former US ambassador to Somalia, and Lieutenant-General Robert Johnston, the UNITAF commander, met Aidid and Ali Mahdi in Mogadishu in the days before UNITAF deployed and gained the militia leaders' agreement that they would not oppose the operation.[39] The naval presence – with its implicit threat of overwhelming firepower – played a significant role in ensuring that this promise was kept. When the first UNITAF-escorted food convoy left Mogadishu on 15 December, three US warships came inshore to offer a deterrent to the militias while USMC AH-1 helicopters provided aerial cover during the convoy's journey. The following day, jets from *Ranger* flew over the inland centre of Baidoa before dawn to intimidate the militia and announce the imminent arrival of US Marines.[40]

Naval units could also cover gaps in Somali infrastructure and assist UNITAF forces. Before the landing, F-14 jets equipped with the Tactical Air Reconnaissance Pod System carried out infra-red surveys of Mogadishu to gather intelligence on the location of militia heavy weapons. Aegis cruisers provided air traffic control for Mogadishu airport until ground-based radars arrived.[41] *Preserver* hosted the Canadian Joint Task Force Headquarters until a headquarters could be established ashore, enabling the Canadian battalion to become operational almost immediately.[42] *Tobruk* provided the Australian battalion with communications with Australia for 20 days in March 1993 when frequency restrictions meant their Medport High Frequency system could only be used for four out of every 24 hours.[43]

The experiences of the Canadian and Australian contingents to UNITAF show the need for joint command structures to ensure effective use of naval assets in littoral operations. The Canadian national commander, Colonel Serge Labbé, had command of all Canadian units in theatre, including *Preserver* and its three Sea King helicopters. Labbé could therefore use the helicopters in a range of roles to support his troops serving ashore. In the first four weeks of operations the Sea Kings airlifted over 300 tonnes of supplies, as well as providing casualty evacuation. They also acted as transport for Labbé and his staff so they could visit detachments scattered across the Canadian sector at Belet Weyne and travel to Mogadishu for meetings. The Canadian Sea Kings were the only helicopters in theatre with Forward Looking Infra-Red capabilities, which enabled their crews to keep track of militia movements at night.[44]

The Australian national commander, Colonel Bill Mellor, did not have a joint command. Appointed by Land Headquarters, he did not have command of *Tobruk* and its one Sea King helicopter. These remained under Maritime Headquarters, which kept tight control over the movements of the Sea King and prevented Mellor using the helicopter for transport, reconnaissance or casualty evacuation.[45] The ADF did not remedy this problem until the establishment of a Deployable Joint Force Headquarters in 1997.[46]

As occurred during the 1990-91 Gulf War and the operations off Former Yugoslavia,[47] the arrangements devised to enable interoperability in UNITAF were ad hoc but generally effective. *Preserver* found itself assuming the role of a communications gateway between the US Navy and those navies whose systems were incompatible with the Americans.[48] Logistic tasks for the operation were allocated to ships regardless of nationality. On one voyage to Mombasa in February 1993, *Tobruk* loaded timber, cement, water and fuel for US forces in Mogadishu as well as 69 pallets of frozen food which were transferred to *Preserver* at sea.[49]

Relations between the military and humanitarian relief organisations are vitally important during peace operations, yet are often problematic, for two reasons. The first is the clash of cultures between military personnel and humanitarian workers. The second is the relief organisations' belief that working with military forces will undermine their political neutrality, at best making it impossible to work in certain places, and at worst putting the lives of the humanitarian workers and the people they are trying to help at risk. The relationship between UNITAF and the humanitarian relief organisations was much more collaborative than normal. This was because most of the humanitarian workers realised that the situation in Somalia had become so violent that they could not have remained in the country without military intervention. Representatives of US relief organisations went so far as to visit Central Command Headquarters in November 1992 where they were briefed on the deployment plan and provided information to assist with military planning.[50]

Naval assistance to humanitarian organisations in Somalia was organised either through UNITAF's Civil Military Operations Center and Humanitarian Operations Centers or in an ad hoc manner.[51] The 30th Naval Mobile Construction Regiment drilled wells for relief agencies, repaired 800 km of roads and rebuilt schools, orphanages and soccer fields.[52] *Preserver* offered rest and recreation facilities for humanitarian personnel such as *Médecins sans Frontières*. The ship's doctor provided medical supplies and assistance at a Mogadishu hospital, while *Preserver*'s technicians formed a 'mobile repair unit' to repair generators, electrical equipment and fax machines for relief organisations.[53]

Improvised arrangements did not always achieve the desired results, however. On 5 March 1993 UNICEF employees in Mombasa approached the captain of *Tobruk*, Commander Ken Taylor, RAN, and asked him to carry a load of 300 tonnes of baby formula and 50 tonnes of general cargo to Mogadishu. Taylor agreed provided that loading the UNICEF cargo did not interfere with his primary task of carrying stores for the Australian and Canadian contingents. The loading of the relief supplies on *Tobruk* was delayed by a lack of communication between

HMAS *Tobruk* unloading containers alongside at Mogadishu. (RAN)

UNICEF's shipping agents and *Tobruk*, and by Mombasa dock workers calling a demarcation dispute about loading a naval ship. This meant that when the ship left harbour on 9 March it was carrying only 30 tons of baby formula and only two tons of general UNICEF cargo.[54]

UNITAF can be considered a successful operation because its deployment enabled humanitarian relief organisations to distribute food and end the Somali famine. During each month of the operation, UNITAF escorted an average of 70 convoys, transporting 9000 tonnes of aid. Only one convoy came under fire. The ability to transport food by road lowered costs for humanitarian relief organisations. Prior to UNITAF, the security situation was so bad that supplies to inland areas had to be carried by air. When UNITAF began escorting aid convoys, the International Committee of the Red Cross decided against taking part on the grounds of keeping political neutrality. The Red Cross, however, found that the militias kept attacking their convoys, and in late January 1993 the organisation decided to accept UNITAF escorts. By early April, the food supply had improved to such an extent that the Red Cross could close 700 of 900 food kitchens which were no longer needed. Families were issued with ration kits so they could begin the journey back to their farms.[55]

On 4 May 1993, UNITAF handed over to a new UN operation, UNOSOM II. Whereas UNITAF had had a narrow and well-defined mandate to assist in the distribution of humanitarian relief, UNOSOM II was given the broader and more vague mandate to re-establish the Somali state. Such a task was beyond the force's capability, especially when it was decided that reconstruction required removing Aidid from power. US Army Rangers and Delta Force were sent to Mogadishu to capture Aidid, but this force lost 18 men on 3 October 1993 while arresting some of Aidid's associates. UNOSOM II ultimately ended in failure, and the last UN troops left the country in March 1995.[56] Somalia has experienced anarchy ever since.

The UNITAF operation illustrates the many ways in which navies can contribute to peace operations. Sea power enabled UNITAF to be deployed to Somalia promptly and continued to support it once ashore. Much of Somalia's infrastructure had been destroyed during the civil war, but naval forces were able to provide vital services ranging from the running of Mogadishu port to the repairing of hospitals. There is much debate about whether organising for peace operations means that armies will be unsuited to taking part in high intensity warfare.[57] But navies have never found it as difficult to transition between roles. Navies have a long history of performing a range of duties from the benign to the coercive across the spectrum of operations. Peace operations in their various forms fit easily into this wide repertoire of naval capabilities.

18. The Challenges for Medium-Sized Navies in the Post-Cold War World

Andrew Dorman

THE WORLD HAS ALTERED dramatically since the end of the Cold War. Benjamin Disraeli, 1st Earl of Beaconsfield and Prime Minister of Great Britain during the 19th century, once commented: 'What we anticipate seldom occurs; what we least expect generally happens'. In many ways this encapsulates the challenges before today's navies. A decade ago few could have imagined the use of passenger aircraft as weapons to destroy the World Trade Center in New York, or envisaged a NATO deployment to Afghanistan. Yet both these events have happened. Predicting the future is always fraught with difficulties. Bill Gates, the famous co-founder of Microsoft, once remarked '640k ought to be enough for anybody.' Times change and perhaps the only guarantee is that whatever is predicted will be wrong. But governments have to attempt to predict what type of capabilities they need from their armed forces in order to prepare them for the future.

Nevertheless, some trends are clear. For many the pace of life seems to have speeded up as the effects of globalisation are felt across the globe.[1] We are all conscious of constant technological change. Communications have been transformed, whether through television, the Internet or intercontinental air travel. We no longer hear about events long after they have happened but often actually as they are happening in real time. The contrast between the reporting of the Falklands War in 1982 and the war in Iraq in 2003 could not be starker. In 1982 every single film report that came back to the UK from the British Task Force had to come via the Ministry of Defence. In Iraq a suitcase-sized satellite dish or even a suitable mobile telephone allowed reporters to transmit moving pictures instantly across the world. Technology has made the world smaller and these processes evolving towards globalisation look set to continue.

Many Western economies are changing away from an emphasis on manufacturing to one based more on the service sector. In their 1995 book *War and Anti-War: Making Sense of Today's Global Chaos* the Futurologists Alvin and Heidi Toffler referred to three levels of economic development: agrarian-based, industrial-based and the emerging knowledge-based societies. They suggest that what we are witnessing is a fundamental adjustment in the nature of economic interaction and, as an offshoot, of warfare.[2] Nation states have again become only one of a number of types of actors on the international stage. Other actors have the ability today to influence events and the decisions of governments at home and abroad to a far greater extent than we have ever witnessed before. The effects of the bombings in Bali in 2002 and London in 2005 were felt across the globe.

In the immediate euphoria that followed the fall of the Berlin Wall and the end of the Cold War commentators talked of the 'end of history' and the victory of the liberal democratic model,[3] while President George Bush Senior spoke of a 'New World Order' in the immediate aftermath of the 1991 Gulf War.[4] Many assumed that with globalisation and the increasing interdependence between states, war and conflict would be reduced. Such assumptions have, however, been increasingly questioned. In 1993 Samuel Huntington wrote of 'The Clash of Civilisations' as the future for the international system.[5] He argued that conflict would no longer be confined to state against state but would increasingly be between civilisations, or within states, along certain so-called faultlines such as that in the Balkans (where Islam, Orthodox Christendom and Western Christendom meet) or between North and Sub-Saharan Africa (Islam and Sub-Saharan Christendom). The experience of the Balkans throughout the 1990s did much to give this view credibility. President Clinton used similar language in a speech during the Kosovo War:

> *Kosovo is a very small place on a very large fault line; on the borderlands of central and eastern Europe; at the meeting place of the Islamic world and the Western and Orthodox branches of Christianity; where people have settled in a complex patchwork of ethnic and religious groups; and where countless wars have been fought over faith, land and power.*[6]

On a related theme, William Pfaff argued that the end of the Cold War had brought an end to the suppression of ancient hatreds typified by the divisions in Yugoslavia.[7]

Others focused their attention on the rise of Islam. In the Spring 1993 edition of *Foreign Affairs* Leon Hadar and Judith Miller argued over the 'Green Peril' and 'The Challenge of Radical Islam',[8] whilst others suggested that Islam would be the new 'ism' for the West to confront following on from Fascism and Communism. The bombings in Bali, Madrid and London imply that what we are now seeing is a clash of ideas between the Liberal Democratic model and those opposed to it. This clash is less geographically oriented than Samuel Huntington inferred in his 'Clash of Civilisations', but one in which we see clashes within existing communities rather than along the civilisation boundaries he identified.

What is clear in this forest of ideas is that the world is changing and the relative certainty that defence planners enjoyed during the Cold War is no longer evident. New threats and new challenges are likely to emerge alongside older problems, often suppressed by the Cold War. The mechanisms and means to deal with these are not necessarily the same as those used during the Cold War. New challenges and old problems need new thinking and ideas if they are to be resolved.

Changing Types of Conflict

Warfare is changing significantly through constant technological development and the increasing complexity of operations. The technological dimension was caught by John Keegan's evocative statement that: 'there is a new date to fix on the calendar: June 3, 1999, when the capitulation of President Miloseviç proved that a war can be won by air power alone'.[9] This statement represented a remarkable about turn for *The Daily Telegraph*'s defence correspondent, and reflected a growing view that technological dominance is vital to winning the wars and conflicts of the future.[10]

More recently, the war in Iraq has seen asymmetries on both sides.[11] The US-led Coalition was able to advance rapidly on Baghdad, which fell with comparatively little opposition. Instead of a war involving a succession of battles that finally resulted in one side being defeated, we saw a series of parallel conflicts leading to the rapid collapse of the Iraqi regime. Yet, since President Bush's proclamation on 1 May 2003 that the war was over, the Coalition has suffered more casualties in stabilisation operations than in warfighting. All this confounds traditional thinking about war as a series of battles eventually producing both a winner and a loser.

Throughout much of the 1990s there was considerable debate among military and defence policy analysts as to whether these changes constituted a so-called 'Revolution in Military Affairs' (RMA).[12] The origins of this debate can, in part, be traced to the experiences of the 1991 Gulf War when there was a remarkable disparity in the conflict.[13] In terms of sheer numbers the Iraqi forces had significantly more tanks and men yet they were comprehensively defeated by an American-led Coalition, which suffered relatively few casualties.[14] The key to the Coalition's success lay in their significant technological superiority over a combat experienced opponent. Speed and precision overcame mass allowing the Coalition to dominate.[15]

A decade or so later, commentators on the 2003 Iraq War frequently referred to what was called a new strategy of 'Shock and Awe',[16] much to the irritation of the Coalition's air commander who claimed to have no knowledge of such a concept.[17] The most recent war in Iraq provides an excellent illustration of how technology is changing warfare. During World War II the Allies used large numbers of heavy bombers in attempting to destroy important strategic targets. In the build up to D-Day, the French railway network was targeted to reduce the Germans' ability to bring reinforcements and munitions to the battlefield. Individual French marshalling yards were targeted for attack by hundreds of aircraft dropping unguided bombs.[18] Many, if not most, went astray with the loss of thousands of civilian lives. In Iraq, the accuracy of the new precision-guided munitions meant that a single aircraft could achieve the same effect against a target that in World War II would have required hundreds of aircraft. Moreover, modern aircraft, such as the B-2A stealth bomber, are now capable of hitting a number of targets successfully on a single mission, rather than requiring a number of aircraft to hit a single target.

The destroyer USS *Milius* fires a Tomahawk cruise missile at a target ashore in Iraq, March 2003. (RAN)

It is not simply pinpoint accuracy that has made the difference; it is accuracy combined with the ability to process information rapidly. On one occasion a US B-1B bomber, while engaged in a pre-planned operation, was diverted to launch a strike against the suspected location of Saddam Hussein within 20 minutes of the information first being received; it then went on to hit 15 additional targets.[19]

The nature of modern conflict has also altered away from the more traditional state versus state war towards conflicts within states, such as those seen in the Balkans. Certain basic elements of warfare have changed. There is now rarely a clear frontline between the opposing sides. The various sides (there may be more than two) will increasingly seek to maximise their advantages and nullify their opponent's strengths through various asymmetric strategies.[20] Armed forces may well find themselves involved in different types of operation on the same battlefield. So-called three-block warfare has come to mean a commander having to oversee war-fighting, support for humanitarian operations and nation-building tasks all occurring within close proximity. Each task has its own challenges and it is not surprising that much of the literature on the subject now refers to the management of 'Complex Emergencies'.[21]

As the nature of warfare has changed, the requirement for future armed forces has had to be adjusted accordingly. The need for transformation in defence management and the construction of armed forces is not new. What changed in the 1990s was the pace. There are pressures on defence organisations that have encouraged the adoption of new management and acquisition strategies. As defence becomes increasingly complex the armed forces have to compete for talented personnel within a more limited pool. Moreover, those skill sets in most demand in the civilian sector also tend to be those most required for complex modern operations: those of logisticians, engineers and communicators. As the armed forces increasingly operate in a complex world they must maintain asymmetric advantages in original technology, which is always costly to acquire. In addition, most defence budgets have been in relative decline since the end of the Cold War and are less able to compete with the inflation associated with the new technologies. Moreover, such technologies are now frequently civilian rather than militarily led and thus the military has little control. For example, the new Merlin helicopter that has recently entered service with the Italian and Royal Navies still uses 286 rather than Pentium microchips, simply because these were the highest performance microchips when the helicopter was designed.

The defence debate is now about 'defence transformation' and encompasses all aspects of defence including acquisition, logistics and management. We are seeing the increasing use of contractors and of new business practices such as 'just-in-time' supply.[22] A single statistic sums this up: just one per cent of those personnel deployed to the Gulf in 1990–91 by the US were contractors. In 2003 this figure had risen to 10 per cent. This had significant consequences in the Second Gulf War in Iraq as the rapidly moving military frontline exposed contractors to attack and capture. Historically, the support for military forces was behind the frontline. Now supporting civilian forces are exposed to risk by the sheer pace of advance. Seizing the opportunity, Iraqi irregular forces targeted supporting elements, with cooks and vehicle repairers among those taken prisoner.

It can be argued that from the end of World War II to the end of the Cold War underlying Western defence policies were subjected to relatively little change in terms of their aims or their measures of success. In contrast defence policy now seems subject to constant change as the environment in which it operates continually alters. One television image from the fighting in Afghanistan caught this dilemma. It showed US Special Forces riding on horseback, using state of the art equipment to direct precision air strikes against Taliban and al-Qaeda positions in support of the Northern Alliance. Sophisticated and highly trained British forces in Sierra Leone found themselves confronted by militias armed with knives who believed in the 'protection' of ju-ju. This mixture of old and new, horses and computers, militia and Special Forces, highlights the challenges for those overseeing defence transformation.

Change is not new, however. A century ago Admiral Sir John Fisher was appointed First Sea Lord – the professional head of the Royal Navy – and in the six years that followed that Navy was transformed. Fisher 'was at the forefront of many innovations: electric lights, large calibre guns, long-range gunnery, the transition from coal to oil, torpedo boat destroyers, and the submarine'.[23] While he is mainly remembered for HMS *Dreadnought*,[24] the first of the all-large calibre gun battleships, he began with educational and personnel reform (such as the Selborne Scheme) to meet the changing requirements of naval warfare. He also attempted to transform the Navy's approach to war with a radical redistribution of resources. This involved a significant reduction in the overall size of the fleet, including the scrapping of approximately 250 ships, and the redistribution of the remainder to bring them into line with Britain's strategic requirements.

It can be too easy, however, to take the idea of change for granted.[25] The deficiencies of the British fleet at Jutland in 1916, where it failed to destroy a smaller German force, have been attributed in part to Fisher's failings, to his fascination with the battlecruiser and his failure to revolutionise the command and control system or maintain the Nelsonian tradition.[26] Writing prior to the World War II, Heinz Guderian wrote in his introduction to *Achtung-Panzer* of the problems of achieving change while vested interests within the German Army sought to oppose it. 'It is a love of comfort, not to say sluggishness, that characterizes those who protest against revolutionary innovations that happen to demand fresh efforts in the way of intellect, physical striving and revolution'.[27] In contrast, Douglas Bader was highly critical of the various 'Fighter Attacks' developed by the United Kingdom's Fighter Command between the world wars because they ignored many of the lessons of World War I.[28]

Applying new technology effectively and adopting a culture of change and innovation in an inherently conservative organisation will always be problematic. A balance is required between the human operator and the technology being used. In his book *Rules of the Game*, naval historian Andrew Gordon highlighted the human dynamic. He showed how the Royal Navy had suffered from an intellectual malaise during the long period between Trafalgar in 1805 and the confrontation with the German Navy during World War I.[29] The period saw massive technological change from wooden-walled sailing ships engaging in close quarter combat to battleships powered by turbines and capable of engaging each other at ranges of over ten miles. The underlying thinking, however, had not changed. Yet Professor Air Vice Marshal Tony Mason has rightly warned against the military's tendency to favour all things technological:

> *Concentration on high technology should not lead to the disparagement of the simpler or even obsolescent weapons. The ultimate measure of a weapon's effectiveness is its value as a political instrument, which may not equate to its operational impact.*[30]

The decision in 1981 to withdraw HMS *Endurance* from its duties in the South Atlantic has been used to illustrate the force of this argument. Although she possessed relatively little military capability the value of *Endurance* lay in her political significance – she was a symbol of Britain's commitment to the defence of the Falklands.

What these examples highlight are the dilemmas surrounding defence transformation. Managing transformation or change is challenging and risky. Forecasting future global trends in an increasingly unpredictable world is an inherently uncertain process. It can prove an expensive one if the investment in new equipment proves unnecessary as a result of change in the strategic climate or, on the other hand if the failure to adapt results in defeat. In the short term it is always easier to maintain the status quo. That is why the flexible use of equipment becomes a priority.

The Role of Maritime Forces

So what is the role for the armed forces? The 19th century military strategist Carl von Clausewitz argued rightly that armed forces are created to achieve a political effect. What has changed is the movement of the use of the military away from last resort, to the position of a strand within overall government policy. This corresponds with the ideas of Julian Corbett who viewed navies as a permanent instrument of statecraft.[31] Reflecting on the experience of 11 September one year on Geoff Hoon, the then British Defence Secretary, concluded:

> *We need to use a range of techniques to win – political, diplomatic, humanitarian, economic, financial, intelligence and law enforcement, as well as military. That is what we have been doing over the last year. That is what we will continue to do. We have to deal with the symptoms and causes in order to help establish the conditions that can deliver long term solutions.*[32]

This means that, as Corbett argued, national navies will be linked directly to their armies and air counterparts with the mission of influencing events on land.[33] Recent operations have confirmed this. The Australian commander of the International Force East Timor (INTERFET), Major-General Peter Cosgrove, remarked that the naval contribution was not an optional extra, 'it was an important indicator of international resolve and most reassuring to all of us who relied on sea lifelines.'[34]

Reflections on Recent Operations

The Value of Partners

East Timor, Kosovo, Afghanistan and Iraq have shown both the benefits and limitations of operating within a coalition. No country wants to be isolated when it undertakes military operations and the benefits of membership of an alliance or coalition are clear. The weight of a number of states coming together and calling for or undertaking action is far more significant than that of an individual state, even a superpower. Moreover, there are times when the world community cannot achieve consensus, as was the case with both Kosovo and Iraq, when a coalition or regional organisation can provide the unifying element. Such agreements are particularly important both in the build up to and following a conflict when the diplomatic and military dimensions are brought together. Working in a coalition brings political, diplomatic and military advantages, including the aggregation of capabilities, flexible war-fighting options and the sharing of intelligence and risk.[35]

From a military perspective the value of additional partners can involve the provision of additional or supplementary capabilities. The Canadian Navy's provision of a tanker for INTERFET, for example, helped to sustain the operation as well as signal Canada's commitment. It must be remembered that modern technology is increasingly expensive, and as result most states inevitably become dependent on one another for the provision of capability. Post-9/11 even America accepted the deployment of NATO AWACS (Airborne Warning and Control System) aircraft to the US to help supplement its capability as war began in Afghanistan.

Impact of Speed and Tempo

Recent operations have shown the advantage that tempo and speed of response provide. The rapid East Timor deployment in 1999 ensured that the warring factions acquiesced to the INTERFET mission. The May 2000 deployment of British forces to Sierra Leone was spectacularly successful because it headed off a crisis caused by the Revolutionary United Front moving against UN forces. Although the deployment of light forces, such as 1st Battalion the Parachute Regiment (1 PARA) to Sierra Leone, can produce a serious effect, there will still be a need for heavier forces which have traditionally taken much longer to deploy. 1 PARA's operation was predicated on the accompanying deployment of the UK's Amphibious Ready Group. In August 1990, in response to the Iraqi invasion of Kuwait, the US deployed the 82nd Airborne Division to Saudi Arabia to act as a deterrent against further Iraqi action. This force would have struggled to cope against an armoured attack by Iraqi forces, and it was only the arrival of the heavy 1st US Cavalry Division in September, with its main battle tanks, which meant that Saudi security could be assured.

Value of Escalation Dominance

In the INTERFET and Sierra Leone deployments the availability of maritime forces just offshore provided a significant deterrent capability. The ground commanders knew that they could run certain risks based on the ability to employ further forces rapidly if need be. The deployment of Royal Navy frigates up the Sierra Leone River highlighted the ability of British forces to provide direct artillery support to the land component. This was backed by a display of airpower in the form of RAF Harriers deployed aboard the carrier HMS *Illustrious*.

The Application of Mission Command

The ability to react rapidly involves the possession of three key elements of capability: sensors capable of identifying targets; a communications network able to transfer this information to commanders who decide on a response; and the strike assets capable of hitting the target accurately within the requisite timeframe.[36] This explains the emphasis on the control of the OODA (Observe-Orientate-Decide-Act) loop or, in current British jargon, Network-Enabled Capability (NEC).[37]

The emphasis on NEC has a further element. Rapid reaction requires rapid decision-making processes, both military and civilian. It is no use having the ability to observe, operate and act if you cannot decide within the requisite time scale.[38] This will place increasing pressure on both political and military leaders. The former will have to make decisions based on the best information available but without necessarily engaging in full and detailed consultation. The latter may well find that authority for action is increasingly delegated based on rules of engagement approved in advance.

This reinforces the emphasis upon Mission Command, with the military commander knowing and understanding the political intent. In Sierra Leone Brigadier Richards was aware that the UN's situation there was collapsing; and while the formal position of the British government had still to be fully articulated, he understood the political goals and could act accordingly. In Kosovo Lieutenant General Mike Jackson was well aware of the dangers of confrontation with the Russians over Pristina Airpor. Rather than provoking the Serbs, he chose a more indirect method of negotiating with the Russians while developing their dependence on him for supplies.

This is not a unique or new situation. In his study of the Battle of Jutland, Andrew Gordon concluded that while it is not an absolute principle of war that the commander ensures that all subordinates understand his or her orders, he or she should appreciate that even distant units may become engaged. They need to know, therefore, what the commander is trying to achieve and where they fit into the overall plan. In the kind of new parallel warfare, in which the simultaneous levels of conflict are becoming increasingly blurred, it is absolutely necessary that all appreciate the overall goals, not least as the civilian leadership becomes increasingly reliant upon military commanders to make appropriate political as well as military decisions.[39]

There are obvious risks involved in such a shift towards increased mission command and faster political decision making. There will be errors which could have severe political consequences, such as the incorrect targeting of the Chinese Embassy in Belgrade during the Kosovo conflict. Yet as Andrew Gordon concludes, experience has shown that nothing can be more fatal than waiting for orders.[40]

War and Diplomacy

The Kosovo and Iraq operations highlight the challenge of being able to develop diplomacy alongside the threat of force. The military dimension can serve as a useful diplomatic lever, but it will also bring its own limitations in terms of subsequent action. The British Ministry of Defence acknowledged this dilemma in *Operations in Iraq: First Reflections*:

> *Supporting diplomacy, while preparing to use force if that diplomacy failed, presented difficult choices and required a fine balance to be struck. We did not wish to prejudice the UN process by making over military preparations too soon. On the other hand, delaying planning until diplomacy had been exhausted would have given Saddam's regime months to prepare for war, or even allowed him to*

> *conduct pre-emptive strikes against Iraq's neighbours and/or coalition forces in the region. Moreover, we hoped that measured military preparations would strengthen diplomatic efforts by demonstrating the resolve of the international community…*[41]

This could be seen as a modern day version of the historian A.J.P. Taylor's 'War by Timetable'.[42] In the immediate crisis which preceded the outbreak of the World War I, he argued, the German war plan finally tipped the continent into war. The German Schlieffen plan was based on the fear of a two front war against France and Russia.[43] The plan sought to hold the Russians in the East temporarily whilst achieving a rapid victory in the West. Thereafter, all German forces were to be devoted to the defeat of Russia. Timing was therefore critical, including the deployment of troops into Belgium as part of the mobilisation process. It followed that once begun there was no turning back. As the crisis of 1914 developed, and mobilisation began across Europe, war was being seen as the inevitable result.

In Kosovo in 1999 and in Iraq in 2003 the Coalition was constrained by the military limitations of time and weather. In 1999 one of the key issues driving NATO to commit ground forces was the number of refugees and the predictable impact of the onset of winter. Given the time required to build up ground capability, conduct a successful land war and return the refugees to their homes before winter, NATO's leaders had to commit to this escalation by early June 1999.

In Iraq the size of the land force deployment meant that a factor encouraging the movement to war in the spring of 2003 was the requirement either to use the forces or to withdraw them. The United Kingdom simply did not have the option of replacing the forces completely as a means of indefinitely extending the diplomatic option. Both 3 Commando and 16 Air Assault Brigades were unique assets and either had to be used or withdrawn. This is one of the reasons for the restructuring of the British Army to provide an additional light brigade, as well as for the need to develop a medium-weight force of three brigades capable of faster deployment than their heavier armoured equivalents.[44]

Forward Presence Versus Surge Capability

A balance will have to be struck between peacetime presence and the ability to deliver a surge capability – particularly one involving rapid response. There are various arguments in favour of maintaining a forward presence. These include the role of military assets in support of defence diplomacy through training missions, as well as visits and the diplomatic signals that are sent by having units in a particular region. A forward presence, moreover, allows a more rapid response to crises.

The presence of the British Amphibious Ready Group in the Mediterranean and of the *Illustrious* carrier group in the Eastern Atlantic was vital to the rapid deployment to Sierra Leone in 2000. The British maritime deployment to Iraq in 2003 was built on the back of a previously planned deployment of a carrier group and Amphibious Ready Group to the Indian Ocean and Pacific region scheduled for the first half of 2003. The surface ships deployed in the Northern Atlantic could support humanitarian operations in the wake of the various hurricanes which struck the Caribbean during the summer of 2004 while those in the Indian Ocean supported Sri Lanka in the immediate aftermath of the Tsunami in December 2004.

At a more tactical level, the US can launch long range bombers, such as the B2 stealth bomber, from the continental US and strike anywhere in the world with the assistance of air-to-air refuelling. But such strikes still take time to reach their targets. Having units permanently in theatre can reduce this response time. In Afghanistan, the UK and US could use their aircraft carriers as mobile bases. This gave them far greater freedom to conduct operations at times of their choosing. Similarly, the deployment of the Amphibious Ready Group to Sierra Leone in 2000 gave the British government the choice of increasing or decreasing the force deployment ashore as it saw fit.

The maintenance of a permanent presence ties up considerable forces. Various estimates have been made of the number of ships required to maintain one deployed. For a permanent commitment it is a minimum of three, because of the need for a ship on station, a ship preparing to deploy and a ship returning from deployment or in refit. The maintenance of a number of such standing commitments will inevitably tie up assets.

The danger of overcommitting to forward presence lies in the reduced capability available for unforeseen operations. This was the lesson of the 1956 Suez Crisis. British forces were so committed elsewhere that there was little capability to respond immediately to President Nasser's nationalisation of the Suez Canal. As a result, the UK and France had to embark on a three month mobilisation of resources and engineer a reason for military action with perhaps predictably dire results.[45] Moreover, the maintenance of a forward presence can distort the overall force balance. The US aims to maintain a carrier battle group in each of the Mediterranean, the Western Pacific and the Persian Gulf. But it cannot do so even with twelve carriers in service. Yet when it comes to surge capacity its plans require only three or four carriers for a major operation. It is arguable, therefore, that it has too few carriers for its peacetime presence tasking and too many for its surge requirements.

Royal Marines from 42 Commando hit Mayoko Beach, Sierra Leone, from Sea King helicopters of 846 Naval Air Squadron in a demonstration of amphibious power. (ROYAL NAVY)

Maintaining significant forward presence also means maintaining an appropriate balance between deployments for service personnel. The retention of personnel has been one of the biggest challenges over recent years as they and their families have sought, like many others, to maintain an appropriate work-life balance. Forward deployments, when combined with a surge capability, require significant personnel resources. In defence terms, the more money spent on personnel the less is available for equipment. The challenge for defence is therefore to manage both forward presence and surge capacities. There are various innovative ideas being trialled to maximise both, including for example partial and full crew rotations of warships, moving right away from the traditional concept of a permanent ship's company.[46]

War is No Longer a Two Player Game

Clausewitz identified war as an essentially political act.[47] The focus, he argued, should be on defeating an enemy's army. All four of our recent case studies, however, show that war is not simply about defeating a particular opponent. Kosovo and Iraq, especially, have shown the importance of third party opinion. The requirements of the modern media involve filling 24 hour news channels around the world. If the 1991 Gulf War was the CNN War, then the 2003 conflict was the *Al Jazeera* War.

The impact of the war in Iraq was perhaps most significant on the streets of the Middle East. This was compounded by the nature of the media coverage.[48] A significant number of journalists were embedded in Coalition combat units and could give an instant tactical view of what was happening in their immediate vicinity. At the central news desks, a few defence analysts of uneven quality attempted to provide some strategic oversight. What was almost completely lost was the operational picture from the Coalition's perspective. The news from the Iraqi side was limited both by the regime and the level of media assets deployed with Iraqi forces. The *Al Jazeera* office in Baghdad, for example, had only three people to deal with the Iraqi view and was therefore inevitably limited in what it could cover and almost wholly dependent on the regime for assistance.

The fear for NATO in Kosovo was the turmoil that might be caused in the states bordering Serbia-Kosovo. This might have led them to prevent further NATO action by closing their airspace. In a meeting between the British Prime Minister, Tony Blair, and the leader of the Liberal Democrats, Paddy Ashdown, the latter drew on his previous experience as a Royal Marine to emphasise the idea of close, rear and deep battle, identifying those states bordering Kosovo as part of the rear battle.[49] The British Government diverted resources not only to help overcome the weakness in NATO's media operations but also to support the neighbouring states.

Overseas Bases and Access

No military operation can be undertaken without the necessary accompanying infrastructure. Aircraft need bases to re-fuel, load ordnance, unload cargo and change crews. Land forces need somewhere to assemble before the conduct of operations and a logistical tail running from home to the frontline. Without bases in Kuwait no operation against Iraq would have been possible in 2003. To stay at sea, ships need to be replenished with fuel and provisions. Even if warships do not enter another state's territorial waters in the course of operations, the accompanying supply ships often will. In other words, it is exceedingly rare for an operation of any size to take place without some form of host nation support. Such support also lies in the political commitment of that state to the conflict, even in what may be limited form. Although a number of Iraq's neighbours may not have been formally part of the Coalition their provision of facilities indicated their tacit support.

Land locked territories such as Afghanistan or Kosovo are impossible to reach without traversing another state's airspace. In the case of Afghanistan, bordering states allowed Coalition aircraft and cruise missiles to pass over their territory. The creation and support of the International Security Assistance Force (ISAF) commitment has been undertaken entirely by air, and without these transit rights the operation would not have been possible. Even where a country is not land locked, as in the case of Sierra Leone or Iraq, there is frequently a requirement to transit over another state's territory to get there quickly as well as the requirement for a fallback should aircraft need to divert. During the initial deployment to Sierra Leone permission to transit other states' territory was being obtained whilst the helicopters and aircraft were in the air because of the speed of the military response. Turkey's decision not to allow American and British forces to pass through its territory caused a major re-adjustment of the war plan against Iraq in 2003.

Recent operations have shown the need to maintain forward bases with or without host nation support. One's own overseas bases offer the benefits without many of the conditions that may be attached elsewhere. For operations over Afghanistan, the US base on the British territory of Diego Garcia was hugely significant. British bases in Cyprus likewise continue to provide an invaluable resource for the UK. They can act as forward operating bases for operations in and around the Mediterranean or provide a useful staging point on the way to the Middle East and beyond.

But the fixed nature of all such bases has encouraged the US in particular to look at more mobile options. The case of Iraq, for example, has shown that while the deployment of support ashore can greatly ease the logistic burden, that support may also become a tempting target for the enemy. Thus various forms of offshore basing may increasingly prove attractive as part of risk reduction. Moreover, maintaining such facilities in a mobile base increases a commander's operational flexibility, in terms of both location and security.

The Way Ahead for Maritime Forces

Overall what we are seeing is the increasing role of maritime forces in influencing events ashore. This is not a new concept and relates directly to the writings of Julian Corbett and others. While traditional maritime roles will still remain at sea there is likely to be an ever greater trend towards supporting forces ashore. This may not be what navies want, but there is likely to be little choice in the matter. Navies which choose only to maintain traditional naval roles such as sea control and sea denial, while ignoring the landward dimension, will increasingly be less relevant to the needs of their countries – the first target when it comes to defence cuts and the last choice when it comes to new investment.

This will have a number of implications. First, the maritime dimension will increasingly be seen as a supporting rather than supported command and as such, when defence budgets are tight which is so often the case, will be likely to be perceived as more eligible for salami-slicing in various forms. Offsetting this vulnerability is the ability of maritime forces to project power ashore from secure bases, offering potential savings to a deployed land force and a much smaller deployment footprint.

Second, there will need to be a change in attitudes within the respective navies. For those which already have some form of marine corps this process should be easier because of the expertise they possess. For others it implies a skill set they will need to develop and will involve close cooperation with their sister services. There is no reason why a joint commander cannot be from the maritime sphere, but he or she will need a thorough understanding of the other environments. It is interesting to note that the Royal Navy has deployed a task group to sea under the command of a Royal Marine two star officer as a way of testing this concept. The Royal Marines also provide one of the UK's three divisional headquarters.

Third, the new strategic environment will almost certainly change the balance of forces within navies and how resources are allocated. In the Royal Navy, for example, one in four personnel is now a Royal Marine – contrasting with fewer than one in nine 25 years ago. This is inevitably impacting upon the service and poses a challenge for the Navy's senior leadership. In terms of capability, there is likely to be significant change in investment emphasis. The shift from blue water to green and brown water operations requires different skills and equipment. Some traditional areas may well have to be sacrificed, such as the ability of submarines to attack surface combatants, in favour of the ability to launch cruise missile strikes ashore. Change has always been part of naval warfare, however, whether it is the abandonment of sail or the introduction of carrier aviation.

19. Triumph or Tragedy? Reflections on the History of Joint Power Projection Ashore

John Reeve

> *A military, naval, littoral war, when wisely prepared and discreetly conducted, is a terrible sort of war. Happy for that people who are sovereigns enough of the sea to put it into execution! For it comes like thunder and lightning to some unprepared part of the world.*
>
> Thomas More Molyneux, *Conjunct Expeditions*, 1759

SINCE ANCIENT TIMES, THE direction of sea power against the land has been a most demanding business. In particular, amphibious warfare (joint power projection in modern parlance) – involving the combined operations of sea and land forces – has been fraught with peril. Perhaps more than any other dimension of warfare, this category of strategy and operations has had a propensity for brilliance and heartbreak in fairly even measure. From Julius Caesar's landing in Britain in 54 BCE to the heights above Quebec in 1759 and the port of Inchon in 1950, joint power projection from the sea has profoundly affected the course of history. But the annals of amphibious success have a parallel story. It runs from the epic disaster of the Athenian expedition to Sicily in 415 – 13 BCE to the defeat of the Spanish Armada in 1588 and the shores of Gallipoli in 1915. That such operations of war—which bring sailors and soldiers and more recently airmen together in a variety of different ways— can produce consummate strategic success was demonstrated as recently as 1982 in the Falklands War. When they have failed, however, the cost has sometimes been terrible. This historical duality, a yawning gulf between possibilities and risks, lies at the heart of the challenge of joint power projection. Why has it tempted many commanders, rewarded fewer, and indeed punished some of the greatest? Why has it caused elite forces to fail? What are its prospects in the complicated strategic world of the early 21st century? Such questions are highly topical. Many maritime powers, since the end of the Cold War, have increasingly committed themselves to joint power projection capabilities, doctrines, and operations.

The History and Theory of Sea Power Ashore

This chapter offers a reflective overview of some of the salient issues raised in the book. As the preceding chapters make clear, there is a wide variety of ways in which sea power can have effect ashore. They range from diplomacy and deterrence, through blockade, to joint operations, and to the actual presence of sailors on land – on occasion as fighting troops. There is thus a substantial grey area which takes in certain offshore capabilities of maritime power, amphibious warfare as traditionally understood (which Geoffrey Till has termed 'the main subset of maritime power projection'[1]), and more current concepts of expeditionary warfare, as well as the many roles which naval people can perform ashore. Today's evolving doctrines of seamless joint operations in the littoral environment underline the interconnectedness of these themes. As much as sailors love the sea, and as powerful as the idea of fleet battle remains, navies are military forces with political purposes, and their utility lies in their ability to influence human societies – on land.[2] The many dimensions of 'sea power ashore' are but the more direct manifestation of this ultimate strategic aim of all naval and maritime operations.

Sea power ashore has both a theoretical and an operational history. The latter has tended to predominate, but it has not been a mainstream aspect of naval lore. For this we can, in part, blame the posthumous cult of Nelson and the uses to which navies have put it politically in the past. Nelson, however, as Colin White has pointed out, received two grave wounds while fighting on land: to his eye and arm at Corsica and Tenerife respectively. Until his mid-30s, that is for most of his career, he was best known as a specialist in joint operations.[3] (Amongst other examples, one might add that Beatty, famous in fleet actions in the 20th century, was twice wounded in land operations at Tientsin during the Boxer Rebellion.) Alfred Thayer Mahan's ideas can also be brought to book as obscuring the theory of maritime power projection. While Mahan has been crudely applied, he was less an advocate of sea power than of a certain kind of sea power revolving around battlefleet capability. Sarah Paine has observed how his view of joint operations was limited.[4] But his influence built upon the Nelsonian legend of naval battle. Conversely, the failure until the late 20th century to appreciate fully the contribution of Julian Corbett – the most distinguished historian and theorist of maritime power projection – had a complementary effect.[5] There have been other barriers to the appreciation of power projection historically. Ironically, the rise of air power – probably the most influential force for the fusion of military operations in the land and sea environments – was accompanied by a body of theory antithetical to jointness. We might also cite the continental commitments made by the great maritime powers during both world wars, as well as US inter-service rivalry during the Cold War.[6] One outcome is that amphibious warfare, in particular, still lacks a substantial and thematic historical literature.[7] But with the passing of the great 20th century conflicts into history, a new strategic environment has served to highlight the littoral regions in all their political and military operational complexity.

Sea Power Ashore: A Panoply of Purposes

History shows how sea power can serve a whole spectrum of purposes directed towards influencing events ashore.

Statesmen have long been aware of the diplomatic and deterrent value of naval power. British intervention in Syria in 1840, for example, with amphibious operations as well as naval blockade and bombardment, brought about the settlement sought by London of the Ottoman civil war. Palmerston concluded that 'Every country that has towns within cannon shot of deep water will remember…whenever such country has any differences with us.'[8] Andrew Lambert has increasingly made clear how such deterrent power lay at the basis of the 19th century Pax Britannica.[9] Blockade has been a staple of naval influence ashore since at least the early 17th century and the era of Anglo-Dutch warfare against Spain. Bruce Elleman has shown how the Nationalist blockade of Communist China was an effective form of political power projection.[10] The Gulf blockade against Iraq in the 1990s was in this tradition.

Amphibious operations in themselves are a remarkably versatile form of power projection ashore. They come under the four principal headings of assault, raid, withdrawal, and feints and demonstrations.[11] As such, they have been one of history's great force multipliers. Inchon stands as a classic example of the strategic value of an assault. MacArthur characterised the operation as the placing of a five dollar bet for the chance to win $50,000.[12] Raids also have a long history of operational value. Drake's raid on Cadiz in 1587 impeded the preparations of the Spanish Armada.[13] Four centuries later, the SAS (Special Air Service) raid on Pebble Island in the Falklands destroyed almost a dozen Argentine aircraft at a critical time.[14]

Amphibious withdrawals reflect the inherent value of sea power to forces deployed ashore. A fleet, in such situations, provides an army with interior lines of communication over water – lines both of supply and of retreat.[15] There is an established Royal Naval tradition of rescuing British armies: Corunna in 1809 and Dunkirk in 1940 are cases in point. In the latter case, at least, the withdrawal changed history. This was a tradition which Admiral Sir Andrew Cunningham was loathe to jettison, despite terrible naval losses, during the evacuation of Crete in 1941.

The capacity for feints and demonstrations is intrinsically amphibious. Corbett wrote of 'the power of confusing' possessed by amphibious operations, observing that while such enterprises are difficult to conceal their objectives are easily disguised.[16] Napoleon, for example, was well aware of the strategic implications: a whole enemy army could be tied down, in this case his own.[17] This has been an enduring capability. The amphibious threat to Norway tied down 18 German divisions in the last stage of World War II, which were therefore useless on either the eastern or western fronts.[18] The Argentinians, in 1982, were distracted by the possibility of a more direct British approach to Port Stanley.[19] The effect of an offshore Marine deployment in tying down five Iraqi divisions in Kuwait in 1991 is well known.

There is a history of sailors themselves serving effectively ashore. British seamen hauled naval guns up the cliffs at Quebec to provide Wolfe's army with artillery. An army, especially a small army configured for power projection, can find the support of naval personnel invaluable.[20] Sailors are traditionally disciplined and have a work ethic indispensable to life at sea. They are also expert in the use of guns and explosives, trained in engineering and logistics, and used to heavy labour.[21] These are qualities readily employable in military operations on land. For generations they have also given soldiers capable medical care on the battlefield or close offshore. With an ancient history of hand-to-hand combat at sea, sailors have frequently not been found wanting in action on land, even at the point of the bayonet. Naval brigades served with distinction in the British colonial conflicts of the late 19th and early 20th centuries.[22] The Australian Navy has always been part of this tradition – from the attack on the German wireless station at Rabaul in 1914 to the clearance of unexploded ordnance ashore in southern Iraq in 2003.[23] Landward service by sailors has also taken the form of riverine warfare, from the gunboat penetration of continental Africa and China during the nineteenth century to the Mekong Delta of Vietnam during the Indo-China war. The US Navy today is consciously drawing upon the latter experience.[24] The requirement of navies for bases has not

A member of Australian Clearance Diving Team 3 keeps watch following the discovery of a weapons cache in 2003. During their Iraq deployment the team travelled more than 200 kilometres in two tactical lodgements and located and destroyed more than 4000 items of ordnance. (RAN)

infrequently put navies and amphibious forces into the business of territorial acquisition. Some such acquisitions, such as Gibraltar (taken in 1704), still retain significant strategic value.

Power projection ashore is not exclusively the tool of great maritime states. Smaller sea powers, from the days of Drake to the 21st century 'coalitions of the willing', have also used power projection to serve a variety of strategic purposes ranging from national survival to the maintenance of alliances. A vast continental power such as the Soviet Union during World War II used amphibious warfare to facilitate land campaigns at the tactical level. Continental powers such as Prussia and Russia have, moreover, benefitted historically from the diversionary power projection capabilities of their maritime allies. The ability of sea power to create effects ashore is – in principle – generally accessible, and can be on the smallest or the grandest scale. This is a result of the deeply practical and elemental bases on which it rests. Maritime power projection ashore combines transport and firepower capabilities within the world's most extensive strategic environment. These elements together tend to have an effect greater than the sum of their parts. The effect has been enhanced by the coming of air power – as Chris Clark has indicated – with its progressive ability to observe, strike, and deploy.[25] Little wonder that so many commanders and political leaders have historically coveted the enabling effects of maritime power projection ashore.

Amphibious Operations: An Audit of Failure and Success

Can history teach us anything about the reasons why joint maritime power projection has succeeded or failed? What do the salient episodes of the past suggest is required? Colin Gray has posited credibly that there are eight essential principles of amphibious operations: inter-service understanding, unity of command structure, command of the sea and air, careful attention to hydrographical and weather conditions, specialist expertise, tactical surprise, and preferably the ability to attack at a weak point.[26] To these we might add a further five, arguably implied by the earlier eight: the analysis of hard-won experience (as notably by the US Marine Corps in the wake of Gallipoli), actionable military intelligence, audacious and judicious command, the ability to improvise, and of course adequate resources.

It is tempting, however, amidst all the complexity of such operations, to isolate three factors. These factors, in different combination, have been fundamental to the outcomes of a host of attempts at joint power projection ashore over the last four centuries – the age of modern sea power.

Command of the sea in the traditional sense—that is of the relevant maritime surface area – has always been a pre-requisite.[27] It was so for the capture of Quebec in 1759, during Wellington's campaign in the Iberian Peninsula (recently called 'the quintessential combined operation'[28]), during the Japanese conquest of Southeast Asia and the Allied Pacific counter-offensives, and at the time of the invasion of Normandy, of the Inchon operation in Korea, and of the British victory in the

Falklands. Failure to achieve such command has always spelt disappointment, as it did for Napoleon's plan to invade England in 1805. Even disputed command of the sea has been sufficient, history suggests, to scuttle the amphibious efforts of great powers. The English fleet could do no more than dispute command against the Armada. Britain lost its American colonies following a tactically indecisive fleet action at the Chesapeake in 1781, leading to the surrender of Cornwallis's army at Yorktown. It has long been established, and indeed was well known at the time, that Hitler's plan for Operation SEA LION in 1940 stood no real chance of success given the ability of the British Royal Navy to contest the issue.[29]

The second factor is successful inter-service co-operation, or the converse failure of it. There was no strategic benefit for Spain in having, in 1588, the services of the greatest land commander of the age, the Duke of Parma, and the most formidable land force, the Army of Flanders – nor in assembling the greatest sea armada ever seen in the Christian world, when poor communication between army and fleet was a material cause of failure.[30] By contrast, the excellent co-operation between Wolfe and the British fleet commander, Vice Admiral Charles Saunders, at Quebec is legendary.[31] There were similar episodes in the 20th century, notably the Dunkirk evacuation in 1940 and the Normandy campaign of 1944. In both cases, the role of British Admiral Sir Bertram Ramsey as operational or naval commander was crucial.

The third factor is the choice of a pivotal operational objective – which, if seized, can yield overwhelming strategic advantage. The Gallipoli Peninsula was supposed to be such a pivotal objective in 1915, but with well-known tragic results. Successful such points of attack have been the Heights of Abraham in 1759, the beaches of Normandy in 1944, the port at Inchon, and San Carlos Bay in the Falklands. The mobility of maritime power can be the key (as in all of the above cases) to an indirect approach and the achievement of surprise. Corbett observed that stealthiness is one of a fleet's most powerful capabilities: 'Secrecy is the soul of combined expeditions'.[32]

Here we may pose a question predicated on the dichotomy of possibility and risk with which we began. Have the dramatic amphibious failures been concomitants of the strategic prizes in prospect, and just the reverse of the coin? Is it necessary to risk such catastrophe to achieve spectacular success? Or is it the case that failure in these operations is caused by acting in the wrong circumstances, by incoherent planning and execution, or by selecting the wrong target? History suggests the latter: that the failures have had the odds stacked against them by historical context, in which the errors of high commanders have played critical roles. The successes, by contrast, have been the work of commanders who have been, in the best sense, 'lucky' – who have painstakingly made their own luck while accepting carefully calculated but inescapable risks. And their relief has been palpable. Wolfe expired praising God that he could die in peace. Eisenhower drafted a statement of personal responsibility in case D-Day became a disaster.[33]

MacArthur threw up when he heard that the Marines had taken Wolmi-Do Island, the gateway to Inchon. Moral courage may be the ultimate lesson of the history of joint power projection from the sea.

Amphibious Operations: People and Skills

Joint maritime power projection has traditionally been manoeuvre warfare, combining the mobility of navies with the striking power of armies. It is thus inherently complex in relying upon professional military expertise in two – and more recently three – environments. Great specialist amphibious commanders, such as Ramsey and MacArthur, have therefore been rare. A survey of commanders who have excelled in the amphibious area – from Wolfe and Saunders at Quebec, through the Second World War and Korea, to the British commanders in the Falklands – suggests that they have been sailors or soldiers in fairly even distribution. The point is that they have generally thought jointly in terms of having respect for the operational complexity of both the maritime and land environments. The question of how amphibious command arrangements are best structured is therefore perhaps redundant, and dependent upon particular circumstances. The point is mutual respect and co-operation on the basis of discrete expertise. Attempts to create joint service cultures, supposedly in the interests of joint operations, must therefore be dubious. Sailors and soldiers, for example, belong to ancient and individual traditions in which skills and values are intrinsically linked, and such traditions must be co-opted rather than fused. This is the means, history suggests, to ensure the expert professional judgement upon which joint operations depend. What can loosely be termed 'soldiers' plans' for campaigns in the maritime environment – such as those for Japan's Pacific perimeter defence after the death of Yamamoto in 1943 – have, therefore, generally failed.[34]

Even more fatal have been amphibious campaigns driven by politicians and warlords, especially those trained as soldiers in continental warfare. Examples come readily to mind: Philip II of Spain in 1588, Napoleon in 1805, Churchill in 1915, and Hitler in 1940. Such situations can be compounded by inadequate professional advice, as Robin Prior has suggested in relation to the Dardanelles operation of 1915.[35] This is not to negate the importance of political will in executing amphibious operations. In the modern democratic age, from Normandy to the Falklands, this has been pivotal. Nor is it to suggest that political leaders can never have the kind of strategic vision required in joint warfare. The role of William Pitt the Elder in pursuing a maritime strategy during the Seven Years' War is a case in point.[36]

Finally, high command in amphibious operations can be complicated by the diplomacy of coalition warfare. But the effort is often worthwhile: coalitions can be political as well as military force multipliers. The Allied campaign in Normandy could thus more legitimately be characterised as a democratic crusade. But managing coalitions indeed requires significant diplomatic skills – such as those of Eisenhower – employed in the fullest measure.

Landing Ships Tank land invasion supplies on Omaha Beach shortly after the 6 June 1944 'D-Day' assault. (NHC 80-G-46817)

The requirement for specialist expertise in amphibious warfare exists at every level. On the naval side, for example, there is no substitute for the navigational and hydrographical skills which can mean safe deployment of troops ashore. The taking of Quebec depended upon the naval personnel (including the young James Cook) who charted the treacherous waters of the St Lawrence. The selection of the British landing area in the Falklands in 1982 is a case study in complex amphibious planning, involving not only hydrography and geography but air, surface and sub-surface warfare capabilities, the requirements of merchant shipping and logistics, the needs of operations on land, and the wider strategic picture.[37] One could employ here the description by US Marine General Oliver Smith of the planning for Inchon: 'It looked simple because it was done by experts.'[38]

Just as fundamental to amphibious operations is the ability to learn. The Seven Years' War and the American War of Independence saw something of a quantum leap in British amphibious capability on the part of both the Army and Navy. A whole variety of equipment and techniques was developed, involving everything from signals to landing craft, and encompassing every stage of operations from embarking troops to supporting them ashore.[39] In the early 20th century, the amphibious forces of Japan, the US, and Britain responded to the challenge of more powerful coastal defences with new equipment and doctrine, such that the Allies, by the closing years of the Second World War, had a new and essential operational capability.[40] The price of inability to learn was evident on Omaha Beach on D-Day, where the lessons of naval gunfire support from the Pacific theatre had not been applied, resulting in death and near failure.[41]

Joint Power Projection Ashore – Today and Tomorrow

Amphibious warfare has survived the great wars and political vicissitudes of the last hundred years. Following the disaster of Gallipoli, it became a war-winning factor in the 1940s in both the European and Pacific theatres. Joint power projection, in the broader sense of combined operations, had what was probably its finest hour in North Africa in 1940–41, when it ensured that the war would continue despite Britain standing alone. After General Omar Bradley's prediction in 1949 that, with atomic weaponry, amphibious landings were obsolete, such a landing the following year at Inchon retrieved the UN position in Korea. Under the shadow of the 1982 UK Defence Review, with its focus on NATO area operations, the Falklands campaign highlighted the abiding importance of expeditionary capabilities. Under the umbrella of the US Maritime Strategy of the 1980s, the US Marine Corps retrieved its amphibious character – set aside during the Vietnam War – with planned attacks on the Soviet flanks in the event of war in Europe and Asia.[42] With the Cold War over, and the Soviet Navy no longer a threat, two US Navy White Papers in the early 1990s set the scene for a new era of power projection from the sea.[43]

Today, 15 years later, we are witnessing a flourishing of joint power projection activities. Within the traditional maritime states such as the US, Australia, and Britain, military services – aided by concepts such as network-centric warfare – are becoming doctrinally conjoined. Established notions of amphibious warfare are now evolving into holistic concepts of expeditionary operations. This has been accompanied, especially in recent years, by a hectic series of coalition expeditionary campaigns, particularly in Afghanistan and Iraq. The US Marine Corps developed the 'vertical envelopment' approach, so as to counter the atomic threat to amphibious assaults, following the Second World War.[44] By the beginning of the 21st century, 'over the beach' deployment had become part of the related US concepts of 'Operational Manoeuvre from the Sea' and 'Ship-to-Objective Manoeuvre'. These, with the idea of sea basing, rest upon the mobility, lethality, and information capability of contemporary forces.[45] In so doing they build upon the traditional operational and strategic advantages of joint maritime power projection, including tactical surprise and naval gunfire support. They are also a response to the danger of the modern littoral environment, with such threats as diesel submarines, mines, and missiles, as well as terrorism and even weapons of mass destruction. The US Navy has recently established a new integrated Expeditionary Combat Command (ECC), with the purpose of enhancing its capacity to pursue littoral and onshore riverine missions within a joint and coalition context. The ECC has a broad mandate in terms of graduated force, ranging from irregular warfare to policing and humanitarian roles.[46] US Naval strike capability now rests upon both carrier groups and independent amphibious strike groups – each of the latter built around Marines on three amphibious ships.[47] British maritime doctrine has moved strongly in the joint operational direction since the late 1990s. This reflects the national strategy emerging from the UK Strategic Defence Review of 1997–98.[48] Australia has had its own history of amphibious operations, particularly during World War II, on which to draw.[49] Today joint power projection is an intrinsic element of Australian maritime doctrine.[50]

What, therefore, are the prospects for joint power projection from the sea in the foreseeable future? All the signs are that it will remain an invaluable operational capability and strategic asset. In a global political environment characterised by failed states, resource conflicts, and terrorism, not to mention climate change, there will be an ongoing need for crisis response in the form of littoral intervention by relevant states, coalitions, and the international community. Recent operations have shown the versatility and indeed graduated force of joint forces deployed and supported from the sea. Deep power projection into Afghanistan in 2002 and an indispensable logistic lifeline into Iraq in 2003 both reflect the traditional relevance of amphibious operations to high intensity conflict. But essentially maritime-enabled forces, such as those of INTERFET in East Timor in 1999–2000, have also demonstrated the potential of joint power projection in terms of peacekeeping. John Connor's study of sea power ashore in Somalia, for example, illustrates the point.[51] There is also wide scope for the use of such forces in counter-terrorist operations and campaigns, forming part of a longer naval policing tradition against piracy and slavery.[52]

As amphibious forces became increasingly professional at the turn of the 19th century, new sea denial weapons such as mines and torpedos made the littoral more dangerous. These two parallel trends continue today. In terms of joint power projection, there are still technical and doctrinal problems to be solved. Mine countermeasures vessels and operations (for example) are slow, as Norman Friedman points out, and the element of surprise could be lost.[53] But maritime forces in particular have been leveraging technological developments for centuries in the interests of power projection: from the gunpowder revolution at sea to long range guns, submarines, air power, missiles, and networked capabilities, and there is no reason to believe that they will not continue to do so. Air warfare destroyers – as part of a networked architecture – have the potential, as Mark Schweikert suggests, to protect and to strengthen joint forces in the littorals.[54] Allies in such operations can also be technical assets, bringing niche capabilities. During the maritime campaign against southern Iraq in 2003, for example, the Australian and British Navies provided local knowledge, maritime interception skills, shallow-draught warships, and mine clearance vessels – all of which were indispensable to a successful outcome.

The need is for more than technology and doctrine, however. In the future the right kind of commanders will also be required. An admiral such as Cunningham, sympathetic to joint operations during World War II, had cut some of his teeth in close concert with the Army during the Boer War. Here the contemporary record is also encouraging. A generation of service officers today, those who have come of age during the post-Cold War period, will provide the leadership for the campaigns of the next quarter century. Amongst such people, historical literacy will also be at a premium. Only the longer look afforded by the study of the past will provide the perspective necessary in negotiating strategic and technological change. MacArthur, it should be remembered, used the example of the taking of Quebec in 1759 in persuading the Joint Chiefs to approve the landing at Inchon.[55] Finally – and perhaps above all – it is important that navies, especially, do not forget that joint operations and expeditionary warfare are but one element of the wider spectrum of sea power missions – not least because the success of such operations depends upon their remembering it.

HMAS *Kanimbla*, seen here with HMA Ships *Darwin* and *Anzac*, earned a reputation as the Queen of the North Arabian Gulf Chessboard during Operation IRAQI FREEDOM in early 2003. In addition to her heavy lift and aviation capabilities she acted as a command platform and mother ship for Coalition patrol craft and boarding parties. (RAN)

Notes

Foreword

[1] P. Cosgrove, 'The ANZAC Lecture at Georgetown University', in *Journal of the Australian Naval Institute*, April/June 2000, p. 9.
[2] N. Friedman, *Seapower as Strategy: Navies and National Interests*, Naval Institute Press, Annapolis, 2001, p. 3.
[3] A. Stephens, submission to 'Inquiry into Australian Defence Force Regional Air Superiority', http://www.aph.gov.au/house/committee/jfadt/adfair/subs/sub1.pdf accessed 21 June 2006.
[4] Royal Australian Navy, *Australian Maritime Doctrine*, Sea Power Centre - Australia, 2000, pp. 48–51

1. Seapower Ashore and in the Air

[1] M.A. Palmer, *Command at Sea: Naval Command and Control Since the Sixteenth Century*, Cambridge USA, Harvard University Press, 2005. p. 243.
[2] For the meaning of this process see Anselm Kiefer's painting *The Ways or Worldly Wisdom – Arminius's Battle of 1978–80*, Art Institute of Chicago; and P.S. Wells, *The Battle that stopped Rome: Emperor Augustus, Arminius, and the Slaughter of the Legions in the Teutoburg Forest*, New York, W.W. Norton and Company, 2003, esp. ch. 2.
[3] F. Nietzsche, *The Use and Abuse of History*, 1873 transl. A. Collins, Indianapolis, Bobbs-Merrill, 1976, p. 40.
[4] Speech by Senator Robert Hill, Australian Minister of Defence, 21 June 2005.
[5] See: www.musee-marine.fr
[6] M. Evans, The *Tyranny of Dissonance: Australia's Strategic Culture and Way of War 1901–2005*, Canberra, Land Warfare Studies Centre, 2005, offers a perceptive insight into the problems that can arise when declared strategy and strategic culture are out of sync. However, I would take issue with some of his conclusions on the merits of inter-war naval planning in London and Canberra.
[7] The nuclear cruise missile submarine *Kursk* was lost with all hands when it sank in the Barents Sea on 12 August 2000.
[8] The Chinese *Luhu* class destroyers were designed with American Gas Turbines, but exports were embargoed after the Tiananmen Square massacre. See J. Bussert, 'New Misssile Destroyers Deploy for Blue-Water Operations', *Signal*, November 1999. My thanks to Professor Bruce Elleman for this information, and much sound advice on the current state of the PLAN.
[9] China announced a successful SSBN test firing at the end of June 2005.
[10] C. McConnaughy, 'China's Undersea Nuclear Deterrent: Will the U.S. Navy Be Ready?' in L.J. Goldstein, *China's Nuclear Force Modernisation*, Newport Papers 22, Newport, Naval War College Press, 2005 pp. 23–48.
[11] The Chinese reportedly paid more than double the scrap value of the *Varyag*. See *AMI International*, May 2005. It has been argued that the imperative to recover Taiwan before it achieves a period of 60 years separate from the mainland, surpassing the longest period of separation hitherto is pushing development at a far faster pace. I am indebted to Bruce Elleman for this important point.
[12] P. Earle, *The Pirate Wars*, London, Methuen, 2003, is very good on the links between legal, quasi-legal and wholly illegal uses of force at sea, and their connections with larger conflicts.

[13] J.B. Kelly, *Britain and the Persian Gulf 1795–1880*, Oxford, Oxford University Press, 1968. However, Sultan M. Al-Qasimi, *The Myth of Arab Piracy in the Gulf*, London, Croom Helm, 1986, disagrees, while C.RE. Davies, *The Blood Red Arab Flag*, Exeter, University of Exeter Press, 2004, attempts to be judicious. The lesson for those making policy today is simple and clear. Do not rush to judgement, and make sure your history is secure, and broadly based. Read widely.
[14] *The Strategic Defence Review*, 2 vols, London, HMSO, 1998. *The Strategic Defence Review: How Strategic? How Much of a Review?*, London, Brassey's, 1998.
[15] J. S. Corbett, *Some Principles of Maritime Strategy*, London, Longman, 1911. See the edition edited by E. Grove, Annapolis, Naval Institute Press, 1988.
[16] *The Fundamentals of British Maritime Doctrine*, 2nd Edition, London, HMSO, 1998. Since 1989 Britain has engaged in a number of expeditionary operations, ranging from high end warfare to disaster relief. These include the Gulf Wars, and longer term Gulf region deployments, the former Yugoslavia, Belize, East Timor, Sierra Leone and Afghanistan.
[17] Command formed on 15 July 1994, by consolidating the Naval Amphibious Warfare School, Coronado, and the Landing Force Training Command Pacific at San Diego: ewtgpac.mil/history.htm. Accessed 26 May 2005.
[18] See: defenseindustrydaily.com/images/SHIP_LCS-GD_cutaw. Accessed 26 May 2005.
[19] *Naval Strategic Plan: The Next Fifteen Years*, UK Ministry of Defence, no date (c. 2000). 'The Future Navy', Admiralty Board Paper 2/00, UK Ministry of Defence, no date (c. 2000).
[20] N. Rich, *Why the Crimean War?*, London, McGraw-Hill, 1985, is only one example, being written as a Cold War diplomatic morality tale. A.D. Lambert, *The Crimean War: British Grand Strategy against Russia 1853–1856*, Manchester, Manchester University Press, 1990, uses a Clausewitzian framework to present the war as a precursor of the 'New Maritime Strategy'.
[21] Corbett, *Some Principles of Maritime Strategy*, p. 14.
[22] A. Gordon, *The Rules of the Game: Jutland and British Naval Command*, London, John Murray, 1996.
[23] P. Colomb, *Naval Warfare*, 2nd edn., London, W.H. Allen and Co., 1895, p. 404.
[24] Corbett, *Some Principles of Maritime Strategy*, remains the most profound enunciation of this strategy.
[25] A.D. Lambert, 'Australia, the "*Trent* Crisis" of 1861 and the Strategy of Imperial Defence', in D. Stevens and J. Reeve (eds), *Southern Trident: Strategy, History and the Rise of Australian Naval Power*, Sydney, Allen and Unwin, 2001, pp. 99–118.
[26] A.D. Lambert '"Good, while it lasts": Great Britain and the Crimean War Coalition, 1854–1856', in D.E. Showalter (ed.), *Future Wars: Coalition Operations in Global Strategy*, Chicago, Imprint Publications, 2002; Lambert, *The Crimean War*.
[27] A. D. Lambert, 'Under the Heel of Britannia' in P. Hore (ed.), *Seapower Ashore*, Chatham, London, 2000, pp. 96–129.
[28] A.D. Lambert, 'Winning without Fighting: British Grand Strategy and its Application to the United States, 1815–1865', in B. Lee and K. Walling (eds), *Strategic Logic and Political Rationality: Essays in Honour of Michael J. Handel*, Newport, Naval War College, 2003.
[29] The two major studies of Imperial Russian Strategy, Fuller and LeDonne, both focus on military issues, devoting little time to the naval policy, and almost none to the coastal fortresses.
[30] A.D. Lambert, *The Foundations of Naval History; Sir John Laughton, the Royal Navy and the Historical Profession*, London, Chatham, 1998; A.D. Lambert, 'Fisher's Formative Years: The Education of a Strategist', paper delivered at the Fisher Conference, Joint Services Command and Staff College Shrivenham, 2004.

[31] Colomb, *Naval Warfare*, pp. v–vi.
[32] Lambert, *Foundations of Naval History*.
[33] Colomb, *Naval Warfare*, pp. 203–430.
[34] R.W. Herrick, *Soviet Naval Theory and Practice*, Annapolis, Naval Institute Press, 1988, pp. 192–200.
[35] Sato Tetsutaro, *Teikoku koubo shiron "On the History of Imperial Defence"*, Tokyo 1908, 1910, 2 vols. I am indebted to Bruce Elleman for this important point.
[36] Colomb, *Naval Warfare*, 1895, p. 430.
[37] E. Hamley, *The Operations of War: Explained and Illustrated,* revised edn., London, W. Blackwood and Sons, 1909. pp. 44, 58–9. The editor, General Kiggell, added a section on the Russo-Japanese War to Hamley's original text, and discussed the subject of amphibious warfare, if only in terms of logistical implications.
[38] C. Callwell, *Small Wars*, London, HMSO, 1896, has nothing to say about this most obvious aspect of the subject.
[39] C. Callwell, *The Effect of Maritime Command on Land Campaigns since Waterloo*, Edinburgh, W. Blackwood and Sons, 1897.
[40] G.G. Aston, *Memories of a Marine: An Amphibiography*, London, John Murray, 1919, pp. 243–7.
[41] Unless otherwise noted career details are taken from: J. Luvass, *The Education of an Army: British Military Thought 1815–1940*, Chicago, University of Chicago Press, 1964, 'The Unfinished Synthesis: Colonel G.F.R. Henderson', pp. 216–247.
[42] Luvass, *The Education of an Army*, p. 227.
[43] Luvass, *The Education of an Army*, p. 244, from Henderson's unsigned article: 'War in South Africa', Edinburgh Review, CXCI (1900), pp. 251–2.
[44] Luvass, *The Education of an Army*, pp. 244–5.
[45] G.F.R. Henderson, *The Science of War: A Collection of Essays and Lectures*, London, Longmans, Green and Co., 1906, 'War', p.28
[46] Fisher to King Edward, 4 October 1907, in A.J. Marder (ed.), *Fear God and Dread Nought: The Correspondence of Admiral of the Fleet Lord Fisher of Kilverstone*, Vol. II, London, Jonathan Cape, p. 143. Fisher claimed he was quoting Foreign Secretary Sir Edward Grey, but Fisher often made up suitable quotes.
[47] A.T. Mahan, *The Influence of Sea Power upon the French Revolution and Empire*, 2 vols, London, Sampson Low, 1892.
[48] R. Mackay, *Fisher of Kilverstone*, Oxford, Oxford University Press, 1973, p. 288
[49] Fisher to E. Goulding, November 1908, in Marder, *Fear God and Dread Nought*, p. 201.
[50] E.S. May, *The Principles and Problems of Imperial Defence*, London, 1903. For a successful career officer, Staff College educator and author Major General Sir Edward Sinclair May (1856–1936) has been given surprisingly little attention. *Who Was Who 1930–1940* contains the only published summary of his career.
[51] B. Bond, *The Victorian Army and the Staff College, 1854–1914*, London, Eyre Methuen, 1972, pp. 197–8.
[52] Report 'Overseas Expeditions', National Archives UK: WO 33/344
[53] J.S. Corbett, *The Successors of Drake*, London, Longmans, Green, and Co., 1900, p. vii.
[54] The title of the published version, which appeared in 1904.
[55] Corbett: 'War Course Lecture Notes October 1903': Liddell-Hart Centre for Military Archives, King's College, London.
[56] Slade to Corbett 22 August 1905 (Royal Naval College): Liddell-Hart Centre.
[57] Rawlinson to Corbett 25 August 1905 and Corbett notes on; Rawlinson to Corbett, 30 August 1905: Liddell-Hart Centre.

[58] Corbett's Trafalgar lecture includes a letter from Barham to Cornwallis, in Laughton's hand: Liddell-Hart Centre.
[59] A.D. Lambert,'The Principle Source of Understanding: Navies and the Educational Role of the Past', in P. Hore (ed.), *The Hudson Papers Volume 1*, London, Ministry of Defence, 2001, pp. 35–66.
[60] Corbett, *Some Principles of Maritime Strategy*, p. 304.
[61] Aston, *Memories of a Marine*, p. 249
[62] Bond, *The Victorian Army and the Staff College*, 1854-1914, pp. 264–7.
[63] J. Gooch, *The Plans of War: The General Staff and British Military Strategy c. 1900– 1916*, London, 1974, esp. pp. 279–295; and N. d'Ombrain, *War Machinery and High Policy: Defence Administration in Peacetime Britain 1902–1914*, Oxford, 1973. An incisive examination of the failure of the CID as a policy making body. A. Offer, *The First World War: An Agrarian Interpretation*, Oxford, 1989, adopts a more positive image of the CID under Hankey.
[64] Offer, *The First World War: An Agrarian Interpretation*, p. 226
[65] C. Callwell, *Military Operations and Maritime Preponderance: Their Relations and Interdependence*, Edinburgh and London, 1905. See Colin Gray's 1997 Edition, Annapolis, USNIP.
[66] P. Simpkins, 'The Four Armies, 1914–1918', in D. Chandler, and I.F.W. Beckett (eds), *The Oxford History of the British Army*, Oxford, Oxford University Press, 1994, p. 242
[67] E.M. Spiers, *Haldane: An Army Reformer*, Edinburgh, 1980, esp. pp. 53–64, 72–3, 76–91, 157–8 and 186–195. J. McDermott, 'The Revolution in British Military Thinking from the Boer war to the Moroccan Crisis' in P.M. Kennedy, (ed.), *The War Plans of the Great Powers 1880–1914*, London, 1979, pp. 99–117
[68] M. Howard, *The Continental Commitment: The Dilemma of British Defence Policy in the Era of Two World Wars*, London, 1972, based on the 1971 Ford Lectures at Oxford. This highly influential text opens with a clear statement of contemporary purpose. For a counter argument see D. Baugh, 'British Strategy during the First World War in the Context of Four Centuries: Blue-Water versus Continental Commitment' in D.M. Masterson, (ed.), *Naval History*, Annapolis, USNIP, 1986, pp. 85–110. See pp. 85 and 95.
[69] D'Ombrain, *War Machinery and High Policy*, p. 113
[70] See Aston, *Memories of a Marine*, pp. 101–5.
[71] Aston, *Memories of a Marine*, p. 28.
[72] The failure of the Naval offensive, which sparked the military operation, was not inevitable. The Navy had not prepared for such operations before the war, did not exercise the gunnery problems, sent the reserve fleet, with magazines half full, of the wrong ammunition. Success was impossible for the force sent, but a well-prepared, fully-equipped force with HE ammunition would have stood a very good chance of getting through.
[73] The same fate befell the US Marines in Vietnam, and is happening again in Iraq. I am indebted to Dr. John Reeve for this point.
[74] Lambert, *Crimean War*.
[75] C. Hall, *Wellington's Navy: Sea Power and the Peninsular War 1807–1814*, London, Chatham, 2004, is a commendable recent attempt to reverse this misleading and dangerous separation.
[76] J. Corbett, 'The Teaching of Naval and Military History', in *History*, April 1916, pp. 12–19
[77] When Michael Howard, Paul Kennedy and Correlli Barnett were writing.

2. The Weihaiwei Blockade and Japanese Hegemony in the Far East

[1] J.P. LeDonne, *The Russian Empire and the World 1700–1924*, New York, Oxford University Press, 1997, p. 337.

[2] B.F. Cooling, *Gray Steel and Blue Water Navy: The Formative Years of America's Military Industrial Complex 1881–1917*, Hamden, Archon Books, 1979, p. 136. Cooling provides a list of the top European navies of 1896. By capacity, China and Japan would have been in the top ten in 1895.
[3] D. Slade, *Queer Things about Japan*, London, Anthony Treherne and Co., 1903, p. xiii. For a similar view, see D. Murray, *Japan*, 6th edn., London, T. Fisher Unwin, 1906, pp. 397–8.
[4] For a recent account of the Battle of the Yalu, see B. Elleman, 'Western Advisors and Chinese Sailors in the 1894–95 Sino-Japanese War', in J. Reeve and D. Stevens (eds), *The Face of Naval Battle*, Sydney, Allen and Unwin, 2003, pp. 55–69.
[5] J.L. Rawlinson, *China's Struggle for Naval Development 1839–1895*, Cambridge, Harvard University Press, 1967, pp. 171–3, 187–8.
[6] Sung-ping Kuo, 'Chinese Reaction to Foreign Encroachment with Special Reference to the First Sino-Japanese War and its Immediate Aftermath', Ph.D. dissertation, Columbia University, 1953, p. 54.
[7] 'War News', *The Japan Weekly Mail* (Yokohama), 3 November 1894, p. 509.
[8] 'War News', *The Japan Weekly Mail* (Yokohama), 24 November 1894, p. 584.
[9] 'China's Apathy', *The North-China Herald* (Shanghai), 25 January 1895, p. 110.
[10] 'The Contending Fleets', *The North-China Herald* (Shanghai), 15 February 1895, p. 214.
[11] H.A. Herbert, 'Military Lessons of the Chino-Japanese War', in *The North American Review*, 160, no. 463 (June 1895), p. 689.
[12] Herbert, 'Military Lessons of the Chino-Japanese War', p. 691.
[13] M. Elliott, 'Resident Aliens: The Manchu Experience in China, 1644–1760', Ph.D. dissertation, University of California, Berkeley, 1993, p. 489.
[14] Citing Der Ostasiatische in 'China's Armies', *The North-China Herald* (Shanghai), 27 July 1894, p. 150.
[15] G. Boulais, *Manuel du Code Chinois* (Chinese Legal Code Manual), Shanghai, Imprimerie de la Mission Catholique, 1924, pp. 415, 418.
[16] Boulais, *Manuel du Code Chinois*, p. 420.
[17] Boulais, *Manuel du Code Chinois*, p. 418.
[18] Rawlinson, *China's Struggle for Naval Development*, pp. 171–3, 187–8.
[19] D.C. Evans and M.R. Peattie, *Kaigun: Strategy, Tactics, and Technology in the Imperial Japanese Navy 1887–1941*, Annapolis, Naval Institute Press, 1997, p. 40.
[20] General Staff Headquarters, *Meiji nijiushichihachinen Nisshin sensō* (The Sino-Japanese War of 1894–5), vol. 6, Tokyo, Tokyo insatsu kabushiki kaisha, 1907, pp. 33, 36–7.
[21] R.H. Chow, *China and Japan: The History of Chinese Diplomatic Missions in Japan 1877–1911*, Singapore, Chopmen Enterprises, 1975, p. 106; Mutsu Munemitsu, *Kenkenryoku: A Diplomatic Record of the Sino-Japanese War*, 1894–95, transl. G.M. Berger, Princeton, Princeton University Press, 1982, pp. 266 n.2, 58, 109, 172, 191.
[22] Sun Kefu and Guan Jia (eds), *Zhongri jiawu luzhan shi* (History of the Land Engagements of the Sino-Japanese War), Harbin, Heilongjiang renmin chubanshe, 1984, pp. 293–4.
[23] Chang Ch'i-yun, *Zhongguo lishi ditu* (Historical Atlas of China), vol. 2, Taipei, Zhongguo wenhuadaxue chubuyinxing, 1984, p. 117; W. Eastlake and Yamada Yoshi-aki, *Heroic Japan: A History of the War between Japan and China 1897*, repr.; Washington, University Publications of America, 1979, pp. 298–9.
[24] 'The Tungchow Affair', *The Japan Weekly Mail* (Yokohama), 16 February 1895, p. 196.
[25] G.A. Ballard, *The Influence of the Sea on the Political History of Japan*, New York, E.P. Dutton, 1921, p. 161.
[26] J.E. Hunter, *Concise Dictionary of Modern Japanese History*, Berkeley, University of California Press, 1984, p. 162; Miwa Kai and P.B. Yampolsky, 'Political Chronology of Japan

1885–1957', Columbia University East Asian Institute Studies, no. 5 (Nov. 1957), p. 5.
[27] Evans and Peattie, *Kaigun*, pp. 36, 40, 524–5.
[28] Qi Qizhang, *Jiawu zhanzheng* (Sino-Japanese War), Beijing: Renmin chubanshe, 1990, p. 426; Rawlinson, *China's Struggle for Naval Development*, pp. 174–5; W.F. Tyler, Pulling Strings in China, London, Constable, 1929, pp. 47, 89.
[29] 'The Wei-hai-wei Campaign', N.W.H. Du Boulay, 'The Chino-Japanese War, 1894–1895', London, typescript, c. 1903, pp. 2–4; Eastlake and Yamada, *Heroic Japan*, p. 306.
[30] S. Spector, *Li Hung-chang and the Huai Army: A Study in Nineteenth Century Regionalism*, Seattle, University of Washington Press, 1964, p. 292; Qi Qizhang, Jiawu zhanzheng, p. 296.
[31] Qi Qizhang, *Jiawu zhanzheng*, pp. 343–4, 353, 366.
[32] 'The Wei-hai-wei Campaign', Du Boulay, 'The Chino-Japanese War, 1894–1895', pp. 2-4; Eastlake and Yamada, *Heroic Japan*, p. 306.
[33] 'The Progress of the War', *The North-China Herald* (Shanghai), 25 January 1895, p. 107.
[34] Qi Qizhang, *Jiawu zhanzheng*, p. 396.
[35] A.T. Mahan, *The Influence of Sea Power upon History, 1660–1783*, 1890, repr.; New York, Hill and Wang, 1957, pp. 357, 366–70, 452–82.
[36] Other spellings include Tungchow and Tungchou.
[37] *The Japan Weekly Mail* (Yokohama), 9 February 1895, pp. 167–8.
[38] 'The Attack on Weihaiwei', *The North-China Herald* (Shanghai), 1 February 1895, p. 156; 'The Capture of Weihaiwei', *The North-China Herald* (Shanghai), 1 March 1895, p. 321; 'The Wei-hai-wei Campaign', Du Boulay, 'The Chino-Japanese War, 1894–1895', pp. 2–4.
[39] For information about traditional Chinese views on fate, auspicious and inauspicious days, and omens, see R.J. Smith, *Fortune-tellers and Philosophers: Divination in Traditional Chinese Society*, Boulder, Westview, 1991; R.J. Smith, *China's Cultural Heritage: The Qing Dynasty, 1644-1912*, 2nd edn., Boulder, Westview, 1994.
[40] 'War News', *The Japan Weekly Mail* (Yokohama), 9 February 1895, p. 157; 'The Weihaiwei Campaign', *The Japan Weekly Mail* (Yokohama), 23 February 1895, pp. 228–31; 'The Lost Navy', *The North-China Herald* (Shanghai), 15 February 1895, p. 242; 'The Capture of Weihaiwei', *The North-China Herald* (Shanghai), 1 March 1895, p. 321; Tyler, *Pulling Strings in China* p. 71; Jukichi Inouye, *The Fall of Weihaiwei*, Yokohama, Kelly and Walsh, 1895; Eastlake and Yamada, *Heroic Japan*, pp. 305–10.
[41] Eastlake and Yamada, *Heroic Japan*, p. 310.
[42] Tyler, *Pulling Strings in China*, pp. 78, 80.
[43] 'The Capture of Weihaiwei', *The North-China Herald* (Shanghai), 1 March 1895, p. 321; 'The Surrender at Weihaiwei', *The Japan Weekly Mail* (Yokohama), 16 February 1895, p. 198.
[44] 'Letter from Hushan', *The Japan Weekly Mail* (Yokohama), 2 March 1895, p. 253; 'War News,' *The Japan Weekly Mail* (Yokohama), 23 March 1895, pp. 335–6; Rawlinson, *China's Struggle for Naval Development*, p. 18.
[45] 'The Capture of Weihaiwei', *The North-China Herald* (Shanghai), 1 March 1895, p. 321.
[46] Rawlinson, *China's Struggle for Naval Development*, p. 189.
[47] Evans and Peattie, *Kaigun*, p. 47; Qi Qizhang, *Jiawu zhanzheng*, p. 421.
[48] 'Weihaiwei Surrenders', *The North-China Herald* (Shanghai), 15 February 1895, p. 215; 'Weihaiwei Occupied', *Asahi shinbun* (Tokyo), 2 February 1895, p. 1; 'Weihaiwei Gun Emplacements All Taken, Current Situation of Enemy Fleet', *Asahi shinbun* (Tokyo), 6 February 1895, supplement, p. 1; 'Complete Destruction of Beiyang Squadron', *Asahi shinbun* (Tokyo), 10 February 1895, p. 1; 'Concerning the Attack on Weihaiwei', *Kokumin shinbun* (Tokyo), 2 March 1895, p. 2.
[49] 'War News', *The Japan Weekly Mail* (Yokohama), 23 February 1895, p. 225. The correspondence between Itÿ and Ding is reprinted in Appendix F of Vladimir [Zenone

Volpicelli], *The China-Japan War Compiled from Japanese, Chinese, and Foreign Sources*, Kansas City, Franklin Hudson Publishing, 1905, pp. 251–4.
[50] Rawlinson, *China's Struggle for Naval Development*, pp. 174–5; Tyler, Pulling Strings in China, pp. 47, 89.
[51] S.C. Chu, 'Sino-Japanese War of 1894: A Preliminary Assessment for the USA,' *Jindaishi yanjiusuo jikan* (Proceedings of the Institute for Modern History) 14 (June 1985)', p. 359; 'Ding Ruchang' in *Zhongguo lishi dacidian* (Comprehensive Chinese Historical Dictionary), vol. 1, Zhongguo lishi dacidian bianzuan weihuanhui, Shanghai, Shanghai cishu chubanshe, 2000, p. 29.
[52] Tyler, *Pulling Strings in China*, p. 79.
[53] Rawlinson, *China's Struggle for Naval Development*, p. 190; Tyler, *Pulling Strings in China*, p. 78.
[54] 'The Surrender at Weihaiwei', *The Japan Weekly Mail* (Yokohama), 16 March 1895, pp. 313–4; 'The Chino-Japanese War', *The Pall Mall Gazette* (London), 19 February 1895, p. 7.
[55] 'Admiral Ding Degraded', *The New York Times*, 10 September 1894, p. 5.
[56] 'Yamagi's Army Advancing', *The New York Times*, 24 November 1894, p. 5; 'Allies against Europe', *The New York Times*, 25 November 1894, p. 5.
[57] 'Count Katsu on Admiral Ting', *The Japan Weekly Mail* (Yokohama), 9 March 1895, p. 285; G.A. Lensen, *Balance of Intrigue: International Rivalry in Korea and Manchuria, 1884–1899*, vol. 1, Tallahassee, University Presses of Florida, 1982, p. 393 n.49; H.B. Morse, *The International Relations of the Chinese Empire*, vol. 3, Shanghai, Kelly and Walsh, 1918, p. 40; D. Keene, 'The Sino-Japanese War of 1894–95 and its Cultural Effects in Japan', in *Tradition and Modernization in Japanese Culture*, D.H. Shively (ed.), Princeton, Princeton University Press, 1971, p. 140; Noriko Kamachi, *Reform in China: Huang Tsun-hsien and the Japanese Model*, Cambridge, Harvard University Press, 1981, p. 193; M. Jansen, Japan and China from War to Peace 1894–1972, Chicago, Rand McNally College Publishing Co., 1975, p. 10; 'A Tribute to the Late Admiral Ting,' *The Japan Weekly Mail* (Yokohama), 9 November 1895, p. 490.
[58] Yamamoto Tsunetomo, Hagakure: *The Book of the Samurai*, trans. W.S. Wilson, Tokyo, Kodansha International, 1979, p. 17.
[59] 'Complete Story of the Surrender of the Beiyang Squadron, Ding, Liu, and Zhang Die for their Country', *Asahi shinbun* (Tokyo), 19 February 1895, p. 2; 'Remains of Admiral Ding', *Asahi shinbun* (Tokyo), 20 February 1895, p. 1.
[60] 'China's Apathy', *The North-China Herald* (Shanghai), 25 January 1895, p. 110.
[61] Rawlinson, *China's Struggle for Naval Development*, p. 190. A contemporary article provides different figures ('The Chino-Japanese War', *The Pall Mall Gazette* [London], 19 February 1895, p. 7).
[62] Rawlinson, *China's Struggle for Naval Development*, pp. 189–90.
[63] Ballard, *The Influence of the Sea on the Political History of Japan*, p. 168; 'The Chino-Japanese War,' *The Pall Mall Gazette* (London), 27 February 1895, p. 7.
[64] Boulais, *Manuel du Code Chinois*, p. 428.
[65] US Adjutant-General's Office, Military Information Division, *Notes on the War between China and Japan*, Washington, Government Printing Office, 1896, p. 22.
[66] 'La guerre en Chine' (The War in China), *Le Temps* (The Times) (Paris), 5 March 1895.
[67] 'Peking', *The North-China Herald* (Shanghai), 10 May 1895, p. 683.
[68] 'China', *The Japan Weekly Mail* (Yokohama), 23 March 1895, p. 344. See also 'Die Lage in Ostasien' (Situation in East Asia), *National Zeitung* (National Newspaper) (Berlin), morning edition, 10 February 1895, p. 1.
[69] 'The Chino-Japanese War', *The Pall Mall Gazette* (London), 1 December 1894, 4th edn., p. 7.
[70] G. Gareau, 'Les Japonais Wei-Hai-Wei' (The Japanese at Weihaiwei), Le Siècle (Century)

(Paris), 3 February 1895, p. 1. Gareau was actually jumping the gun because Weihaiwei would not fall for another week and a half.
[71] 'Chine et Japon' (China and Japan), *Journal de St-Pétersbourg* (St. Petersburg Journal), no. 345, 6 January 1895, p. 2.
[72] 'The War in the East', *The Times* (London), 2 January 1895, p. 7.
[73] 'Planning to Save Pekin', *The New York Times*, 11 February 1895, p. 1.
[74] 'The Problem of China's Fall,' *The North-China Herald* (Shanghai), 5 April 1895, p. 504.
[75] J.S. Corbett, *Some Principles of Maritime Strategy*, New York, Longmans, Green and Co., 1911.
[76] Mutsu, *Kenkenryoku*, pp. 27–8.
[77] S. Lone, *Japan's First Modern War: Army and Society in the Conflict with China, 1894–1895*, London, St. Martin's Press, 1994, pp. 59–69.
[78] 'Count Okuma on the Situation', *The Japan Weekly Mail* (Yokohama), 1 September 1894, p.256.
[79] O. Borisov, *The Soviet Union and the Manchurian Revolutionary Base*, transl. D. Fidlan, Moscow, Progress Publishers, 1975, p. 61.

3. Maritime Aspects of the Gallipoli Operation

[1] This chapter draws upon the author's research for a forthcoming book on the Dardanelles campaign to be published by Yale University Press.
[2] Churchill to Asquith, 29 December 1914, Asquith Papers 13/242–3, Bodleian Library, Oxford.
[3] Buchanan (British Ambassador in Russia) to Grey, 1 January 1915, in M. Gilbert, *Winston S. Churchill, vol.3, Companion Documents*, London, Heinemann, 1972, p. 381.
[4] Churchill to Admiral Carden, 2 January 1915, the National Archives, Kew (TNA): ADM 137/96.
[5] Gilbert, *Winston S. Churchill*, pp.360–1.
[6] Carden to Churchill, 5 January 1915, TNA: ADM 137/96.
[7] Churchill to Jellicoe, 4 January 1915, Jellicoe Papers, Add/Mss 48890, British Library.
[8] Churchill to Fisher 4 January 1915, in Gilbert, *Winston S. Churchill*, p.71.
[9] War Council Minutes 13 January 1915, TNA: CAB 42/1/16.
[10] Richmond, 'Remarks on Present Strategy', Richmond Papers, RIC14/3, National Maritime Museum, Greenwich.
[11] Oliver, 'Draft Autobiography', Oliver Papers OLV12, National Maritime Museum, Greenwich.
[12] Note by Admiral Jackson 15 January 1915, TNA: ADM 116/3491.
[13] Note by Admiral Jackson 15 January 1915.
[14] Fisher to Jellicoe, 19 January 1915, in Gilbert, *Winston S. Churchill*, pp.429–30.
[15] War Council Minutes 28 January 1915, TNA: CAB 42/1/26.

4. The Royal Naval Division at Gallipoli

[1] For a more detailed study of RND operations during World War I see C. Page, 'The Royal Naval Division 1914–19', in P. Hore (ed.), *Seapower Ashore: 200 Years of Royal Navy Operations on Land*, Chatham Publishing, 2001.
[2] C. Page, *Command in the Royal Naval Division. A Military Biography of Brigadier A.M. Asquith DSO*, Staplehurst, Spellmount, 1999.
[3] D. Jerrold, *The Royal Naval Division*, Hutchison, London, 1923, p. xiii.

5. Naval Aviation at Gallipoli

[1] For instance, in July 1915 three Short 827 seaplanes were sent to Mombasa, in modern Kenya, for use by the RNAS against German East Africa, but upon arrival, these were diverted to Mesopotamia and reached Basra in early September. Also used in Mesopotamia were Short 184 and Wight two seater seaplanes originally built for the RNAS.
[2] R. Jackson, *Strike from the Sea*, London, Barker, 1970, p. 20.
[3] H.A. Jones, *The War in the Air*, vol. 2, Oxford, Clarendon Press, 1928, p. 10.
[4] R.D. Layman, 'Over the Wine-Dark Sea: Aerial Aspects of the Dardanelles/Gallipoli Campaign', in *Over the Front*, vol. 9, no. 1, 1994, p. 26.
[5] Interestingly, one of the five observers later serving with 3 Wing, Flight Commander A.H. Keith-Jopp, was identified as an Australian. See C.R. Samson, *Fights and Flights*, London, Ernest Benn, 1930, pp. 225, 228.
[6] S.W. Roskill (ed.), *Documents Relating to the Naval Air Service*, vol. 1: 1908–1918, London, Navy Records Society, 1969, p. 223.
[7] Roskill, *Documents Relating to the Naval Air Service*, p. 242
[8] Roskill, *Documents Relating to the Naval Air Service*, p. 201
[9] O. Thetford, *British Naval Aircraft since 1912*, 6th edn, London, Putnam Aeronautical, 1991, p.12.
[10] Roskill, *Documents Relating to the Naval Air Service*, p. 202; see also R. Hayward, *The Fleet Air Arm in Camera 1912–1996*, Stroud, Sutton Publishing and Public Record Office, 1996, pp. 6, 8–9.
[11] Roskill, *Documents Relating to the Naval Air Service*, p. 258.
[12] R.R. James, *Gallipoli*, London, Pan Books, 1974, p. 159.
[13] T. Henshaw, *The Sky their Battlefield*, London, Grub Street, 1995, pp. 493–4.
[14] Jones, *The War in the Air*, pp. 11–12.
[15] James, *Gallipoli*, p. 46.
[16] James, *Gallipoli*, p. 47.
[17] Jones, *The War in the Air*, p. 18.
[18] P. Young (ed.), *Marshall Cavendish Illustrated Encyclopedia of World War I*, vol. 4: 1915–16, New York, Marshall Cavendish, 1984, p. 724.
[19] C.E.W. Bean, *The Story of Anzac*, vol. 1, Sydney: Angus and Robertson, 1921, p. 393.
[20] C.E.W. Bean, *The Story of Anzac*, vol. 2, Sydney: Angus and Robertson, 1924, pp. 137–8.
[21] F.H. Sykes, *Aviation in Peace and War*, London, Edward Arnold, 1922, p. 74.
[22] James, *Gallipoli*, p. 211.
[23] James, *Gallipoli*, p. 264.
[24] See the oblique views reproduced in Captain J.B. Sterndale Bennett, 'Gallipoli Interlude', *War in the Air*, pt. 3, London, 21 November 1935, p. 93.
[25] Sykes, *Aviation in Peace and War*, p. 59.
[26] Jones, *The War in the Air*, p. 64.
[27] Malone to Director of the Air Department, 14 August 1915, quoted in Roskill, *Documents Relating to the Naval Air Service*, p. 221.
[28] Jones, *The War in the Air*, p. 65.
[29] Author's conversation with Mr Kenan Çelik, Çanakkale, Turkey, 14 September 2002.
[30] Jones, *The War in the Air*, p. 65.
[31] Jones, *The War in the Air*, p. 65.
[32] Samson, *Fights and Flights*, p. 245.
[33] D. Hamer, *Bombers versus Battleships*, Sydney, Allen and Unwin, 1998, p. 9.
[34] Jones, *The War in the Air*, pp. 30–1; Samson, *Fights and Flights*, p. 228.
[35] Sykes, *Aviation in Peace and War*, pp. 62–3.

[36] James, *Gallipoli*, p. 204.
[37] James, *Gallipoli*, p. 206.
[38] Roskill, *Documents Relating to the Naval Air Service*, p. 258.
[39] Jones, *The War in the Air*, p. 69; Samson, *Fights and Flights*, p. 230.
[40] Jones, *The War in the Air*, p. 71, states that this incident occurred near Turshun Keui, but Samson, *Fights and Flights*, p. 268, gives the location as near Selvili.
[41] Roskill, *Documents Relating to the Naval Air Service*, p. 204; Jackson, *Strike from the Sea*, p. 29.
[42] Thetford, *British Naval Aircraft since 1912*, p. 39.
[43] C. Bowyer, *For Valour: The Air VCs*, London, William Kimber, 1978, pp. 59–60; Stephen Snelling, *VCs of the First World War: Gallipoli*, Stroud, Alan Sutton, 1995, pp. 230 7.
[44] Thetford, *British Naval Aircraft since 1912*, p. 258.
[45] Cenk Avci, *The Skies of Gallipoli*, Istanbul, Nart Yayincilik, 2003, pp. 7–9.
[46] Ole Nikolajsen and B. Yilmazer, *Ottoman Aviation* 1909-1919, CD-ROM obtainable from http://www.ole-nikolajsen.com.
[47] B.P. Flanagan, 'The History of the Ottoman Air Force in the Great War. The Reports of Major Erich Serno', in *Cross and Cockade*, no. 2–4, 1970.
[48] Roskill, *Documents Relating to the Naval Air Service*, p. 204.
[49] James, *Gallipoli*, p. 159.
[50] N. Steel, *Defeat at Gallipoli*, London, Macmillan, 1994, p. 346; R.J. Brownell, *From Khaki to Blue*, Canberra, Military Historical Society of Australia, 1978, p. 35.
[51] Avci, *The Skies of Gallipoli*, p. 137.
[52] Jones, *The War in the Air*, p. 55.
[53] Bean, *The Story of Anzac*, vol. 2, p. 485.
[54] This unit had arrived during May; see Jones, *The War in the Air*, p. 28, Samson, *Fights and Flights*, p. 242.
[55] E.W.A. von Hoeppner, *Germany's War in the Air*, Nashville, Battery Press, 1994, pp. 47–8.
[56] Henshaw, *The Sky their Battlefield*, p. 496.
[57] Sykes, *Aviation in Peace and War*, p. 47.
[58] Roskill, *Documents Relating to the Naval Air Service*, p. 258.
[59] Sykes, *Aviation in Peace and War*, p. 48.
[60] Sykes, *Aviation in Peace and War*, p. 49.
[61] Layman, 'Over the Wine-Dark Sea', p. 37.
[62] R.D. Layman, *Naval Aviation in the First World War: Its Impact and Influence*, London, Chatham Publishing, 1996, pp. 152-3.

6. An Inferior Naval Power Ashore: German Navy Baltic, Mediterranean and Black Sea operations

[1] R.A. Bowling, 'Mahan's Principles and the Battle of the Atlantic,' in T. Runyon and J. Copes (eds), *To Die Gallantly: The Battle of the Atlantic*, Boulder, Westview, 1994, p. 239.
[2] G. Weinberg, *A World at Arms: A Global History of World War II*, Cambridge, Cambridge University Press, 1994, pp. 721, 773.
[3] On the command mess the German military created in World War II see G. Megargee, *Inside Hitler's High Command*, Lawrence, Kansas, 2000, especially pp. 234–5.
[4] For the German ships committed to BARBAROSSA see B.R. 1840 (2) *German Naval Operations in the Baltic Against Russia, 1941–1945*, London, Ministry of Defence: Naval Historical Branch, December 1965, p. 10. J. Meister, *Der Seekrieg in den osteuropäischen Gewässern, 1941–1945*, München, J.F. Lehmanns, 1958, describes the 'massacre' of Soviet U-boats, pp. 39–40. For a slightly more balanced view see F. Ruge, *The Soviets as Naval*

Opponents, 1941–1945, Annapolis, Naval Institute Press, 1979, pp. 12, 15–20.
[5] H. Boog et al., *Das Deutsche Reich und der Zweite Weltkrieg, Bd. 4: Der Angriff auf die Sowjetunion*, Stuttgart, Deutsche Verlags-Anstalt, 1983, p. 560.
[6] M. Salewski, *Die Deutsche Seekriegsleitung, 1935–1945*, Bd. 2, München, Bernard & Graefe, 1975, pp. 383–4, Ruge, *The Soviets as Naval Opponents*, pp. 21–2.
[7] 1.Skl. Iop 262/41 Gkdos. An Skl. U. 'Weisung Nr.1 für Fall 'Barbarossa,' 6 March 1941, PG 33347, Roll 2096, Microfilm Publication T1022 (German Naval Records), National Archives and Records Administration (NARA).
[8] In June 1941 the combined total of training and sea trials' U-boats was almost 100, with half a dozen committed to support BARBAROSSA. See E. Grove (ed.), *The Defeat of the Enemy Attack on Shipping*, Aldershot, Ashgate for the Naval Records Society, 1997, vol. 1B, Plan 7.
[9] The marginal notation '15.3.!' suggests so in the appendix (*Anlage*) to 'Weisung…', 'Vorläufige Zeittafel für Vorbereitung von Fall "Barbarossa".'
[10] In Latvian, their respective names are Saaremaa, Hiiumaa, and Muhu.
[11] HG Nord, Ia 'Notizen des Herrn O.B. zu anl.Studie,' 6 May 1941, first frame 7063053, Roll 51, Microfilm Publication T311 (German Army Group Records), NARA.
[12] *German Naval Operations in the Baltic*, p. 14. Carls's title was Commander, Naval Group North.
[13] OKH zu HG Nord, Ia 'Unternehmen "Beowulf," 31 May 1941, frame 7063137, roll 51, T311, NARA.
[14] On the Siebelfähren see S. Roskill, *The War at Sea*, 1939–1945, vol. III, pt. 1, London, HMSO, 1960, p. 145, n. 1.
[15] Teletype message to HG Nord, Ia, 4 June 1941, frames 7063147–7063148, roll 51, T311, NARA. As an aside, the Latvian name for the Dvina River is the Daugava, while the German name is Duna.
[16] HG Nord, Ia, 'Regelung der Befehlsverhältnisse für Operation Beowulf', 31 May 1941, frames 7063127–7063129, roll 51, T311, NARA.
[17] HG Nord, Ia, 'Erfahrungsbericht über die Landungsunternehmen gegen die Baltischen Inseln', 8 October 1941, frames 7063557–7063566, roll 51, T311, NARA.
[18] The two ports are in Estonia and Finland, respectively. *German Naval Operations in the Baltic*, pp. 11–12 and 19–21, Meister, *Der Seekrieg in den Osteuropäischen Gewässern*, pp. 11–17, and E. Withol, 'Mine Warfare in the Baltic Sea and the Gulf of Finland, 1939–1941', in W. Roberts and J. Sweetman (eds), *New Interpretations in Naval History: Selected Papers from the 9th Naval History Symposium*, Annapolis, Naval Institute Press, 1991. The German 'official' history criticises the lack of surface ship participation but characterises the mining campaign as offensive in character, sinking over a dozen ships evacuating troops from Talinn. See Boog et al., *Das Deutsche Reich und der Zweite Weltkrieg*, pp. 560–62.
[19] HG Nord, Ia (untitled), 16 June 1941, frames 7063157–7063158, T311, roll 51, NARA.
[20] HG Ia/M.V.O. 'Beowulf,' 18 June 1941, frames 7063159–7063160, T311, roll 51, NARA. On the crossing of the Dvina in Riga by 30 June see map volume in Boog et al., *Das Deutsche Reich und der Zweite Weltkrieg*, 'Frontveränderungen der H. Gr. Mitte und Nord vom 24.6. bis 6.7.41.'
[21] HG Nord, Ia, 'Divisionsbefehl für die Durchführung des Unternehmens Beowulf II,' 7 September 1941, frames 7063490–7063495, roll 51, T311, NARA. On the naval attacks see *German Naval Operations in the Baltic*, pp. 14-16; for the air threat see Meister, *Der Seekrieg in den Osteuropäischen Gewässern*, p. 26, and W. Hubatsch, 61. *Infanterie-Division: Kampf und Opfer Ostpreussicher Soldaten*, Bad Nauheim, 1961, p. 74.
[22] R. Hofmann, 'Die Operationen des XXXXII A.K. zur Eroberung der Baltischen Inseln im

September/Oktober 1941', (no date), Foreign Military Study #P-143b, pp. 53–54. Copy at the US Army Military History Institute, Carlisle, PA.
[23] Meister, *Der Seekrieg in den Osteuropäischen Gewässern*, p. 26.
[24] Ruge, *The Soviets as Naval Opponents*, p. 21. On the Finnish *Ilmarinen* (lost) and *Väinämöinen* coastal defence ships see A. Bredt, *Weyers Taschenbuch der Kriegsflotten*, München, J.F. Lehmanns, 1936, repr. 1972, pp. 58–9.
[25] H. Spaeter, *Die Brandenburger: Eine Deutsche Kommandotruppe*, zbV 800, München, Walter Angerer, 1978, pp. 181–5.
[26] Spaeter, *Die Brandenburger*, see also *German Naval Operations in the Baltic*, p. 22, Hofmann, 'Die Operationen des XXXXII A.K. zur Eroberung der Baltischen Inseln im September/Oktober 1941', pp. 27–8, and Ruge, *The Soviets as Naval Opponents*, p. 21.
[27] 'Erfahrungsbericht über die Landungsunternehmen . . .', frame 7063557, roll 51, T311, NARA.
[28] Hubatsch, 61. *Infanterie-Division*, p. 77.
[29] Meister, *Der Seekrieg in den Osteuropäischen Gewässern*, p. 31.
[30] *German Naval Operations in the Baltic*, p. 24.
[31] A.O.K. 18 Ia, 'Fragen des A.O.K. und Beantwortung durch das XXXXII A.K.', frames 7063287–7063290, roll 51, T311, NARA. *German Naval Operations in the Baltic*, p. 24, notes that the passengers on these craft were captured.
[32] Meister, *Der Seekrieg in den Osteuropäischen Gewässern*, p. 32
[33] F.H. Hinsley et al., *British Intelligence in the Second World War: Its Influence on Strategy and Operations* Vol. 3, pt. I, London, HMSO, 1984, p. 87.
[34] See Grove, *The Defeat of the Enemy Attack on Shipping*, vol. 1A, pp. 134–40, and vol. 1B, Plans 65 and 66.
[35] Roskill, *The War at Sea*, p. 121.
[36] W. Murray, *Strategy for Defeat: The Luftwaffe, 1933–1945*, Maxwell AFB, USGPO, January 1983, p. 165. See also *Hinsley, British Intelligence in the Second World War*, pp. 82–4.
[37] Hinsley, *British Intelligence in the Second World War*, pp. 88–91.
[38] G. Wagner (ed.), *Lagevorträge des Oberbefehlshabers der Kriegsmarine vor Hitler*, Munich, Lehmanns, 1972, pp. 528–30.
[39] P. Schramm (ed.), *Kriegstagebuch des Oberkommandos der Wehrmacht. 1943: Teilband II*, Munich, Bernard & Graefe, 1982, pp. 837–9, entry for 29 July. Achse translates as 'Axis.' This work is the published OKW war diary.
[40] Schramm, *Kriegstagebuch des Oberkommandos der Wehrmacht*, pp. 896-7.
[41] W. Rahn et al., *Kriegstagebuch der Seekriegsleitung*, 1939-1945, Bd.48: August 1943, Berlin, E.S. Mittler & Sohn, 1994, p. 95. The published War Diary of the German Naval Staff.
[42] Schramm, *Kriegstagebuch des Oberkommandos der Wehrmacht*, p. 919.
[43] For the bulk of the debate see Wagner, *Lagevorträge des Oberbefehlshabers der Kriegsmarine vor Hitler*, p. 534, Salewski, *Die Deutsche Seekriegsleitung*, p. 370, and C. Thomas, *The German Navy in the Nazi Era*, Annapolis, Naval Institute Press, 1990, p. 230. For Dönitz's own view, see K. Dönitz, *10 Jahre und 20 Tage*, Frankfurt, Athenäum, 1963, pp. 355–6.
[44] B.B. Tomblin, *With Utmost Spirit: Allied Naval Operations in the Mediterranean, 1942–1945*, Lexington, University Press of Kentucky, 2004, p. 226. On HMS *Warspite* see S. Roskill, *White Ensign: The Royal Navy at War, 1939–1945*, Annapolis, Naval Institute Press, 1966, p. 293. For examples of German reports of Allied fire support see Rahn et al., *Kriegstagebuch der Seekriegsleitung*, entry for 11 August 1943, p. 195, and entry for 12 August 1943, p. 214.
[45] On Allied air attacks see OKW war diary entries for 10 and 12 August, Schramm, *Kriegstagebuch des Oberkommandos der Wehrmacht*, pp. 925, 933.
[46] Schramm, *Kriegstagebuch des Oberkommandos der Wehrmacht*, p. 896. As an aside, the XIV

Panzer Corps was a headquarters reconstituted after its loss with the balance of the German 6th Army at Stalingrad.

[47] The Italians did not work with German planners, and began extracting the bulk of their troops on 9 August; see Hinsley, *British Intelligence in the Second World War*, p. 95. *Lehrgang* means 'school year'.

[48] Schramm, *Kriegstagebuch des Oberkommandos der Wehrmacht*, p. 929. 'Auf Sizilien verlief das Absetzen auf die letzte Brückenkopfstellung planmässig'. Emphasis in original. See also Salewski, *Die Deutsche Seekriegsleitung*, pp. 370–71.

[49] Hinsley, *British Intelligence in the Second World War*, p. 105, notes that Allied signals intelligence identified four other divisions in Italy, including Sardinia.

[50] See the map in Roskill, *The War at Sea*, facing p. 145. Italian troops employed a different set of four evacuation points.

[51] An estimate composed by Oberbefehlshaber Süd, that is Field Marshal Albert Kesselring's headquarters, for the *Wehrmacht Führungsstab* in Berlin. See Rahn et al., *Kriegstagebuch der Seekriegsleitung*, p. 46.

[52] Roskill, *The War at Sea*, p. 148.

[53] Numbers vary. German contemporary sources – the OKW and Naval Staff war diaries – cite 60,000 Germans and a like number of Italians, which seems high. See Schramm, *Kriegstagebuch des Oberkommandos der Wehrmacht*, entry 18 August 1943, pp. 963–4 and Rahn et al., *Kriegstagebuch der Seekriegsleitung*, entry for 18 August, p. 347. Tomblin and Roskill, using postwar German records, report the numbers as cited here. See Tomblin, *With Utmost Spirit*, p. 230, Roskill, *The War at Sea*, p. 150.

[54] S. Morison, *History of United States Naval Operation in World War II, Vol. Nine: Sicily-Salerno-Anzio*, New York, Little, Brown, 1954, p. 212, although n. 16 on that page offers higher numbers. Hinsley, *British Intelligence in the Second World War*, pp. 97–8 notes 75 ferries reportedly used by Allied intelligence. On 14 May Dönitz told Hitler (Wagner, *Lagevorträge des Oberbefehlshabers der Kriegsmarine vor Hitler*, p. 484) the Axis had just 18 MFPs in theatre, six of them German. For the characteristics of the 220-ton MFP, larger than an Allied LCVP, E. Gröner, *Die Schiffe der Deutschen Kriegsmarine und Luftwaffe und ihr Verbleib*, München, J.F. Lehmanns, 1972, pp. 42–3, 120.

[55] Roskill, *The War at Sea*, pp. 144–5.

[56] Roskill, *The War at Sea*, p. 150. Italian forces lost a further eight transport craft, and a further five damaged and scuttled.

[57] Hinsley, *British Intelligence in the Second World War*, p. 99 and V. Orange, *Coningham: A Biography of Air Marshal Sir Arthur Coningham*, London, Methuen, 1992, p. 168.

[58] At 10:10 pm. See Orange, *Coningham*, p. 166.

[59] M. Simpson, *A Life of Admiral of the Fleet Andrew Cunningham: A Twentieth Century Naval Leader*, London, Frank Cass, 2004, p. 164. 40 Commando, Royal Marines, did land at Scaletta on 16 August, but German troops had already departed, see Tomblin, *With Utmost Spirit*, p. 230.

[60] Orange, *Coningham*, p. 167.

[61] Orange, *Coningham*. In contrast, the RCAF official history suggests 'the cost was high . . . The Canadians alone lost five machines (and twenty-five airmen).' B. Greenhous et al., *The Official History of the Royal Canadian Air Force, Vol. III: The Crucible of War, 1939–1945*, Toronto, University of Toronto Press, 1994, p. 651.

[62] Simpson, *A Life of Admiral of the Fleet Andrew Cunningham*, p. 164.

[63] Tomblin, *With Utmost Spirit*, p. 228.

[64] Morison, *Sicily-Salerno-Anzio*, p. 217.

[65] As quoted in Simpson, *A Life of Admiral of the Fleet Andrew Cunningham*, p. 164.

[66] See Simpson, *A Life of Admiral of the Fleet Andrew Cunningham*, p. 150. See also M. Simpson, *The Cunningham Papers, Vol. II: The Triumph of Allied Sea Power 1942–1946*, Aldershot, Ashgate for the Naval Records Society, 2006, pp. 25, 75–6.
[67] Rahn, et al., *Kriegstagebuch der Seekriegsleitung, 1939–1945*, Bd. 49: September 1943, Berlin, E.S. Mittler & Sohn, 1994, entry for 9 September, pp. 175–7.
[68] Salewski, *Die Deutsche Seekriegsleitung.*
[69] E. Ziemke, *Stalingrad to Berlin: The German Defeat in the East*, Washington, USGPO, 1987, p. 18.
[70] E. Marolda, 'The Failure of German World War II Strategy in the Black Sea', *Naval War College Review*, Summer 1975, pp. 39–54. German naval forces built up slowly in the Black Sea over the first year of the Soviet-German war; see pp. 42–4.
[71] Marolda, 'The Failure of German World War II Strategy in the Black Sea', pp. 46–9.
[72] Marolda, 'The Failure of German World War II Strategy in the Black Sea', p. 52.
[73] Ziemke, *Stalingrad to Berlin*, pp. 81–9. Some accounts refer to it as the Taman Peninsula.
[74] Ruge, *The Soviets as Naval Opponents*, p. 95. The Germans had built much of this MFP and MAL force over the winter of 1941–1942 using a captured shipyard in the Ukrainian port of Nicolayev; see Ruge, *The Soviets as Naval Opponents*, p. 79.
[75] On the limited Romanian capabilities shaping their commitment east of Sevastopol, see Marolda, 'The Failure of German World War II Strategy in the Black Sea', p. 43; for the Italian, see pp. 45, 48.
[76] Ruge, T*he Soviets as Naval Opponents*, pp. 79, 95–96. Salewski, *Die Deutsche Seekriegsleitung*, pp. 370–1.
[77] Marolda, 'The Failure of German World War II Strategy in the Black Sea', p. 41.
[78] J. Hayward, *Stopped at Stalingrad: The Luftwaffe and Hitler's Defeat in the East 1942–1943*, Lawrence, University Press of Kansas, 1998, p. 316. Another historian characterises air support for the Black Sea naval effort as minimal; see Marolda, 'The Failure of German World War II Strategy in the Black Sea', p. 51.
[79] Hayward, *Stopped at Stalingrad*, p. 317, Ruge, *The Soviets as Naval Opponents*, p. 79. On the Soviets gaining air superiority see D. Glantz & J. House, *When Titans Clashed: How the Red Army Stopped Hitler*, Lawrence, Kansas, 1995, p. 151.
[80] Both as quoted in Salewski, *Die deutsche Seekriegsleitung*, p. 384.
[81] Salewski, *Die Deutsche Seekriegsleitung*, pp. 383–4.
[82] Ruge, *The Soviets as Naval Opponents*, p. 96, Meister, *Der Seekrieg in den Osteuropäischen Gewässern*, p. 262.
[83] J. Rohwer & G. Hümmelchen, *Chronology of the War at Sea 1939–1945, The Naval History of World War Two*, Annapolis, Naval Institute Press, 1992, p. 196. The authors note a total of 99 'Kleiner Bär' ('smaller bear') MFP convoys, with loss of just three landing craft. See also Meister, *Der Seekrieg in den Osteuropäischen Gewässern*, p. 268.
[84] 'War Diary of Admiral, Black Sea,' 16 May 1943 – 31 May 1943, p. 79, copy in the Navy Department Library. This is a translation of captured German document PG 31532 prepared by the US Office of Naval Intelligence.
[85] Meister, *Der Seekrieg in den Osteuropäischen Gewässern*, p. 262, n. 1.
[86] Dönitz, *10 Jahre und 20 Tage*, p. 380. The evacuees were in addition to the Kuban garrison – code-named *Gotenkopf* – remaining in the bridgehead and noted earlier.
[87] Meister, *Der Seekrieg in den Osteuropäischen Gewässern*, p. 261 and Ruge, *The Soviets as Naval Opponents*, p. 96. The Soviet light attack craft had been dragged overland to the Sea of Azov.
[88] Wagner, *Lagevorträge des Oberbefehlshabers der Kriegsmarine vor Hitler*, p. 544, and Meister, *Der Seekrieg in den Osteuropäischen Gewässern*, pp. 263, 266.

[89] On German aspirations for a 1943 Caucasus offensive see Glantz and House, *When Titans Clashed*, pp.141–2. For the summer campaign see Chapter 11.
[90] Excluding Crimean forces, 17th Army possessed six German infantry divisions and six Romanian divisions. Schramm, *Kriegstagebuch des Oberkommandos der Wehrmacht*, p. 1155.
[91] Rahn et al., *Kriegstagebuch der Seekriegsleitung*, Bd. 49, entry for 16 September, p. 339.
[92] Rahn et al., *Kriegstagebuch der Seekriegsleitung*, Bd. 49, entry for 17 September, p. 357.
[93] Rahn et al., *Kriegstagebuch der Seekriegsleitung*, Bd. 49, entries for 21 and 25 September, pp. 436, 518.
[94] Rahn et al., *Kriegstagebuch der Seekriegsleitung*, Bd. 49, entry for 27 September, p. 548.
[95] Rahn et al., *Kriegstagebuch der Seekriegsleitung*, Bd. 49, entries for 29 and 30 September, pp. 585, 603.
[96] W. Rahn et al., *Kriegstagebuch der Seekriegsleitung*, 1939–1945, Bd. 50: October 1943, Berlin, E.S. Mittler & Sohn, 1994, entry for 6 October, pp. 117–18. See also Meister, *Der Seekrieg in den Osteuropäischen Gewässern*, pp. 270–71.
[97] Rohwer & Hümmelchen, *Chronology of the War at Sea 1939-1945*, p. 239, list the ships as flotilla leader *Kharkov*, and destroyers *Besposhchadny* and *Sposobny*.
[98] Meister, *Der Seekrieg in den Osteuropäischen Gewässern*, p. 264. Rohwer & Hümmelchen, *Chronology of the War at Sea 1939–1945*, p. 234, cite 239,669 troops, 16,311 wounded, 27,456 civilians, 74,657 horses, 21,230 vehicles, 27,741 wagons, 1815 guns, 74 tanks, and 115,477 tons of supplies; Ruge's numbers concur with these (*The Soviets as Naval Opponents*, p. 112). The Navy Staff War Diary cites a final total only for naval lift. Rahn et al., *Kriegstagebuch der Seekriegsleitung*, 1939–1945, Bd. 50, entry for 10 October, p. 218.
[99] Ruge, *The Soviets as Naval Opponents*, p. 113.
[100] Salewski, *Die Deutsche Seekriegsleitung*, pp. 383–400, especially 386.

7. Sea Power Ashore: North Africa to Normandy

[1] See for example N. Barr, *Pendulum of War*, London, Jonathon Cape, 2004; S. Hart *Montgomery and Colossal Cracks*, London, Praeger, 2000; R. Hart, *Clash of Arms*, London, Lynne Rienner, 2001.
[2] See R. Woodman, *The Victory of Seapower*, London, Chatham Publishing, 1998; P. Hore, *Seapower Ashore*, London, Chatham Publishing, 2001.
[3] G. Sheffield, *Forgotten Victory*, London, Headline, 2001, p.79.
[4] E. Tauber, 'Alexandretta: Three Plans Blocked', in *Army Quarterly and Defence Journal*, Volume 122, No. 3, July 1992; D. Massam, *British Maritime Strategy and Amphibious Capability*, PhD Thesis, Oxford University, 1995.
[5] The Manuals are available at the National Archives (TNA), Kew as Defence (DEFE) 2/708 and 709.
[6] I. Speller, *The Role of Amphibious Warfare in British Defence Policy, 1945–56*, Basingstoke, Palgrave, 2001, pp. 17–21.
[7] See D. Grinnell-Milne, *The Silent Victory*, London, The Bodley Head, 1958; C. Page 'The Planned Invasion of Great Britain', in T. Lovering (ed.), *Amphibious Assault*, Portsmouth, Royal Navy, 2005.
[8] See for instance J. Levy, *The Royal Navy's Home Fleet in World War II*, Basingstoke, Palgrave, 2003, p.160.
[9] R. Overy, *Why the Allies Won*, London, Jonathon Cape, 1995, p.228.
[10] See M. Brice, *Axis Blockade Runners of World War II*, London, Batsford, 1981.
[11] See R. DiNardo, *Mechanized Juggernaught or Military Anachronism*, New York, Greenwood, 1991.

[12] S. Roskill, *The War at Sea, Volume 1,* London, HMSO, 1954, pp. 239, 436, 444–6.
[13] Cited in J. Greene and J. Massignani, *The Naval War in the Mediterranean 1940–1943*, London, Chatham, 1998, p. 169. This quote is sometimes reported as 'It takes two years to build a battleship; it takes two hundred to build a naval tradition'.
[14] S. Hart, *Montgomery and Colossal Cracks*, p.151.
[15] Quoted in G.Gill, *Royal Australian Navy 1939–1942*, Canberra, Australian War Memorial, 1957, p. 297.
[16] R. Westphal, *The German Army in the West*, London, Cassel and Co., 1950, p. 116.
[17] G. Kerr, *Business in Great Waters*, London, Faber and Faber, 1951, p. 83 and Naval Historical Branch (Portsmouth) Compilation C174, 'Troopship Losses'.
[18] I. Playfair, *The Mediterranean and the Middle East*, Volume III, London, HMSO, 1960, pp. 371–2.
[19] See Playfair, *The Mediterranean and the Middle East, Volume III*, pp. 325–9 and *Volume IV*, pp. 202–10 and H. Boog et al, *Germany and The Second World War, Volume VI*, Oxford, Clarenden Press, pp.748–83 and 834–9.
[20] Quoted in F. Hinsley *British Intelligence in the Second World War, Volume 2*, London, HMSO, 1981, p. 443. See Playfair, *The Mediterranean and the Middle East, Volume III*, pp. 325–9 and *Volume IV*, pp. 202–10 and H. Boog et al, *Germany and The Second World War, Volume VI*, Oxford, Clarenden Press, pp.748–83 and 834–9.
[21] Book of Reference (BR) 1736(51) *The Defeat of the Enemy Attack on Shipping, 1939–1945*, London, Admiralty (internal use only) 1957. Published by the Naval Records Society in 1997, Para.32.
[22] BR 1736(26) *Battle Summary No.33 Raid on Dieppe*, London, Admiralty (internal use only), 1944, Revised 1959, Appendix C.
[23] P. Ziegler, *Mountbatten The Official Biography*, London, Book Club Associates, 1985, p. 189
[24] See for instance W. Buckingham, *D-Day, The First 72 Hours*, Stroud, Tempus, 2004, p. 19.
[25] J. Campbell, *Dieppe Revisited: A Documentary Investigation*, London, Frank Cass, 1993; S. Prince, 'The Raids on St Nazaire and Dieppe' in Lovering, *Amphibious Assault*.
[26] See G. Weinberg, *A World At Arms*, Cambridge, Cambridge University Press, 2005.
[27] S. Bidwell and D. Graham, *Tug of War*, New York, St Martins Press, 1986, Ch. 6.
[28] BR 1736(30) *Battle Summary No. 37 Salerno*, London, Admiralty (internal use only) 1946, pp. 57–63; S. Morison, *History of the United States Navy in World War II, Sicily-Salerno-Anzio*, Boston, Little, Brown and Company, 1954, p. 280; J. Gordon, 'Joint Fire Support, The Salerno Experience', in *Military Review*, March 1989.
[29] Westphal, *The German Army in the West*, p.151
[30] C. D'Este, *Fatal Decision*, New York, Harper Collins, 1986; Morison, *Sicily-Salerno-Anzio*.
[31] W. Jackson, *Battle for Rome*, London, Batsford, 1969.
[32] Quoted in S. Roskill. *The War at Sea, Volume III, Part II*, London, HMSO, 1961, p. 38
[33] See R. Hart, *Clash of Arms*; S. Hart, *Montgomery and Colossal Cracks*.
[34] See N. Lewis, *Channel Firing*, London, Penguin, 1990.
[35] For an excellent account of the anti-FAC campaign see J. Tent, *E-Boat Alert, Defending the Normandy Invasion Fleet*, Shrewsbury, Airlife, 1996.
[36] S.R. Fiske 'Amount of Experience of U-Boat Commanders now on Operations', Admiralty Operational Intelligence Centre, SI.1009, 12 July 1944, TNA: ADM 223/172.
[37] *Invasion Europe, Battle Summary No. 39 Operation Neptune, Landings in Normandy, June 1944* (1952, Admiralty, internal use only), London, HMSO, 1994.
[38] C. Goulter, *A Forgotten Offensive*, London, Frank Cass, 1995.
[39] 'Operation Overlord: bombardment of heavily defended coasts, 1943–44'and 'Gunnery and Bombardment, 1944', TNA: ADM 116/5111 and ADM 199/2505.

[40] See D. Brown, 'The Air Spotting Pool', in D. Chandler and J. Collins Jr (eds), *The D-Day Encyclopaedia*, Oxford, Helicon, 1994.
[41] J. Bailey, *Field Artillery and Firepower*, London, The Military Press, 1987.
[42] 'Casualties and the Effects of Fire Support on the British Beaches in Normandy', Army Operation Research Group Report No. 261, p. 13, TNA: DEFE 2/492.
[43] See Buckingham, *D-Day, The First 72 Hours*, pp.157–9 and 175.
[44] G. Harrison, *Cross Channel Assault*, Washington D.C., US Government Printing Office, 1951.
[45] 'Special Tactical Study No.30, German Views of the Normandy Landings', 28 November 1944, TNA: War Office (WO) 232/25 M; Reynolds, *Steel Inferno: I SS Panzer Corps in Normandy*, Staplehurst, Spellmount, 1997, pp. 73–75; Buckingham, *D-Day, The First 72 Hours*, p. 224.
[46] See S. Rowell, *Full Circle*, Melbourne, MUP, 1974.
[47] B. Kennett and J. Tatman, *Craftsmen of the Army*, London, Leo Cooper, 1970, Ch. 12
[48] M. Hastings, *Overlord*, London, Michael Joseph, 1984.
[49] *Invasion Europe, Battle Summary No. 39 Operation Neptune, Landings in Normandy, June 1944* (1952, Admiralty, internal use only), London, HMSO, 1994, Para. 70.
[50] 'Operations of the 1st Canadian Army in NW Europe, 31 July–1October 1944', 17 October 1944, available at www.forces.gc.ca/dhh/downloads/cmhq/cmhq146.pdf. See also C. Stacey, *The Victory Campaign*, Ottawa, Queens Printer, 1960.
[51] J. Moulton *The Battle for Antwerp*, New York, Hippocrene Books, 1978.
[52] See www.navy.gov.au/spc/semaphore/issue3_2005.html; www.operations.mod.uk/telic and National Audit Office 'Operation Telic – UK Military Operations in Iraq', 11 December 2003, available at www.nao.org.uk/publications/nao_reports/03-04/030460.pdf.
[53] See *The Royal Navy Strategic Plan*, available at www.royal-navy.uk/server/show/nav.1242.

8. Together Again for the First Time: The Army, the RAN and Amphibious Warfare 1942-45

[1] DCGS to DCAS, 'Long Range Planning for Offensive Action – Landing Operations', 13 March 1942, National Archives of Australia (NAA) (Vic): B6121 289.
[2] H.V. Evatt to Prime Minister, 19 May 1942, NAA (ACT): A5954 5/8/19.
[3] MacArthur to Prime Minister, 8 June 1942, NAA (ACT): A5954 5/8/19.
[4] DCNS to CNS, 29 April 1942; CNS to Secretary, Defence Committee, 25 May 1942, NAA (Vic): B6121 194B.
[5] Major A.L. Rose, 'Combined Operations Training School, Toorbul Point', 21 March 1943, Australian War Memorial (AWM): AWM54 492/4/31.
[6] N. Friedman, US *Amphibious Ships and Craft: An Illustrated Design History*, Annapolis, Naval Institute Press, 2002, pp. 77–100.
[7] MacArthur to Commader Allied Naval Forces, 11 June 1942, AWM Blamey Papers: 3DRL 6643 2/53; 'History of Combined Operations Training in Australia', NAA (Vic): B6121 194B.
[8] R.M. Leighton and R.W. Coakley, *Global Logistics and Strategy 1940–1943*, Washington DC, Department of the Army, 1954, pp. 408–09.
[9] Naval Attaché, Washington to CNS, 19 February 1942, 9 June 1942, NAA (ACT): A981/4 DEF146.
[10] Colonel N.F. Wellington to Commander Allied Land Forces (CALF), 5 August 1942, Blamey Papers: 3DRL 6643 2/53.
[11] 'History of ALCV Marks I and II', Notes on War Cabinet Agendum 12/1944, NAA (ACT): A1308 776/1/51.

[12] H.J. Casey, *Engineers of the Southwest Pacific 1941–1945: Vol.: Amphibian Engineer Operations*, Washington DC, Government Printing Office, 1959, pp. 32–3, 39–41, 713.
[13] A more orthodox abbreviation would have been CINCUS, but this had an unfortunate pronunciation.
[14] *Memorandum of Agreement of the Chief of Staff, US Army and Commander in Chief, US Fleet and Chief of Naval Operations*, 13 April 1943, US National Archives: RG496 Box 667.
[15] W. Roskill, *The War at Sea, Vol. III: The Offensive, Part I: 1 June 1943–31 May 1944*, London, HMSO, 1960, pp. 61–2.
[16] Director of Plans (DoP) to DCNS, 'Assault Landing Training', 29 May 1942, NAA (Vic): B6121 289.
[17] *War Cabinet Minute* (2402), NAA (ACT): A2684 1080 Part 3.
[18] G. Hermon Gill, *Royal Australian Navy 1942-1945*, Adelaide, AWM, 1968, pp. 277–8; *Command History VII Amphibious Force*, NAA (Vic): B6121 291.
[19] Commander VII Amphibious Force to Commander Seventh Fleet, 4 December 1943, US National Archives: RG496 Box 683.
[20] DoP to DCNS, 'Disposal of LCVPs and LCMs', 17 October 1945, NAA(Vic): B6121 289E.
[21] OC, 3 Landing Ship Detachment, 'Report on the Formation of Aust Landing Ships Dets RAE', 20 September 1944, AWM: AWM54 963/21/7.
[22] G-3 GHQ SWPA to CoS GHQ SWPA, 13 February 1943, US National Archives: RG496 Box 667.
[23] *War Cabinet Agendum No. 389/1942*, NAA(ACT): A5954 518/9.
[24] *War Cabinet Agendum No. 293/1944*, NAA(ACT): A5954 518/9.
[25] HMAS *Assault* to Naval Board, 31 August 1942.
[26] 'HMAS *Assault* (Port Stephens)', NAA (Vic): B6121 194B.
[27] 'History of Combined Operations Training in Australia', NAA (Vic): B6121 194B.
[28] Friedman, *US Amphibious Ships and Craft*, pp. 115–20, 131–4, 140–3.
[29] S.E. Morison, *Vol. II: Operations in North African Waters, October 1942–June 1943*, Boston, Little, Brown and Co., 1947, p. 268.
[30] S.E. Morison, *Vol. VI: Breaking the Bismarcks Barrier, 22 July 1942–1 May 1944*, Boston, Little, Brown and Co., 1950, pp. 131–2.
[31] Chief of Staff GHQ SWPA, 'Amphibious Training', 8 February 1943, War Diary, Adv LHQ DA&QMG Branch, January – June 1943, AWM: AWM52 1/2/6.
[32] CALF to GHQ SWPA, 10 February 1943, Blamey Papers: 3DRL 6643 2/53.
[33] MacArthur to Blamey, 20 February 1943, NACP: RG496 Box 667
[34] Blamey to MacArthur, 22 February 1943, Blamey Papers: 3DRL 6643 2/53 .
[35] *CNS Weekly Progress Report No. 161*, 26 February 1943, NAA(ACT): A5954 518/19.
[36] Barbey to CO Amphibious Training Center Port Stephens, 'Discontinuance of Amphibious Training Center Port Stephens', 1 October 1943, US National Archives: RG496 Box 667.
[37] 'HMAS *Assault* (Port Stephens)', NAA (Vic): B6121 194B.
[38] G-3 GHQ SWPA to Chief of Staff GHQ SWPA, 'Control of Operations of the 2nd Engineer Special Brigade', G-3 Journal, 25 May 1943, AWM: AWM54 589/3/8.
[39] *Report of I Corps on the Operations in New Guinea from 22 Jan 43 to 8 Oct 43*, AWM: AWM54 519/6/32.
[40] Brigadier Hopkins to Blamey, 'Naval Beach Party', 26 October 1943, Blamey Papers, 3DRL 6643 2/43.
[41] GOC I Corps to CinC AMF, [10 September 1943], Herring Papers, State Library of Victoria: MS 11355, Box 10.
[42] DCGS to G-3 GHQ SWPA, 25 September 1943, War Diary, G Branch, New Guinea Force, AWM: AWM52 1/5/51.

[43] CO Shore Battalion to CO 532nd EBSR, 24 September 1943, US National Archives: RG496 Entry 388 Box 2383; Beachmaster Buna to CTF 76.9, 'Report of Landing Operation on Scarlet Beach Eight Miles North of Finschhafen', 7 October 1943, USN Historical Center, Barbey Papers, Box 31 Series 4.
[44] CTF 76, 'Finschhafen Operation – Report On', AWM: AWM54 589/7/27.
[45] CGS to CNS, 'Formation of Beach Group – 6 Div Group', 15 October 1943, NAA (Vic): MP742 240/3/502.
[46] NA2NM, 29 April 1943, NAA (Vic): B6121 289.
[47] Brig Hopkins to Blamey, 'Naval Beach Party', 26 October 1943, Blamey Papers, 3DRL 6643 2/43.
[48] SD Adv LHQ, 10 November 1943, AWM: AWM54 923/3/10; Barbey to Landforces, 25 October 1943, NAA (Vic): MP742 240/3/502.
[49] ACNB to Landforces, 6 November 1943, NAA(Vic): MP742 240/3/502.
[50] AG, 'HQ 1 Aust Beach Group – Raising of', 10 December 1943, AWM: AWM54 923/3/10.
[51] S.H. Kipling, *The Royal Naval Commandos*, www.btinternet.com/~keith.oakley/rncommando/rnhist.htm, accessed 29 January 2005; and A.C. Hampshire, *The Beachhead Commandos*, London, William Kimber and Co., 1983.
[52] Sec Navy, 'Naval Beach Signal Section' and 'Naval Beach Commando', 6 January 1944, AWM: AWM54 721/29/11.
[53] BGS I Corps, 'Formation of 2 Aust Beach Group', 2 March 1944, AWM: AWM54 721/29/11; Sec Army to Sec Navy, 'Second Naval Beach Commando', 12 April 1944, NAA(Vic): MP742 240/3/502.
[54] SNOALS to Sec Naval Board, 'Retention of Nucleus Beach Party in LSI', 14 May 1944, NAA (Vic): B6121 289.
[55] DoP, 'Naval Requirements for AMF Formations in Amphibious Operations', 18 September 1944, NAA (Vic): B6121 289.
[56] ACNB 13, 27 October 1944, NAA (Vic): B6121 289.
[57] Morshead to Berryman, 17 October 1944, AWM: AWM54 505/3/6.
[58] DoP, 'Naval Beach Organisation', 15 November 1944, NAA (Vic): B6121 289.
[59] A.E. Jones, *Sailor and Commando: A Royal Australian Navy Special Beach Commando 1942-1946*, Carlisle, Hesperian Press, 1998, p. 53.
[60] Sub-Lieutenant J.E. Dighton, *Report on Operation Conducted against Sadau Island*, NAA (Vic): B6121 61D.
[61] COHQ Bulletin Y/50, *Tarakan, Borneo, 1st May 1945*, October 1945, US National Archives: RG496 Box 2365.
[62] PBM B, *Report on Operation Oboe One*, NAA (Vic): B6121 61D.
[63] Casey, *Amphibian Engineer Operations*, pp. 660–1.
[64] PBM B, *Report on Operation Oboe One*, NAA (Vic): B6121 61D; *Building the Navy's Bases in World War II: History of the Bureau of Yards and Docks and the Civil Engineer Corps 1940-1946*, Vol. II, Washington, DC, Government Printing Office, 1947, p. 396
[65] CTU 78.1.12 to OC 26th Infantry Brigade, 6 May 1945, NAA (Vic): B6121 61D.
[66] CRE 9th Division to CE I Corps, 'Report on Activities of USN Underwater Demolition Teams in OBOE Six Operations', 17 Jun 1945, AWM: AWM54 619/4/14.
[67] SNOBU, *Report of Proceedings of RAN Beach Unit Operation 'OBOE 6' Landings by 9 Aust Division at Labuan Island and Brunei, North West Borneo*, 21 July 1945, NAA(Vic): B6121 62F.
[68] S.E. Morison, *Vol. XIII: The Liberation of the Philippines: Luzon, Mindanao, the Visayas 1944–1945*, Boston, Little, Brown and Co., 1959, p. 265.
[69] OC 1st Beach Group, 14 June 1945, War Diary, 1st Beach Group, AWM: AWM52 1/11/3.
[70] Morison, *The Liberation of the Philippines*, pp. 270-71.

[71] SNOBU, Report of Proceedings of RAN Beach Unit in Operation 'Oboe Two', 7 Aust Div Landing at Klandasan, Near Balikpapan, South East Borneo, AWM: AWM54 505/10/5; *Building the Navy's Bases in World War II*, Vol. II, p. 396

[72] See M. Evans, 'Maritime Power and the Australian Army', in D. Stevens (ed.), *Maritime Power in the 20th Cenury: The Australian Experience*, Sydney, Allen and Unwin, 1998, p. 234.

9. Tethered to Okinawa: The US Navy's Role in the Okinawa Operation

[1] The views presented in this paper are those of the author and do not necessarily represent those of the US Navy or the US Department of Defense.

[2] R.E. Appleman, J.M. Burns, R.A. Gugeler, and J. Stevens, *Okinawa: The Last Battle, The War in the Pacific Series, United States Army in World War II*, Washington, DC: Historical Division, Department of the Army, 1948, p. 4.

[3] Geographical and topographical information on the Ruykyus, and Okinawa in particular, is taken from a variety of sources, including Appleman, et al., *Okinawa: The Last Battle*, pp. 7–14; Major C.S. Nichols, USMC, and H.I. Shaw, *Okinawa: Victory in the Pacific*, Washington, DC, Historical Branch, G-3 Division, Headquarters, USMC, 1955, pp. 6–8; and S.E. Morison, *History of United States Naval Operations in World War II*, Vol. 14, *Victory in the Pacific, 1945*, Boston, Little, Brown and Company, 1960, pp. 79–80.

[4] *Commander-in-Chief Pacific Ocean Areas Joint Staff Study ICEBERG*, PACFLT/POA serial 000131, 25 October 1944, 1–2, 4; Box 52, World War II Plans Records, Record Group (RG) 38, National Archives and Records Administration, College Park, Maryland (hereafter NARA). This document and others in this series of files were examined while the World War II Plans Records were still located in the US Naval Historical Center's Operational Archives (hereafter OA), Washington Navy Yard, District of Columbia.

[5] J.G. Barlow, 'The Battle for Okinawa', in *Great American Naval Battles*, J. Sweetman (ed.), Annapolis, Naval Institute Press, 1998, pp. 364–5, citing CTF 51 and Commander Amphibious Forces US Pacific Fleet Operation Plan No. A1-45, serial 000120, 9 February 1945, 24, 35, Box 137, World War II Plans Records, RG 38, NARA II; and 'Commander Task Force Fifty-One [and] Commander Amphibious Forces US Pacific Fleet Report on Okinawa Gunto Operation from 17 February to 17 May, 1945 ('L' Day 1 April, 1945),' (I)-1, enclosure to memorandum from COMPHIBSPAC to COMINCH, serial 01400, 25 July 1945, Box 9, Harry W. Hill Papers, OA.

[6] The first *kamikaze* attacks had been initiated during the latter stages of the Battle of Leyte Gulf in October 1944.

[7] *Imperial General Headquarters Army High Command Record Mid-1941–August 1945*, Japanese Monograph 45, Tokyo, Japanese Research Division, Military History Section, Headquarters Far East Command, undated, pp. 172–6, 198–200; and *Okinawa Operations Record*, revised edn., Japanese Monograph 135, Tokyo, First Demobilization Bureau, 1949, Book 1, pp. 32, 45–6; both in Navy Department Library, US Naval Historical Center. See also *Reports of General MacArthur*, Vol. 2, pt. 2 , *Japanese Operations in the Southwest Pacific Area*, Washington, DC, Compiled from Japanese Demobilization Bureau Records by G-2 Section, Headquarters Far East Command, published by the Department of the Army, 1966, pp. 584–96.

[8] T.M. Huber, *Japan's Battle of Okinawa, April-June 1945, Leavenworth Papers* No. 18, Fort Leavenworth, US Army Command and General Staff College, May 1990, p. 13.

[9] Test of message (pink) from COM5THFLT to CTF 58 Info CTF 51, CINCPAC ADV HQ, 020359 March 1945 (GCT); in 'COMMAND SUMMARY [–] FLEET ADMIRAL C. W.

NIMITZ[,] U.S. NAVY [–] BOOK SIX[:] 1 JANUARY 1945 – 1 JULY 1945 [hereafter cited as CINCPAC/POA GRAYBOOK],' 3,030; Chester W. Nimitz Papers, OA.

[10] Barlow, 'The Battle for Okinawa', citing report from CTF 58 to CinC US Fleet, Subj: 'Report of Operations of Task Force FIFTY-EIGHT in Support of Landings at OKINAWA, 14 March through 28 May (East Longitude Dates), including Actions against KYUSHU, NANSEI SHOTO, Japanese Fleet at KURE, the YAMATO, and Operations in Direct Support of Landings at OKINAWA,' serial 00222, 18 June 1945, 3-5, Enclosure A-1, Enclosure B-2, and Appendix Two to Part 2; Box 216, World War II Action Reports, RG 38, NARA II.

[11] Summary of COM5THFLT preliminary report on strikes on Kyushu, 18-19 March 1945, entry for 20 March 1945 (Guam date); in CINCPAC/POA GRAYBOOK, Book Six, 2576.

[12] Summary of amplifying report from Commander Fifth Fleet, entry for 23 March 1945 (Guam date); in CINCPAC/POA GRAYBOOK, Book Six, 2579.

[13] J.G. Barlow, 'The U.S. Navy's Fight against the Kamikazes', in *New Interpretations in Naval History: Selected Papers from the Tenth Naval History Symposium Held at the United States Naval Academy, 11–13 September 1991*, J. Sweetman (ed.), Annapolis, Naval Institute Press, 1993, pp. 408–9; and Barlow, 'The Battle for Okinawa', p. 367.

[14] Because of the extent of her damage, *Franklin*, together with the cruiser USS *Santa Fe*, was being sent back to Pearl Harbor once she reached Ulithi. Summary of Task Force 58 reorganization, entry for 22 March 1945 (Guam date); in CINCPAC/POA GRAYBOOK, Book Six, 2578.

[15] During the OCTAGON Conference at Quebec in September 1944, British Prime Minister Winston Churchill had offered up the British Fleet to take part under American command in main operations against Japan, and President Roosevelt (despite Admiral Ernest King's strong demurrals) had eagerly accepted the offer. In mid-December 1944, the newly selected CinC of the British Pacific Fleet, Admiral Bruce Fraser, had met with Admiral Nimitz at Pearl Harbor to work out the operational and logistic details.

[16] Morison, *Victory in the Pacific*, p. 105.

[17] See the text of the message (pink) from CINCPOA ADV HQ [Nimitz's Advanced Headquarters on Guam] to COM5THFLT Info CTF 113, CTF 58 et al., 150658 March 1945 (GCT); in CINCPAC/POA GRAYBOOK, Book Six, 3052.

[18] Summary of TF 58 strikes on 23 March, entry for 24 March 1945 (Guam date); in CINCPAC/POA GRAYBOOK, Book Six, 2580.

[19] Text of message from CTF 58 to CTF 51 Info COM5THFLT, CTF 52, 252155 March 1945 (GCT); in CINCPAC/POA GRAYBOOK, Book Six, 2801.

[20] Appleman et al., *Okinawa: The Last Battle*, pp. 52-4. See also the summary of the Kerama Retto attacks on 26 March, entry for 26 March 1945 (Guam date); in CINCPAC/POA GRAYBOOK, Book Six, 2582.

[21] Morison, *Victory in the Pacific*, pp. 133-4; and summary of the air attack on the night of 26/27 March, entry for 27 March 1945 (Guam date); in CINCPAC/POA GRAYBOOK, Book Six, 2584.

[22] See the text of the message from CTF 57 to COM5THFLT Info CINCPAC ADV HQ, CINCBPF [Commander-in-Chief British Pacific Fleet], 280503 March 1945 (GCT); in CINCPAC/POA GRAYBOOK, Book Six, 2805.

[23] Morison, *Victory in the Pacific*, p. 105.

[24] Summaries of Kerama Retto operations for 28 and 29 March, entries for 28 March 1945 (Guam date) and 30 March 1945 (Guam date); in CINCPAC/POA GRAYBOOK, Book Six, 2585, 2587; and Appleman et al., *Okinawa: The Last Battle*, p. 56.

[25] Appleman et al., *Okinawa: The Last Battle*, p. 57.

[26] Huber, *Japan's Battle of Okinawa*, pp. 27–8.

[27] Summary of Okinawa landings, entry for 1 April 1945 (Guam date); in CINCPAC/POA GRAYBOOK, Book Six, 2591.
[28] Huber, *Japan's Battle of Okinawa*, p. 32.
[29] Okinawa ground battle summary, entry for 7 April 1945 (Guam date); in CINCPAC/POA GRAYBOOK, Book Six, 2600.
[30] Okinawa ground battle summary, entry for 8 April 1945 (Guam date); in CINCPAC/POA GRAYBOOK, Book Six, 2603.
[31] Barlow, 'The Battle for Okinawa', p. 369; and Barlow, 'The US Navy's Fight against the Kamikazes', pp. 409–10.
[32] *Reports of General MacArthur*, vol. 2, pt. 2, p. 598.
[33] This screen proved its effectiveness, since during the Okinawa fighting Japanese suicide boats sank only one vessel (an LCI [G]) and damaged only three other vessels.
[34] Barlow, 'The Battle for Okinawa', pp. 369-70, citing CTF 51 and COMPHIBSPAC OpPlan A1–45, Annex K, Appendix 2, pp. 1–2, 6–10; and COMPHIBSPAC Report on Okinawa Gunto Operation (III), pp. 11–19.
[35] Barlow, 'The Battle for Okinawa', p. 370.
[36] Barlow, 'The US Navy's Fight against the Kamikazes', p. 411.
[37] Technical Air Intelligence Summary # 31 [–] BAKA, OPNAV 16V T-131, June 1945, 1, 7; 'AIR INTELLIGENCE GRP. /CENTER Intell. Series 'T' Nos. 130-136' Folder, Box 86, World War II Command File, OA.
[38] Barlow, 'The Battle for Okinawa', p. 374, citing NavShips A-4 (424), *Summary of War Damage to US Battleships, Carriers, Cruisers, Destroyers and Destroyer Escorts, 8 December 1944 to 9 October 1945* (Washington, DC, 1946), 10; Box 456, World War II Command File, OA.
[39] Text of message from CTF 51 to TF 51 Info CINCPAC BOTH, COM5THFLT, 130840 April 1945 (GCT); in CINCPAC/POA GRAYBOOK, Book Six, 2847.
[40] R.F. Wallace, *From Dam Neck to Okinawa: A Memoir of Antiaircraft Training in World War II*, J.G. Barlow (ed.), *The US Navy in the Modern World* Series, No. 5, Washington, Naval Historical Center, 2001, p. 41.
[41] Okinawa naval battle summaries, entries for 12 and 13 April 1945 (Guam dates); in CINCPAC/POA GRAYBOOK, Book Six, 2610, pp. 2612–3.
[42] Barlow, 'The US Navy's Fight against the Kamikazes', p. 411.
[43] Text of message from CTF 51 to COM5THFLT, 131040 April 1945 (GCT); in CINCPAC/POA GRAYBOOK, Book Six, 2853.
[44] Quoted in E.B. Potter, *Nimitz*, Annapolis, Naval Institute Press, 1976, p. 375.
[45] Text of message from CTF 57 to COM5THFLT, CTG 52.1 Info CINCPAC BOTH, CINC BPF, CTF 112, CTF 73, 091151 May 1945 (GCT); in CINCPAC/POA GRAYBOOK, Book Six, 2898.
[46] 'Report of Sr. U.S. Naval Liaison Officer, British Pacific Fleet, 20 April–25 May 1945, Forwarding Of,' Serial 005043, 3 June 1945, 2; Box 37, World War II Command File, OA.
[47] Untitled manuscript speech by Captain W.A. Read, USNR, [probably 1946], pp. 16–17; quoted in T. Taylor, *The Magnificent Mitscher*, with a Foreword by Admiral Arthur W. Radford and an Introduction by J.G. Barlow, New York, W. W. Norton and Company, 1954, reprint [with Introduction], Annapolis, Naval Institute Press, 1991, pp. xxx–xxxi.
[48] Taylor, *The Magnificent Mitscher*, p. xxxi.
[49] See the summary of ground fighting, entry for 21 June 1945 (Guam date); in CINCPAC/POA GRAYBOOK, Book Six, 2704.
[50] Barlow, 'The Battle for Okinawa', p. 378.

10. The US Navy and Marines in the Civil War in China 1945-47

[1] Selected List of References.

a) US National Archives: Record Group 59: Records of the Department of State; Record Group 165: Records of General and Special Staffs; Record Group 218: Records of the Joint and Combined Chiefs of Staff; Record Group 226: Records of the Office of Strategic Services.
b) Admiral Daniel E. Barbey Papers, US Naval Historical Center, Washington DC.
c) Albert C. Wedemeyer Papers, Hoover Institution, Stanford California
d) Foreign Relations of the United States 1945: Vol. VI: China
e) Foreign Relations of the United States 1946: Vol. VII: The Far East
f) L. Bland (ed.) George C. *Marshall's Mediation Mission To China*, Lexington VA, George C. Marshall Foundation, 1996.
g) L. Bland and S.R. Stevens (eds.) *The Papers of George C. Marshall, Vol 6: The Finest Soldier*, Baltimore, Johns Hopkins University Press, 2004.
h) O. Cary, (ed.) *Eyewitness To History*, Tokyo, Kodansha, 1991.
i) B.M. Frank and H.I. Shaw *Victory and Occupation: History of US Marine Corps Operations in World War II*, Washington, G-3 Headquarters US Marine Corps, 1968.
j) M. S. Gallichio, *The Cold War Begins in Asia*, New York, Columbia University
k) Press, 1988.
l) D. Gillin and F. Etter, 'Staying On: Japanese Soldiers and Civilians in China 1945' in T*he Journal of Asian Studies*, XLII, May 1983.
m) Captain E. Klein, 'The Situation In North China', Marine Corps Gazette, April 1946.
n) E.B. Sledge, *China Marine*, Tuscaloosa, University of Alabama Press, 2002.
o) O. A. Westad, *Decisive Encounters*, Stanford, Stanford University Press 2003

11. The Shanghai Blockade and British Diplomatic Recognition of the PRC.

[1] For more information on the Nationalist blockade in its entirety, see B.A. Elleman 'The Nationalists' Blockade of the PRC, 1949–58', in B.A. Elleman and S.C.M. Paine (eds), *Naval Blockades and Seapower: Strategies and Counter-Strategies, 1805–2005*, London, Routledge, 2005, pp. 133–43; quoting from SO(I) HK to the Admiralty, 23 June 1949, the National Archives UK(TNA): FO 371/75900.

[2] 'Confidential' Cypher message from the British Embassy, Canton, to Nanking Office, 22 June 1949, TNA: FO 371/75900.

[3] Message from the British Embassy, Canton, to the Nationalist Ministry of Foreign Affairs, 1 July 1949, TNA: FO 371/75901.

[4] Report on the Chinese Blockade by P.D. Coates, 4 July 1949, TNA: FO 371/75901; emphasis in the original.

[5] 'Appreciation of the Ability of the Chinese Nationalist Navy to Effect a Blockade of Communist Territorial Waters (Secret)', Intelligence Division, Naval Staff, Admiralty, 9 July 1949, TNA: FO 371/75902.

[6] CinC FES Afloat to the Admiralty (Secret), 18 July 1949, TNA: FO 371/75903.

[7] CinC FES Afloat to the Admiralty (Secret), 18 July 1949.

[8] 'Chinese Nationalist Blockade and Bombing', 24 February 1950, TNA: FO 371/83425.

[9] Foreign Office telegram No. 7470 to Washington, TNA: FO 371/75903.

[10] Foreign Office telegram No. 7470 to Washington

[11] From the British Consul in Shanghai to the Foreign Office, marked 'Secret', 20 July 1949, TNA: FO 371/75903.

[12] From the British Consul in Shanghai to the Foreign Office, marked 'Secret', 30 July 1949, TNA: FO 371/75904.
[13] 'Chinese Nationalist Blockade and Bombing', 24 February 1950, TNA: FO 371/83425.
[14] CinC FES to Admiralty, 25 August 1949, Rec'd: 0919, TNA: FO 371/75907.
[15] CinC FES to Admiralty, 25 August 1949, Rec'd: 0825, TNA: FO 371/75907.
[16] CP(49) 180 'China', 26 August 1949, TNA: FO 371/75810-75815.
[17] Telegram from Foreign Office to Washington, 7 September 1949, TNA: FO 371/75907.
[18] CP(49) 180 'China', 26 August 1949, TNA: FO 371/75810-75815.
[19] CinC FES to Admiralty, 12 September 1949, National Archives UK: FO 371/75910.
[20] Secretary of State 'Chinese Nationalist Blockade', 23 December 1949, TNA: FO 371/75916.
[21] Message from Butterfield and Swire, Shanghai, to John Swire and Sons, Ltd., London, 22 September 1949, TNA: FO 371/75911.
[22] 'On the Way In', *The China Weekly Review*, 24 September 1949.
[23] CinC FES to Admiralty, 5 October 1949, TNA: FO 371/75911.
[24] 'Chinese Nationalist Blockade and Bombing', 24 February 1950, TNA: FO 371/83425.
[25] 'Another Hole in the Blockade', *The China Weekly Review*, 22 October 1949.
[26] Telegram from Shanghai to Foreign Office, 8 November 1949, TNA: FO 371/75912.
[27] 'Blockade Stalemate', *The China Weekly Review*, 12 November 1949.
[28] Telegram 1866 from Nanking to Foreign Office, 8 November 1949, TNA: FO 371/75912.
[29] Secretary of State 'Chinese Nationalist Blockade', 23 December 1949, TNA: FO 371/75916.
[30] 'What Will America Do?' *The China Weekly Review*, 15 October 1949.
[31] 'The Illegal Blockade', *The China Weekly Review*, 31 December 1949.
[32] British Consulate-General, Canton, to Political Adviser, Hong Kong (Confidential), TNA: FO 371/75810-75815.
[33] Cablegram 198 from Nanking to Australia Department of External Affairs, 4 August 1949, http://www.infor.dfat.gov.au/info/historical/HistDocs.nsf/vVolume.
[34] 'Record of a Meeting Held at the State Department, Washington, on 13th September, 1949', TNA: FO 371/75810-75815.
[35] L.H. Woolsey, 'Editorial Comment: Closure of Ports by the Chinese Nationalist Government', *The American Journal of International Law*, vol. 44 (1950), pp. 350–6; 355.
[36] He Di, 'The Last Campaign to Unify China', M.A. Ryan, D.M. Finkelstein, and M.A. McDevitt (eds), *Chinese Warfighting: The PLA Experience Since 1949*, Armonk, M.E. Sharpe Publishers, 2003, pp, 84, 87–8.
[37] 'China Blockade: How it Works; Ships by the US – Sailors by Chiang Kai-shek', *US News and World Report*, 20 February 1953.
[38] Memorandum by the Secretary of State for Foreign Affairs entitled 'China', C.P. (50) 73, 20 April 1950, TNA: CAB 21/3273.
[39] Letter from A.E.E. Franklin to H.A Graves, 21 January 1950, TNA: FO 371/83424.
[40] Chiefs of Staff Committee Joint Planning Staff, 'Recognition of Chinese Communist Government', 28 November 1949, TNA: CAB 21/3273.
[41] 'Chiang Mines China Coast', *Daily Worker*, 20 December 1949, TNA: FO 371/75917.
[42] Nanking Telegram to Foreign Office, 15 December 1949, TNA: FO 371/75916.
[43] 'Draft Signal' to 'CinC, Far East', undated, but probably mid-to-late December 1949, TNA: FO 371/75917; final policy issued on 9 January 1950, FO 871/83424 with a crucial addition that if the Nationalists were responsible for a hostile action, then 'air reconnaissance must not be carried out without prior approval of HM Government'.
[44] From Shanghai to Foreign Office, 10 January 1950, TNA: FO 371/83424.
[45] 'Chinese Nationalist Blockade and Bombing', 24 February 1950, TNA: FO 371/83425.
[46] Assistant Naval Attaché in Shanghai to Foreign Office, 15 February 1949, TNA: FO 371/83425.

[47] Admiral of the Fleet Lord Fraser, to Sir William Strang, Permanent Under-Secretary of State, 9 February 1950, TNA: FO 371/83425.
[48] CinC FES to Admiralty, 14 February 1950, TNA: FO 371/83425.
49 Brief entitled 'China Blockade', undated, but probably late January 1950, TNA: FO 371/83425.
[50] A.A.E. Franklin, Minutes, 14 February 1950, TNA: FO 371/83425.
[51] Foreign Office Note to the Cabinet Far East (Official) Committee, entitled 'Chinese Nationalist Blockade and Bombing', 17 March 1950, TNA: LO/2/943.
[52] Memorandum by the Secretary of State for Foreign Affairs entitled 'China', C.P. (50) 73, 20 April 1950, TNA: CAB 21/3273.
[53] Mao Zedong, 'Farewell, Leighton Stuart!' *Selected Works of Mao Tse-tung*, Beijing, Foreign Languages Press, 1969, Vol. IV, pp. 433–40.
[54] Memorandum by the Secretary of State for Foreign Affairs entitled 'China', CP (50) 73, 20 April 1950, TNA: CAB 21/3273.
[55] 'The Nationalist Closure of Chinese Ports in Communist Hands and of Certain Territorial Waters', Hong Kong, 12 July 1949, TNA: FO 371/75905.
[56] 'The Southeast China Coast Today', *The ONI Review*, February 1953, pp. 51–60.
[57] 'A New Type of Formosa Warfare: Reds Spot British Cargo Ships for Chiang's Bombers', *US News and World Report*, 18 November 1955.
58 'Report on Visit to Hong Kong, 15–21 February, 1952', (Secret), by D.F. Allen, TNA: ADM 1/23217.
[59] Letter from H.A. Graves, British Embassy in Washington, to P.W.S.Y. Scarlett, Far Eastern Department (Secret), 4 February 1950, TNA: FO 371/83425.
[60] 'Notes on Sino-Soviet Relations', 1958, TNA: FO 371/133366.

12. Korea: the First Challenge for Australian Naval Aviation

[1] D. Horner, *Defence Supremo: Sir Frederick Shedden and the Making of Australian Defence Policy*, St. Leonards, Allen and Unwin, 2000. p. 187.
[2] A. Wright, *Australian Carrier Decisions: The Acquisition of HMA Ships Albatross, Sydney and Melbourne*, Canberra, Department of Defence (Navy) Maritime Studies Program, 1998. p. 69.
[3] Horner, *Defence Supremo*, p. 187.
[4] Horner, *Defence Supremo*, p. 187.
[5] J. Goldrick, 'Carriers for the Commonwealth,' in T.R. Frame, J.V.P. Goldrick and P.D. Jones (eds), *Reflections on the Royal Australian Navy*, Kenthurst, Kangaroo Press, 1991. pp. 224 and 226.
[6] Wright, *Australian Carrier Decisions*, p. 96.
[7] J. McCaffrie, 'Australia's Maritime Strategy and Air Power,' in Wing Commander Keith Brent (ed.) 100 Years of Aviation: The Australian Military Experience, Fairbairn, RAAF Aerospace Centre, 2004, p. 156.
[8] Wright, *Australian Carrier Decisions*, p. 133.
[9] Goldrick, 'Carriers for the Commonwealth,' p. 230.
[10] Goldrick, 'Carriers for the Commonwealth,' p. 232.
[11] Goldrick, 'Carriers for the Commonwealth,' p. 232.
[12] Wright, *Australian Carrier Decisions*, p. 151.
[13] Wright, *Australian Carrier Decisions*, pp. 151 and 157.
[14] The Australian Naval Aviation Museum friends and volunteers, *Flying Stations: A Story of Australian Naval Aviation*, St. Leonards, Allen and Unwin, 1998, p. 53.
[15] The Australian Naval Aviation Museum, *Flying Stations*, p. 54.
[16] The Australian Naval Aviation Museum, *Flying Stations*, p. 59.

[17] The Australian Naval Aviation Museum, *Flying Stations*, pp. 73–4.
[18] The Australian Naval Aviation Museum, *Flying Stations*, p. 75.
[19] The Australian Naval Aviation Museum, *Flying Stations*, p. 76
[20] The Australian Naval Aviation Museum, *Flying Stations*, p. 84.
[21] C. Jones, *Wings and the Navy 1947–1953*, Kenthurst, Kangaroo Press, 1997, p. 75.
[22] R. O'Neill, *Australia in the Korean War 1950–1953, Vol. I: Strategy and Diplomacy*, Canberra, Australian Government Publishing Service (AGPS), 1981, p. 6.
[23] O'Neill, *Australia in the Korean War 1950-1953, Vol. I*, pp. 7–8.
[24] O'Neill, *Australia in the Korean War 1950-1953, Vol. I*, p. 11.
[25] O'Neill, *Australia in the Korean War 1950-1953, Vol.I*, p. 14.
[26] O'Neill, *Australia in the Korean War 1950-1953, Vol.I*, p. 20.
[27] O'Neill, *Australia in the Korean War 1950-1953, Vol.I*, pp. 31–2.
[28] O'Neill, *Australia in the Korean War 1950-1953, Vol. I*, pp. 38 and 41.
[29] R. O'Neill, *Australia in the Korean War 1950–1953, Vol. II: Combat Operations*, Canberra, AGPS, 1985, p. 416.
[30] O'Neill, *Australia in the Korean War 1950–1953, Vol. II*, pp. 417 and 461.
[31] O'Neill, *Australia in the Korean War 1950–1953, Vol. II*, p. 425.
[32] O'Neill, *Australia in the Korean War 1950–1953, Vol. II*, p. 464.
[33] O'Neill, *Australia in the Korean War 1950–1953, Vol. II*, p. 465.
[34] O. Thetford, *British Naval Aircraft Since 1912*, 6th rev.ed., Annapolis, Naval Institute Press, 1991, pp. 185–6
[35] S. Wilson, *Sea Fury, Firefly and Sea Venom in Australian Service*, Weston Creek, Aerospace Publications, 1993, p. 81.
[36] Jones, *Wings and the Navy*, p. 86.
[37] Jones, *Wings and the Navy*, p. 76.
[38] O'Neill, *Australia in the Korean War 1950–1953, Vol. II*, p. 486.
[39] O'Neill, *Australia in the Korean War 1950–1953, Vol. II*, p. 480
[40] O'Neill, *Australia in the Korean War 1950–1953, Vol. II*, p. 477.
[41] O'Neill, *Australia in the Korean War 1950–1953, Vol. II*, p. 471.
[42] O'Neill, *Australia in the Korean War 1950–1953, Vol. II*, p. 485.
[43] O'Neill, *Australia in the Korean War 1950–1953, Vol. II*, p. 350.
[44] Major M.B. Simkin, *H.M.A.S. Sydney in Korean Waters*, Report by the Carrier Borne Ground Liaison Officer, undated, pp. 1–3. Copy held by SPC-A.
[45] Jones, *Wings and the Navy*, p. 89.
[46] Jones, *Wings and the Navy*, p. 86.
[47] F. Lane, 'Naval Aviation in the Korean War,' *Naval Officers Club Newsletter*, No. 60, 1 March 2005, p. 18.
[48] 72 CBAL Section, HMAS SYDNEY, War Diary – Korean Operations, 21 October 1951.
[49] O'Neill, *Australia in the Korean War 1950–1953, Vol. II*, p. 483. The actual number of aircraft lost varies among sources. Taking account of enemy action and all other causes 13 aircraft appear to have been lost – nine Sea Furies and four Fireflies.
[50] The Australian Naval Aviation Museum, *Flying Stations*, p. 87.
[51] O'Neill, *Australia in the Korean War 1950–1953, Vol. II*, p. 484.
[52] The Australian Naval Aviation Museum, *Flying Stations*, p. 89.
[53] O'Neill, *Australia in the Korean War 1950–1953, Vol. II*, p. 472.
[54] Lane, 'Naval Aviation in the Korean War', p. 16.
[55] O'Neill, *Australia in the Korean War 1950–1953, Vol. II*, p. 472.
[56] Jones, *Wings and the Navy*, p. 77.
[57] O'Neill, *Australia in the Korean War 1950–1953, Vol. II*, p. 470.

[58] Jones, *Wings and the Navy*, p. 78. Reference to the 72 CBAL De-briefing Reports shows that in fact 93 aircraft sorties were flown on that day.
[59] 72 CBAL Section, HMAS SYDNEY, *War Diary – Korean Operations*, 31 October 1951.
[60] O'Neill, *Australia in the Korean War 1950–1953, Vol. II*, p. 474.
[61] Jones, *Wings and the Navy*, p. 84.
[62] 72 CBAL Section, HMAS SYDNEY, *War Diary – Korean Operations*, 30 November 1951.
[63] O'Neill, Australia in the *Korean War 1950–1953, Vol. II*, p. 480.
[64] O'Neill, Australia in the *Korean War 1950–1953, Vol. II*, p. 482.
[65] 72 CBAL Section, HMAS SYDNEY, *War Diary – Korean Operations*, 5 September 1951 and 3–4 October 1951.
[66] 72 CBAL Section, HMAS SYDNEY, *War Diary – Korean Operations*, 5 October 1951.
[67] O'Neill, *Australia in the Korean War 1950-1953, Vol. II*, p. 472.
[68] Naval Staff History, *British Commonwealth Naval Operations, Korea, 1950-53*, London, Ministry of Defence, Historical Branch (Naval) 1967, p. 188. Fred Lane also provides this figure, claiming that only one Firefly was lost overboard, together with a motor boat and a forklift. See, Lane, 'Naval Aviation in the Korean War,' p. 18. O'Neill claims that four aircraft were lost, see *Australia in the Korean War 1950–1953*, Vol. II, p. 473.
[69] O'Neill, *Australia in the Korean War 1950–1953, Vol. II*, p. 473.
[70] O'Neill, *Australia in the Korean War 1950–1953, Vol. II*, p. 453.
[71] O'Neill, *Australia in the Korean War 1950–1953, Vol. II*, pp. 478–79.
[72] O'Neill, *Australia in the Korean War 1950–1953, Vol. II*, pp. 479.
[73] O'Neill, *Australia in the Korean War 1950–1953, Vol. II*, pp. 486.
[74] 72 CBAL Section, HMAS SYDNEY, *War Diary – Korean Operations*, 29 October 1951.
[75] O'Neill, *Australia in the Korean War 1950–1953, Vol. II*, pp. 475.
[76] O'Neill, *Australia in the Korean War 1950–1953, Vol. II*, pp. 485.
[77] 72 CBAL Section, HMAS SYDNEY, *War Diary – Korean Operations*, 1 February 1952.
[78] Lane, 'Naval Aviation in the Korean War,' p. 18
[79] The Australian Naval Aviation Museum, *Flying Stations*, p. 86.
[80] Jones, *Wings and the Navy*, p. 89.
[81] The Australian Naval Aviation Museum, *Flying Stations*, p. 84.
[82] O'Neill, *Australia in the Korean War 1950–1953, Vol. II*, pp. 486.
[83] Simkin, *H.M.A.S. Sydney in Korean Waters*, p. 4.
[84] Jones, *Wings and the Navy*, p. 95.
[85] Jones, *Wings and the Navy*, pp. 101–2.

13. Virtual War in the 'Ice Jungle'

[1] Serving a very different purpose, a slightly altered portion of this essay originally appeared in the *Journal of Strategic Studies* (www.informaworld.com) as part of a collection of essays presented at the Cold War at Sea Conference. The US Naval War College hosted this gathering in May of 2004 in collaboration with the Russian Federation Navy. My essay, entitled 'Virtual War in the Ice Jungle: "We don't know how to do this"', (volume 28, no. 2, April 2005) was prepared as part of my regular duties as a federal employee, civil servant, and senior member of the staff at the US Naval Historical Center, attached to the office of the Director of Navy Staff. As a product of my regular tasks as a civil servant the original manuscript version has always resided in the public domain. Any value added, of course, belongs to the publishers, Taylor and Francis. I wish to thank the *Journal of Strategic Studies* for publishing the earlier essay focusing on the American Cold War encounter with the Russians and for their enlightened patience and policy in this matter.

[2] Oral History with Dr. Waldo Lyon, by G.E. Weir, US Naval Historical Center, 16 April 1994, US Navy Operational Archive, Washington DC (OA). USS *Hawkbill* (SSN 666) was a *Sturgeon* class nuclear boat and USS *Seadragon* (SSN 584) belonged to the *Skate* class.
[3] Oral History with Vice Admiral Anatoli I. Shevchenko by G.E. Weir, Moscow, 10 and 13 February 2002. The author also consulted an essay prepared for him by Vice Admiral Shevchenko entitled 'The Arctic Cruises'. The oral history and the essay are in the possession of the author.
[4] W. Østreng, 'The Geostrategic Conditions of Deterrence on the Barents Sea', in L.W. Brigham (ed.), *The Soviet Maritime Arctic*, Annapolis, Naval Institute Press, 1991, pp. 201–14.
[5] A poliyna is an opening in the ice that will provide space for the submarine to surface safely. Very often it can be covered with a relatively thin ice layer that is easily punctured.
[6] The Soviet nuclear submarine *K-21* had made the first Soviet Arctic cruise in 1962. The *K-3* commander, Captain First Rank L. Zhiltsov, his on-board supervising officer, Rear Admiral A. Petelin, and the propulsion officer, R. Timofeev, were elevated to Heroes of the Soviet Union. In 1963 the nuclear submarine *K-115* (Commander I. Dubyaga) made the first cruise from the Barents Sea to the Pacific Ocean under the Arctic ice; the submarine also surfaced at the drifting base 'North Pole – 12'. In September of the same year the ballistic missile submarine *K-178* (Commander A. Mikhailovsky) had cruised from the Pacific Ocean to the Northern Fleet. On 29 September in 1963, the nuclear submarine *K-181* (Commander Y. Sysoev and the cruise supervisor Admiral V. Kasatonov) surfaced at the North Pole. Oral History with Rear Admiral Oleg Chefonov and Captain First Rank Igor Chefonov by G.E. Weir, Moscow, 12 February 2002. The author also consulted an essay prepared for him by the brothers Chefonov entitled 'The Missile Firing Submarines'. This essay and the oral history are in the author's possession.
[7] Oral History with Rear Admiral Oleg Chefonov and Captain First Rank Igor Chefonov; G.P. Steele, *Seadragon: Northwest Under the Ice*, New York, Dutton, 1962. Other works consulted relative to the under ice experience include: W.M. Leary, *Under Ice: Waldo Lyon and the Development of the Arctic Submarine*, College Station, Texas A & M University Press, 1999; G.E. Weir, *An Ocean in Common: American Naval Officers, Scientists, and the Ocean Environment*, College Station, Texas A & M University Press, 2001; W. Anderson with Clay Blair Jr., Nautilus 90 North, Blue Ridge Summit, Tab Books, 1989.
[8] 'Study of Undersea Warfare' (The Low Report), 22 April 1950, Post 1 January 1946 Command File, OA.
[9] Admiral J.D. Watkins, USN, 'The Maritime Strategy', in *US Naval Institute Proceedings*, Vol. 112, (Supplement to January 1986), pp. 3–17. See particularly page 5ff.
[10] O.R. Cote Jr., *The Third Battle: Innovation in the US Navy's Silent Cold War Struggle with Soviet Submarines*, Newport Paper Number 16, Newport, Naval War College, 2003. Nuclear torpedoes made the Cuban missile crisis much more dangerous than US President John F. Kennedy thought at the time. He was unaware of their presence on board the four Soviet Foxtrot submarines on their way to Mariel during the crisis. Oral History with Captain First Rank Nikolai Shumkov by G.E. Weir, Moscow, 14 February 2002. Personal collection of the author. G.E. Weir and W.J. Boyne, *Rising Tide: The Untold Story of the Russian Submarines that Fought the Cold War*, New York, Basic Books, 2003.
[11] Oral history with Vice Admiral Anatoli I. Shevchenko; Shevchenko, *The Arctic Cruises*.
[12] Weir, *An Ocean in Common;* G.E. Weir, *Forged in War: The Naval-Industrial Complex and American Submarine Construction, 1940–1961*, Washington DC, Brassey's, 1998.
[13] The LRAPP complemented SOSUS by addressing, by means of scientific and engineering solutions, the immediate at-sea acoustic detection requirements of the American surface operating forces. SURTASS became operational in 1984 in an at-sea, very mobile

configuration on board dedicated surface ships using many of the same detection technologies developed for SOSUS.

[14] Oral History with Captain First Rank Boris Kolyada by G.E. Weir, St. Petersburg, 17 February 2002. Kolyada, 'The Last Cruise of the K-278 (the *Komsomolets*)'. Both the essay and the oral history are in the author's possesion.

[15] Oral History with Dr. Waldo Lyon.

[16] W.K. Lyon, 'Submarine Combat in the Ice', in *US Naval Institute Proceedings*, Vol. 118, February, 1992, p. 38.

[17] One of the best souces for unclassified details on open basin navigation in the Arctic is T. Stefanik, *Strategic Antisubmarine Warfare and Naval Strategy*, Lexington, Lexington Books, 1987.

[18] Oral History with Vice Admiral Anatoli I. Shevchenko. See also, Weir and Boyne, *Rising Tide*.

[19] Steele, *Seadragon*.

[20] S. Holroyd, 'Canadian and U.S. Defense Planning Toward the Arctic', RAND Corporation, p. 8. Defense Technical Information Service.

[21] Stefanik, *Strategic Antisubmarine Warfare and Naval Strategy*, p. 45.

[22] Oral History with Rear Admiral Oleg Chefonov and Captain First Rank Igor Chefonov; Oral History with Captain First Rank Igor Kurdin by G.E. Weir, St Petersburg, 19 February 2002. Personal collection of the author. The possession and use of ice breaking torpedoes was discussed during these interviews.

[23] D. Glick, 'The Great Thaw', *National Geographic Magazine*, Vol. 206, No. 3, September 2004, p. 16. Note also the graphic images of the Arctic ice retreat on page 20f.

[24] D. MacKenzie, 'Arctic Meltdown', in *New Scientist*, 3 March 2002, http://www.newscientist.com, consulted October 2004.

[25] 'Naval Operations in an Ice Free Arctic Symposium, Handbook', 17–18 April 2001, Sponsored by the Office of Naval Research, Oceanographer of the Navy, and the Arctic Research Commission, p. 3. Contemporary History Branch, US Naval Historical Center.

[26] US Arctic Research Commission, 'The Arctic Ocean and Climate Change: A Scenario for the US Navy', p. 15. Contemporary History Branch, US Naval Historical Center.

14. A Maritime Approach to Joint and Coalition Warfare: Three Case Histories

[1] Naval Staff History, *Middle East Operations*, London, Naval Historical Branch, Ministry of Defence, 1968.

[2] Naval Staff History, *Middle East Operations*.

[3] Flag Officer Second-in-Command Mediterranean Fleet (Vice Admiral Sir Robin Durnford-Slater) ROP, May 1958.

[4] Naval Staff History, *Middle East Operations*.

[5] D. Hobbs, *Aircraft Carriers of the Royal and Commonwealth Navies*, London, Greenhill Books, 1996.

[6] FOAC ROP, 14–24 July 1958, TNA: MII/276/153/58.

[7] N. Polmar, *Aircraft Carriers*, London, Macdonald and Co, 1969.

[8] CinC Mediterranean, ROPs May to October 1958, TNA: MII/282/5/59.

[9] H. Popham, *Into Wind – A History of British Naval Flying*, London, Hamish Hamilton, 1969.

[10] D. Brown, 'Study 1/97', Naval Historical Branch, London.

[11] FOME 356/ME 386/70, 15 September 1961, TNA.

[12] Hobbs, *Aircraft Carriers of the Royal and Commonwealth Navies*.

[13] TNA: JIC(61)53/10.
[14] TNA: MII/276/128/61.
[15] HMS *Victorious*, 130/5, 30 July 1961, the National Archive.
[16] TNA: MII/276/128/61.
[17] FOME, 356/ME/386/70, 15 September 1961, TNA.
[18] 'The Fourth Commission of HMS *Centaur*', published by the ship's company, 1965.
[19] Much of the detail for this section was taken from D.K. Hankinson, 'HMS Centaur at Dar es Salaam', in United States Naval Institute *Proceedings*, November 1969.
[20] CinC East Indies Station, 'Remarks and Recommendations', TNA: MII/288/19/58.
[21] Naval Staff History, *Middle East Operations*.
[22] G. Prins, 'Air Power from the Sea – Alternative Views', a paper read at the 'Air Power from the Sea' conference at the Fleet Air Arm Museum, UK, 1 February 2005.
[23] United Nations Recommendation 55 Article 203.

15. The RAN and Air Mobile Operations: The RAN Helicopter Flight Vietnam

[1] The author has made use of the following sources: D. Halberstam, *Best and the Brightest*, Random House, New York, 1992 and *The Making of a Quagmire*, McGraw-Hill, New York, 1987; B. Fall, *Street Without Joy*, Leo Cooper, London, 2005; J. Grey, *Up Top – The Royal Australian Navy and Southeast Asian Conflicts 1955–1972*, Allen and Unwin, Sydney, 1998; The Australian Naval Aviation Museum, *Flying Stations – A Story of Australian Naval Aviation*, Allen and Unwin, Sydney, 1998; and D. Fairfax, *Royal Australian Navy in Vietnam*, Australian Government Publishing Service, Canberra, 1980.

16. The Type 42 in the Falklands War: The Influence of the Air Warfare Destroyer on Joint Operations

[1] A. Stephens, submission to 'Inquiry into Australian Defence Force Regional Air Superiority', http://www.aph.gov.au/house/committee/jfadt/adfair/subs/sub1.pdf accessed 21 June 2006.
[2] In 1981 a film crew from the British Television network Anglia visited the Falklands to make a documentary about the Falkland Islanders. After meeting and talking to the people the crew produced a documentary which went to air with the title 'More British than the British'. Cited in R. Hunt, *My Falkland Days*, Politico's Publishing, London, 2002, p. 120.
[3] Hunt, *My Falkland Days*, p. 58, 131.
[4] A.W.Grazebrook, 'NOTT the way to defend the West', *Pacific Defence Reporter*, Oct 1981, p. 41.
[5] Hunt, *My Falkland Days*, p. 125.
[6] Lt Col John Marr, *War in the Falklands: Perspectives on British Strategy and the Use of Air Power*, Research Report, Maxwell Airbase, Alabama, April 1988, p. 26.
[7] Nigel West, *The Secret War for the Falklands, The SAS, MI6, and the War Whitehall Nearly Lost*, Little Brown and Company, London, 1997, p. 109.
[8] Preparations seem to have become urgent only when an Argentine 707 transport aircraft found the RN Task Force south of Ascension Island. See Middleton, *Fight for the Malvinas*, p. 54.
[9] Martin Middleton, *Fight for the Malvinas*, Penguin Books, London, 1989, p. 5.
[10] The Argentine conscript training cycle begins in February. Middleton, *Fight for the Malvinas*, p. 49.
[11] Capt John Moore, RN (Rtd), *Jane's Fighting Ships, 1983-84*, Jane's, Surrey, 1983, p. 129.
[12] *The Falklands Campaign, The Lessons*, UK Government White Paper, p. 6.

[13] Peter Morton, *Fire across the Desert, Woomera and the Anglo-Australian Joint Project 1946-1980*, Australian Government, Canberra, 1997, p. 353.
[14] Grove, *Vanguard to Trident – British Naval Policy Since World War II*, The Bodley Head Ltd, London, 1987, p. 316.
[15] Leo Marriott, *TYPE 42*, Ian Allan Ltd, Surrey, 1985 p.16.
[16] Grove, *Vanguard to Trident*, p. 349.
[17] Grove, *Vanguard to Trident*, p. 317.
[18] Marriott, *TYPE 42*, p. 58
[19] Marriott, *TYPE 42*, p. 58
[20] ADM Sandy Woodward, *One Hundred Days, The Memoirs of the Falklands Battle Group Commander*, Blue Jacket Books, Naval Institute Press, Annapolis, 1997, p. 303.
[21] Woodward, *One Hundred Days*, p. 272-273.
[22] Middlebrook, *Fight for the Malvinas*, p. 21.
[23] Middlebrook, *TASK FORCE, The Falklands War 1982*, Penguin books, London, 1987, p. 53.
[24] Middlebrook, *Fight for the Malvinas*, p. 212.
[25] Middlebrook, *Fight for the Malvinas*, p.157.
[26] *The Falklands Campaign, The Lessons*, UK Government White Paper, p. 45.
[27] Ethell, Price, *Air War over the South Atlantic*, p. 153.
[28] Nigel West, *The Secret War for the Falklands*, p. 114.
[29] Woodward, *One Hundred Days*, p. 99.
[30] Grove, *Vanguard to Trident*, p. 369.
[31] Grove, *Vanguard to Trident*, p. 369.
[32] Jeffery Ethell, Alfred Price, *Air War over the South Atlantic*, Sidgwick & Jackson, London, 1983, p. 89.
[33] Burden, Draper, Rough, Smith, Wilton; *Falklands – The Air War*, p. 118.
[34] *The Falklands Campaign, The Lessons*, UK Government White Paper, p. 9.
[35] Burden, Draper, Rough, Smith, Wilton; *Falklands – The Air War*, Arms and Armour Press Limited, London, 1986, p. 118.
[36] Ethell, Price, *Air War over the South Atlantic*, p. 193.
[37] *The Falklands Campaign: The Lessons*, UK Government White Paper, p. 45.
[38] CMDR Sharky Ward, *Sea Harrier over the Falklands,* Cassell & Co, London, 2001, p. 323.
[39] David Brown, *The Royal Navy and the Falklands War*, Arrow Books, London, 1989, p. 154.
[40] Frank Uhlig, *How Navies Fight, The US Navy and its Allies*, Naval Institute Press Annapolis, 1994, p. 369.
[41] *RUSI Journal,* June 2002, RUSI Panel Discussion, 'Twenty years on: The Falklands War in perspective', p. 66.

17. Sea Power and Peace Operations: The Naval Contribution to UNITAF in Somalia, 1992–94

[1] RAN, *Australian Maritime Doctrine*, Canberra, Defence Publishing Service, 2000, p. 65. It is under this wider popular definition that the term 'peacekeeping' is used in the title of the most recent Australian official history series: the *Official History of Australian Peacekeeping and Post-Cold War Operations*.
[2] R. Morris, 'Education, discovery and idealism, 1760–1895', in J.R. Hill (ed.), *The Oxford Illustrated History of the Royal Navy*, Oxford, Oxford University Press, 1995, pp. 242–5.
[3] M. Battesti, 'L'aspect naval des operations de "maintien de la paix" durant le Second Empire', in S. Bernier (ed.), *Peacekeeping 1815 to Today: Proceedings of the XXIst Colloquium of the International Commission of Military History*, Directorate of History, Department of

National Defence, Ottawa, 1995, p. 52. I am indebted to Dr Matthew Glozier, Strategic and Defence Studies Centre (SDSC), Australian National University (ANU), for the translation.

[4] www.un.org/aboutun/charter/.

[5] Quoted in J. Hillen, *Blue Helmets: The Strategy of UN Military Operations*, 2nd ed., Washington, Brassey's, 2000, p. 10.

[6] www.un.org/aboutun/charter/.

[7] Speech by Blair, to the Economic Club of Chicago, Hilton Hotel, Chicago, USA, 22 April 1999, www.globalpolicy.org/globaliz/politics/blair.htm.

[8] N. Chomsky, *The new military humanism: Lessons from Kosovo*, London, Pluto, 1999, p. 13; and *A new generation draws the line: Kosovo, East Timor and the standards of the West*, London, Verso, 2000, p. 107. I am indebted to Mr Daniel Flitton, SDSC, ANU, for these references.

[9] M.C. Smouts, 'Political aspects of peace-keeping operations', in B. Stern (ed.) (D. Boyle trans), *United Nations peace-keeping operations: A guide to French policies*, Tokyo, UN University Press, 1998, pp. 23–5.

[10] *Australian Maritime Doctrine*, pp. 65–7.

[11] G. Swinden, '"The natives appear restless tonight": HMAS *Adelaide* and the punitive expedition to Malaita in 1927', in D. Stevens (ed.), *Maritime Power in the Twentieth Century: The Australian Experience,* Sydney, Allen and Unwin, 1998, pp. 54-67.

[12] P. Londey, *Other people's wars: A history of Australian peacekeeping*, Sydney, Allen and Unwin, 2004, pp. 227-9.

[13] R. Mobley, 'The Beira Patrol: Britain's Broken Blockade against Rhodesia', *Naval War College Review*, Vol. LV, No. 1, Winter 2002, pp. 63-81.

[14] M. Buracchia, 'Italian peacekeeping operations (1950-1995)', in Bernier (ed.), *Peacekeeping 1815 to Today*, pp. 480-483. The three vessels currently deployed, the *Esploratore*, *Sentinella* and *Vedetta*, were constructed specially for service with the MFO. www.mfo.org/site_tree/9/42/base.asp.

[15] A 1948 UN document listing historical precedents for UN peace operations mistakenly refers to this ship as HMS *Carisbrooke* and this error has been repeated in published works. 'Precedents concerning the creation of an international force', UN Department of Public Information press memorandum, 23 January 1948, NAA: A1838, TS852/20/2 Pt 3; Pugh, 'The historical record and the relevance of force thresholds', p. 34.

[16] Telegram Earl Derby, British Ambassador Paris, to Earl Curzon, British Foreign Secretary, 10 January 1920, letters Sir C. Marling, British Ambassador Copenhagen and President of the International Commission for Schleswig, to Curzon, 12 February and 1 March 1920, R. Butler et al. (eds), *Documents on British Foreign Policy 1919-1939*, First Series, Vol. X, London, Her Majesty's Stationery Office, 1960, pp. 563-4, 569, 573-6. I am indebted to Graeme Rymill of the University of Western Australia Library and Nis Maybom of the Royal Archives of Denmark for information on this operation. The USMC formed the 15th Separate Battalion for service in Schleswig, but the US Government decided not to take part in this mission. E.N. McClellan, *The United States Marine Corps in the World War*, Washington DC, Historical Branch, G-3 Division HQ, USMC, 1920, p. 78.

[17] HMAS *Australia* was serving with the Mediterranean Fleet at the time and was tasked with covering the withdrawal of *Glorious* following the attack. J. Sears, '1929-1939: Depression and Rearmament' in D. Stevens (ed.), *The Royal Australian Navy*, Melbourne, Oxford University Press, 2001, p. 95.

[18] Many histories of peace operations (for example W.J. Durch (ed.), *The evolution of UN peacekeeping: Case studies and comparative analysis*, New York, St Martin's Press, 1993, pp. 8-10) have ignored the Indonesian deployment of September 1947 and refer to UN Special Committee on the Balkans formed later in 1947 or the UN Truce Supervision Organization

formed in Palestine in 1948 as the first UN peace operation. For an account of UN operations in Indonesia from 1947 to 1951, see Londey, *Other people's wars*, pp. 13–28.

[19] For a brief introduction to Somali history, see W.J. Durch, 'Introduction to anarchy: Humanitarian intervention and "state-building" in Somalia', in W.J. Durch (ed.), *UN Peacekeeping, American politics, and the uncivil wars of the 1990s*, New York, St Martin's Press, 1996, pp. 312–5.

[20] D.J. Zvijac and K.A.W. McGrady, *Operation Restore Hope: Summary Report*, Alexandria, Center for Naval Analyses, 1994, pp. 22–3, 25.

[21] P. Robinson, *The CNN Effect: The myth of news, foreign policy and intervention*, London and New York, Routledge, 2002; A. Natsios, 'Illusions of Influence: The CNN Effect in Complex Emergencies', in R. Rothberg and T. Weiss (eds), *From Massacres to Genocide: The Media, Public Policy and Humanitarian Crises,* Washington, DC and Cambridge, The Brookings Institution and The World Peace Foundation, 1996, pp. 149–68.

[22] *United States Forces, Somalia After Action Report and Historical Overview: The United States Army in Somalia, 1992–1994*, Washington DC, Center of Military History, United States Army, 2003, p. 61.

[23] I am indebted to Dr Rosalind Hearder, SDSC, ANU, for compiling a list of the ships that took part in UNITAF.

[24] Zvijac and McGrady, *Operation Restore Hope*, pp. 35–6.

[25] *From the Sea: Preparing the naval service for the 21st century*, Washington DC, Department of the Navy, 1992, www.chinfo.navy.mil/navpub/policy/fromsea/fromsea.txt; [US] Navy Times, 21 December 1992.

[26] A. Seigel, *Eastern Exit: The Noncombatant Evacuation Operation (NEO) from Mogadishu, Somalia, in January 1991*, Alexandria, Center for Naval Analyses, Virginia, 1992.

[27] L.J. Higgins, *Canadian Naval Operations in the 1990s: Selected Case Studies,* Maritime Security Occasional Paper No. 12, Halifax, Nova Scotia, Centre for Foreign Policy Studies, Dalhousie University, 2002, pp. 51, 53; C. Oliviero, 'Operation "Deliverance": International success or domestic failure?', *Canadian Military Journal*, Summer 2001, pp. 52–3, 57.

[28] Zvijac and McGrady, *Operation Restore Hope*, p. 38.

[29] *United States Forces, Somalia After Action Report*, p. 55.

[30] Zvijac and McGrady, *Operation Restore Hope*, pp. 40–1; [US] Navy Times, 22 February 1993.

[31] Zvijac and McGrady, *Operation Secure Hope*, pp. 40–1, 28–9.

[32] Zvijac and McGrady, *Operation Secure Hope*, pp. 45, 47, 59.

[33] Higgins, *Canadian Naval Operations in the 1990s*, p. 52.

[34] B. Breen, *A little bit of hope: Australian Force Somalia*, Sydney, Allen and Unwin, 1998, pp. 41–2.

[35] *United States Forces, Somalia After Action Report,* p. 53; HMAS *Tobruk,* Report of Proceedings, January 1993, SPC-A, Canberra.

[36] [US] *Navy Times*, 22 February 1993.

[37] [US] *Navy Times*, 22 February 1993; Zvijac and McGrady, *Operation Restore Hope*, p. 25; J. Dworken, *Military Relations with Humanitarian Relief Organizations: Observations from Restore Hope*, Alexandria, Center for Naval Analyses, 1993, pp. 37–9.

[38] L. De Vos, L. Du Four and R. Paulissen, 'Les opérations de paix et les operations humanitaires de l'Armee Belge', in Bernier (ed.), *Peacekeeping 1815 to Today*, p. 431.

[39] J.L. Hirsch and R.B. Oakley, *Somalia and Operation Restore Hope: Reflections on peacemaking and peacekeeping*, Washington DC, United States Institute of Peace Press, 1995, p. 54.

[40] [US] *Navy Times*, 28 December 1992.

[41] Zvijac and McGrady, *Operation Restore Hope*, p. 49.

[42] Higgins, *Canadian Naval Operations in the 1990s*, p. 53.
[43] Breen, *A little bit of hope*, pp. 267–8.
[44] Higgins, *Canadian Naval Operations in the 1990s*, p. 54.
[45] Breen, *A little bit of hope*, p. 253.
[46] D. Horner, *Making the Australian Defence Force*, Melbourne, Oxford University Press, 2001, p. 128.
[47] J. Ginifer, 'Multinational naval force structures', in Pugh (ed.), *Maritime security and peacekeeping*, p. 114; Grove, 'Navies in Peacekeeping and Enforcement', pp. 467–8.
[48] Higgins, *Canadian Naval Operations in the 1990s*, p. 55.
[49] HMAS *Tobruk,* Report of Proceedings, February, 1993, SPC-A, Canberra.
[50] Dworken, *Military Relations with Humanitarian Relief Organizations*, p. 14.
[51] Dworken, *Military Relations with Humanitarian Relief Organizations*, p. 17.
[52] [US] *Navy Times*, 15 February 1993.
[53] Higgins, *Canadian Naval Operations in the 1990s*, p. 55.
[54] HMAS *Tobruk*, Report of Proceedings, March, 1993, SPC-A, Canberra.
[55] Zvijac and McGrady, *Operation Restore Hope*, pp. 26, 37.
[56] Londey, *Other people's wars*, pp. 179–93.
[57] For an example, see J. Grey, *The Australian Army*, Melbourne, Oxford University Press, 2001, p. 259.

18. The Challenges for Medium-sized Navies in the Post-Cold War World.

[1] There is a mass of literature on globalisation. See, for example, A. de Borchgrave, 'Globalisation: The Bigger Picture', in *Washington Quarterly*, Summer 1996, pp. 159–78; J.N. Rosenau, 'The Dynamics of Globalization: An Operational Formulation', in *Security Dialogue*, Vol. 27, No. 3, September 1996, pp. 247–62; and C. Hill, *The Changing Politics of Foreign Policy*, Basingstoke, Palgrave, 2003.
[2] A. and H. Toffler, *War and Anti-War: Making Sense of Today's Global Chaos*, New York, Mass Market Paperback, 1995.
[3] F. Fukuyama, 'The End of History?', in *The National Interest*, 1989, reprinted in G. Tuathail et al (eds), *The Geopolitics Reader*, London, Routledge, 1998, pp. 114–24.
[4] President George Bush, 'Towards a New World Order', in *Public Papers of the Presidents of the United States* reprinted in Tuathail et.al., *The Geopolitics Reader*, pp. 131–8.
[5] S.P. Huntington, 'The Clash of Civilisations', in *Foreign Affairs*, Vol. 72, No. 3, Summer 1993, pp. 22–49.
[6] 'Transcript of Clinton's Remarks', *ASNE Reporter*, 16 April 1999.
[7] W. Pfaff, 'Invitation to War, *Foreign Affairs*, Vol. 72, No. 3, Summer 1993, pp. 97–109.
[8] L.T. Hadar, 'What Green Peril?'; and J.Miller, 'The Challenge of Radical Islam', in *Foreign Affairs*, Vol.72, No.2, Spring 1993, pp. 27–42 and pp. 43–56.
[9] J. Keegan quoted by R. Grant, in 'Air power made it Work', in *Air Force Magazine*, November 1999, p. 37.
10 See, for example, G. and M. Friedman, T*he Future of War: Power, Technology and American World Dominance in the Twenty-First Century*, New York, St Martins Griffin, 1996; and M. Van Creveld, *Technology and War: From 2000BC to the Present*, London, Brassey's, 1991.
[11] While there is no universally agreed definition for asymmetric warfare the UK Ministry of Defence has defined it as 'the threat of an attack by unconventional methods which could have a disproportionate effect'. See 'The Strategic Defence Review: A New Chapter', *Cm.5,566*, London, The Stationery Office, 2002, p. 7, www.mod.uk/issues/sdr/newchapter.htm.

[12] See, for example, C. Gray, *Strategy for Chaos: Revolutions in Military Affairs and the Evidence of History*, London, Frank Cass, 2002; and B. Berkowitz, *The New Face of War: How War will be Fought in the 21st century*, New York, The Free Press, 2003.
[13] See R. Atkinson, *Crusade: The Untold Story of the Persian Gulf War*, Boston, Houghton Mifflin Co, 1993; and L. Freedman and E. Karsh, *The Gulf Conflict 1990–91*, London, Faber and Faber, 1993.
[14] A.H. Cordesman, *The Iraq War: Strategy, Tactics and Military Lessons*, Westport, Praeger, 2003, pp. 40–1.
[15] R.B. Polk, 'A Critique of the Boyd Theory: Is it Relevant to the Army?', in *Defense Analysis*, Vol. 16, No. 2, December 2000, pp. 257–76.
[16] H.K. Ullman, *Shock and Awe: Achieving Rapid Dominance*, Washington DC, National Defense University, 1996.
[17] Lieutenant-General M. Mosley, 'Coalition Forces Air Component Command Briefing', 5 April 2003, www.defense.mil/transcripts/2003/t04052003_t405mose.html.
[18] T.D. Biddle, 'British and American Approaches to Strategic Bombing: Their Origins and Implementation in the World War II Combined Bomber Offensive', in *Journal of Strategic Studies*, March 1995, pp. 41–144.
[19] Cordesman, *The Iraq War*, p. 302.
[20] S.J. Blank, 'Rethinking Asymmetric Threats', in *Strategic Studies Institute Papers*, Carlisle, US Army War College, 2003, www.comword/rma/fulltext/0309blank.pdf.
[21] See R.A.D. Applegate, 'Observations on the Use of Force in Complex Emergencies', in *The RUSI Journal*, Vol. 147, No. 1, February 2001, pp. 22–7; and J. MacKinlay, 'Beyond the Logjam: A Doctrine for Complex Emergencies', in *Small Wars and Insurgencies*, Vol. 9, No. 1, Spring 1998, pp. 114–31.
[22] See P.W. Singer, *Corporate Warriors: The Rise of the Privatized Military Industry*, New York, Cornell University Press, 2003.
[23] F. Hoffman, 'What can we Learn from Jackie Fisher', in *Proceedings*, April 2004, p.70.
[24] See N.A. Lambert, *Sir John Fisher's Naval Revolution*, Columbia, University of South Carolina Press, 1999.
[25] S. Croft, 'Introduction', in Croft et al, *Britain and Defence, 1945-2000: A Policy Re-Evaluation*, Harlow, Longman, 2001, p. 1.
[26] See Lambert, *Sir John Fisher's Naval Revolution*; A. Gordon, *The Rules of the Game: Jutland and British Naval Command*, London, John Murray, 1996.
[27] H. Guderian, *Achtung-Panzer: the Development of Tank Warfare*, translated by C. Duffy, London, Cassell Military Paperbacks, 1992, p. 24.
[28] See M.G. Burns, *Bader: the Man and his Men*, London, Arms and Armour, 1990, pp. 33–4, 44.
[29] See Gordon, *The Rules of the Game*.
[30] R.A. Mason, *Air Power: A Centennial Appraisal*, London, Brassey's, 1994, p. 160; See also Colonel G.L. Crowder, 'Effects Based Operations Briefing', News Transcript, Department of Defense, 19 March 2003, www.iwar.org.uk/rma/resources/ebo/effects-based-operations-briefing.htm.
[31] G. Till, 'Corbett and the 1990s', in J. Goldrick and J.B. Hattendorf (eds), *Mahan is Not Enough: Proceedings of a Conference on the Works of Sir Julian Corbett and Admiral Sir Herbert Richmond*, Newport, Naval War College Historical Monograph Series, No. 10, 1993, p. 217–8.
[32] G. Hoon, 'One Year On – A UK Perspective', 9 September 2002, University of Louisville, http://news.mod.uk/news/press/news_press_notice.asp?newsItem_id=1982.
[33] G. Till, 'Sir Julian Corbett: Ten Maritime Commandments', in A. Dorman et al (eds), *The Changing Face of Maritime Power*, Basingstoke, Macmillan, 1999, p. 30.
[34] P. Cosgrove, 'The ANZAC lecture at Georgetown University', in *Journal of the Australian*

Naval Institute, April/June 2000, p. 9.
[35] *Operations in Iraq: First Reflections,* London, Ministry of Defence , 2003, www.mod.uk, p. 7.
[36] House of Commons Defence Committee, 'Defence White Paper 2003', Third Report of Session 2003–04, *HC.465-I*, 2004, p. 7.
[37] *Operations in Iraq: First Reflections*, p. 24.
[38] House of Commons Defence Committee, 'Defence White Paper 2003', p. 24.
[39] Gordon, *Rules of the Game*, p. 569; P. Cornish, 'Transformed Warfare: Who's in Charge?', in *World Defence Systems*, Vol. 7, No. 2, pp. 24–7.
[40] Gordon, *Rules of the Game*, p. 570.
[41] *Operations in Iraq: First Reflections*, p. 4.
[42] A.J.P. Taylor, *The Struggle for Mastery in Europe*, Oxford, Oxford University Press, 1954, pp. 528–9.
[43] A. Bucholz, *Moltke, Schlieffen and Prussian War Planning*, New York, Berg Publishers Inc, 1991.
[44] 'Delivering Security in a Changing World: Defence White Paper', *Cm.6,041-I*, London, The Stationery Office, 2003, pp. 12–3.
[45] W.R. Louis and R. Owen (eds), *Suez 1956: the Crisis and its Consequences*, Oxford, Oxford University Press, 1991; J. Pearson, *Sir Anthony Eden and the Suez Crisis: Reluctant Gamble*, Basingstoke, Palgrave Macmillan, 2003.
[46] 'Interview: Admiral Jonathan Band', in *Jane's Navy International*, Vol. 110, No. 3, April 2005, p. 34.
[47] See Carl von Clausewitz, *On War*, London, Penguin, 1968.
[48] C. Paul and J.J. Kim, 'Reporters on the Battlefield: The Embedded Press System in the Historical Context', MG-200-RC, Santa Monica, RAND, 2004, www.rand.org/publications/MG/MG200/.
[49] Lord Ashdown, *The Ashdown Diaries: Volume Two 1997–99*, London, Allen Lane, 2001, p. 436.

19. Triumph or Tragedy? Reflections on the History of Joint Power Projection Ashore

[1] G. Till, *Seapower: A Guide for the Twenty-First Century*, London, Frank Cass, 2004, p. 199.
[2] As Julian Corbett famously observed. J.S. Corbett, *Some Principles of Maritime Strategy*, London, Longmans, Green and Co., 1911, repr. E. Grove (ed.), London, Brassey's, 1988, p. 16.
[3] C. White, 'Nelson Ashore, 1780–97' in P. Hore (ed.), *Seapower Ashore: 200 Years of Royal Navy Operations on Land*, London, Chatham, 2001, pp. 53ff.
4 See Paine, chapter 2 above.
[5] The new appreciation of Corbett has been due principally to the work of Donald Schurman. See in particular D. Schurman, *Julian S. Corbett, 1854–1922: Historian of British Maritime Policy from Drake to Jellicoe*, London, Royal Historical Society, 1981.
[6] This is a theme of G.W. Baer, *One Hundred Years of Sea Power: The US Navy, 1890–1990*, Stanford, CA, Stanford University Press, 1994.
[7] As Ian Speller has observed. See entry on 'Amphibious Warfare' in C. Messenger (ed.), *Reader's Guide to Military History*, London, Fitzroy Dearborn, 2001, p. 17. The notable exception is of course C.E. Callwell, *Military Operations and Maritime Preponderance: Their Relations and Independence*, Edinburgh and London, Blackwood, 1905, repr. C.S. Gray (ed.), Annapolis, Naval Institute Press, 1996.
[8] Quoted in A. Lambert, '"Within Cannon Shot of Deep Water": The Syrian Campaign of 1840' in Hore, *Seapower Ashore*, p. 92.
[9] See also for example A. Lambert, 'Under the Heel of Britannia: The Bombardment of Sweaborg 9-11 August 1855' in Hore, *Seapower Ashore*.

[10] See Elleman, chapter 11 above.
[11] Till, *Seapower: A Guide for the Twenty-First Century*, p. 199.
[12] R.D. Heinl, Jr., 'Inchon, 1950' in M.L. Bartlett (ed.), *Assault from the Sea: Essays on the History of Amphibious Warfare*, Annapolis, Naval Institute Press, 1983, p. 347.
[13] G. Parker, *The Grand Strategy of Philip II*, New Haven and London, Yale University Press, 1998, p. 193.
[14] M. Clapp and E. Southby-Tailyour, *Amphibious Assault Falklands: The Battle of San Carlos Water*, paperback edn, London, Orion, 1997, pp. 141–2.
[15] Corbett made this point about the Quebec campaign. J.S. Corbett, *England in the Seven Years' War: A Study in Combined Strategy*, 2 vols, London, Longmans, Green and Co., 1907, I, pp. 427–9.
[16] Corbett, *England in the Seven Years' War*, I, pp. 193–4, 437.
[17] Corbett, *Some Principles of Maritime Strategy*, p. 69.
[18] I. Speller and C. Tuck, *Amphibious Warfare: The Theory and Practice of Amphibious Operations in the 20th Century*, Staplehurst, Spellmount, 2001, p. 18.
[19] Speller and Tuck, *Amphibious Warfare*, p. 102.
[20] On these issues see M. Duffy, '"Science and Labour". The Naval Contribution to Operations Ashore in the Great Wars with France 1793–1815' in Hore, *Seapower Ashore*, pp. 39ff.
[21] Duffy, 'Science and Labour'.
[22] See in general R. Brooks, *The Long Arm of Empire: Naval Brigades from the Crimea to the Boxer Rebellion*, London, Constable, 1999.
[23] D. Stevens, *The Royal Australian Navy: The Australian Centenary History of Defence*, Volume III, Melbourne, Oxford University Press, 2001, p. 36; G. Nash and D. Stevens, *Australia's Navy in the Gulf: From Countenance to Catalyst, 1941–2006*, Topmill, 2006, p. 72.
[24] See note 44 below.
[25] See Clark, chapter 5 above.
[26] C. Gray, entry on 'Amphibious Warfare' in R. Holmes (ed.), *The Oxford Companion to Military History*, Oxford, Oxford University Press, 2001, pp. 51–2.
[27] How far command of the air has always been an equal pre-requisite is a greyer area, I would suggest, and beyond the scope of this chapter. The British task force in the Falklands, for example, did not command the air at the time of the San Carlos landing. Clapp and Southby-Tailyour, *Amphibious Assault Falklands*, p. 176.
[28] Till, *Seapower: A Guide for the Twenty-First Century*, p. 52. See also C.D. Hall, *Wellington's Navy: Sea Power and the Peninsula War, 1807–1814*, London, Chatham, 2004.
[29] See in general D. Grinnell-Milne, *The Silent Victory: September 1940*, London, The Bodley Head, 1958.
30 See C. Martin and G. Parker, *The Spanish Armada*, London, Hamish Hamilton, 1988. A classic work is G. Mattingley: *The Defeat of the Spanish Armada*, London, Jonathan Cape, 1959.
[31] N.A.M. Rodger, *The Command of the Ocean: A Naval History of Britain, 1649–1815*, London, Allen Lane, 2004, p. 288. On the Quebec campaign see in general C.P. Stacey, *Quebec, 1759: The Siege and the Battle*, Toronto, Macmillan, 1959.
[32] Corbett, *England in the Seven Years' War*, I, pp. 181–2.
[33] S.E. Ambrose, *D-Day June 6, 1944: The Climactic Battle of World War II*, New York, Simon and Schuster, 1994, repr. Touchstone, 1995, p. 190.
[34] Corbett called the French plan for the invasion of England in 1759 a soldier's plan. Corbett, *England in the Seven Years' War*, II, p. 19. For an analysis of the fallacies of the Japanese Pacific perimeter strategy see J. Reeve, *Maritime Strategy and Defence of the Archipelagic Inner Arc*, RAN Sea Power Centre Working Paper No. 5, Canberra, Australian Defence Publishing Service, 2001.

[35] See Prior, chapter 3 above.
[36] See J. Black, *Pitt the Elder*, Cambridge, Cambridge University Press, 1992.
[37] Clapp and Southby-Tailyour, *Amphibious Assault Falklands*, pp. 97ff.
[38] Heinl, 'Inchon, 1950', p. 352.
[39] D. Syrett, 'British Amphibious Operations during the Seven Years and American Wars' in Bartlett (ed.), *Assault from the Sea*, pp. 51ff. On the complexities of British amphibious operations in the eighteenth century see R. Harding, 'Sailors and Gentlemen of Parade: Some Professional and Technical Problems Concerning the Conduct of Combined Operations in the Eighteenth Century", *The Historical Journal*, xxxii, 1989.
[40] Speller and Tuck, *Amphibious Warfare*, pp. 155–6.
[41] W. Murray and A.R. Millett, *A War to be Won: Fighting the Second World War*, Cambridge, Mass., Harvard University Press, 2000, p. 419.
[42] N. Friedman, 'A New Kind of Amphibious Warfare', *Asia-Pacific Defence Reporter*, April, 2006, pp. 22, 29. See also T. Terriff, '"Innovate or Die": Organizational Culture and the Origins of Maneuver Warfare in the United States Marine Corps', *The Journal of Strategic Studies*, 29, 3, June, 2006.
[43] US Navy White Papers: *From the Sea*, Washington, 1992 and *Forward…From the Sea*, Washington, 1994.
[44] Speller and Tuck, *Amphibious Warfare*, p. 157.
[45] Speller and Tuck, *Amphibious Warfare*, pp. 161ff.
[46] In January 2006. The future of this initiative still depends upon Congressional budgetary approval for FY2007. G.I. Petersen, 'The U.S. Navy Expeditionary Combat Command', *Naval Forces*, IV, 2006, pp. 27-35.
[47] Friedman, 'A New Kind of Amphibious Warfare', p. 31.
[48] *BR1806: British Maritime Doctrine*, Second Edition, London, HMSO, 1999; Hore, *Seapower Ashore*, p .23.
[49] See for example G. Wahlert (ed.), *Australian Army Amphibious Operations in the South-West Pacific: 1942–45*, Canberra, Army Doctrine Centre, 1995. A significant text in terms of the evolution of thinking in the Australian Army was: M. Evans, *The Role of the Army in a Maritime Concept of Strategy*, Canberra, Land Warfare Studies Centre, Working Paper No. 101, 1998.
[50] *Australian Maritime Doctrine: RAN Doctrine 1 2000*, Canberra, Department of Defence, 2000.
[51] See Connor, chapter 17 above.
[52] There has been little written on conceptual links between maritime capabilities and counter-terrorism. See for example J. Reeve, 'Maritime Operations and Counter-Terrorism: An Australian View' in J. McCaffrie (ed.), *Positioning Navies for the Future*, Sydney, Halstead Press, 2006. For a regional political case study see J.F. Bradford, 'The Growing Prospects for Maritime Security Cooperation in Southeast Asia', *Naval War College Review*, 58, 3, Summer, 2005.
[53] Friedman, 'A New Kind of Amphibious Warfare', p. 32.
[54] See Schweikert, chapter 16 above.
[55] Heinl, 'Inchon, 1950', p. 346.

Index

D

E

F

G